53.00
100K

FEW–BODY SYSTEMS AND MULTIPARTICLE DYNAMICS

AIP CONFERENCE PROCEEDINGS 162

RITA G. LERNER
SERIES EDITOR

FEW–BODY SYSTEMS AND MULTIPARTICLE DYNAMICS

CRYSTAL CITY, VA 1987

EDITOR:
DAVID A. MICHA
UNIVERSITY OF FLORIDA

AMERICAN INSTITUTE OF PHYSICS   NEW YORK 1987

Authorization to photocopy items for internal or personal use, beyond the free copying permitted under the 1978 US Copyright Law (see statement below), is granted by the American Institute of Physics for users registered with the Copyright Clearance Center (CCC) Transactional Reporting Service, provided that the base fee of $3.00 per copy is paid directly to CCC, 27 Congress St., Salem, MA 01970. For those organizations that have been granted a photocopy license by CCC, a separate system of payment has been arranged. The fee code for users of the Transactional Reporting Service is: 0094-243X/87 $3.00.

Copyright 1987 American Institute of Physics

Individual readers of this volume and non-profit libraries, acting for them, are permitted to make fair use of the material in it, such as copying an article for use in teaching or research. Permission is granted to quote from this volume in scientific work with the customary acknowledgment of the source. To reprint a figure, table or other excerpt requires the consent of one of the original authors and notification to AIP. Republication or systematic or multiple reproduction of any material in this volume is permitted only under license from AIP. Address inquiries to Series Editor, AIP Conference Proceedings, AIP, 335 E. 45th St., New York, NY 10017.

L.C. Catalog Card No. 87-72594
ISBN 0-88318-362-5
DOE CONF-8704193

Printed in the United States of America

Contents

Introduction to the Symposia on Few-Body Systems and Multiparticle Dynamics ix
    D. A. Micha

PART I: BOUND STATES OF FEW-BODY SYSTEMS

The Trinucleon Bound-State System 1
    G. L. Payne

Monte Carlo Techniques for Ground-State Properties 19
    K. E. Schmidt

Quarks and Nucleon–Nucleon Correlations 30
    G. A. Miller

Eigenfunctions and Periodic Orbits in Classically Chaotic Systems: Order in Chaos? 43
    E. J. Heller, P. W. O'Connor, and J. N. Gehlen

The Quark Model and the Nucleon–Baryon Interaction 60
    A. Faessler

PART II: COLLISION DYNAMICS OF FEW-BODY SYSTEMS

Time Dependent Approach to Photodetachment in E, B, and Parallel E and B Fields 94
    W. P. Reinhardt

Deuteron–Nucleus Collisions: Reduction of a Many-Fermion Collision System to an Equivalent Three-Body Model 111
    R. Kozack and F. S. Levin

Coincidence Reactions and the Three-Body Structure of ^{6}Li 131
    D. R. Lehman

Three-Body Theory of Electron Capture 145
    J. H. Macek

The Time-Dependent Hartree-Fock Method and Chaos in Nuclear Dynamics 157
    M. R. Strayer

PART III: FEW-BODY DYNAMICS IN LARGER SYSTEMS

Microscopic Cluster Theory in Nuclear Physics 174
    Y. C. Tang

Simulations of Structure, Dynamics and Transformations in Finite Aggregrates 200
    U. Landman, R. N. Barnett, C. L. Cleveland, J. Luo, D. Scharf, and J. Jortner

Reactive Flux Correlation Functions and Monte Carlo Evaluation of Real Time Path Integrals 245
    W. H. Miller

Solid and Liquid Molecules and Microclusters 261
    R. S. Berry

Foreword

This volume contains all the invited contributions presented at the symposia of the Topical Group on Few-Body Systems and Multiparticle Dynamics, on April 20th and 21st 1987, in Crystal City, Virginia, the site of the 1987 Spring meeting of the American Physical Society.

Few-body systems and multiparticle dynamics have been the focus of research by many scientists working in different areas of physics and chemistry. The American Physical Society has responded to the interest in these topics and the need to bring together people from the different divisions of the APS by the 1985 approval of the creation of a new topical group on few-body systems and multiparticle dynamics. It is charged with organizing sessions at APS meetings, organizing topical conferences, workshops and satellite meetings, acting as a liaison with similar groups in other countries and as a bridge among the APS divisions, and generally with keeping interested scientists informed on related activities. At the time of this writing our topical group already includes more than 700 members.

Three symposia were organized within the framework of the National APS Meeting in Crystal City. The invited speakers were chosen from suggestions of the topical group members, channelled through its program committee. This was constituted by:

> H.E. Conzett, Lawrence Berkeley Laboratory
> B.F. Gibson, Los Alamos National Laboratory
> M.D. Girardeau, Jr., University of Oregon
> E.J. Heller, University of Washington
> D.A. Micha, University of Florida (Chairman)
> W.W. Smith, University of Connecticut
> A.W. Thomas, University of Adelaide, Australia

Because the suggestions covered a wide range of topics, the talks were grouped into the following three symposia:

> – Bound States of Few-Body Systems
> – Collision Dynamics of Few-Body Systems
> – Few-Body Dynamics in Larger Systems.

Each symposium consisted of presentations lasting thirty minutes each, with five minutes for questions and answers. Each presentation introduced its subject in a way suitable to an interdisciplinary audience. The symposia were very well attended, even while competing with more established divisions and groups of the APS, and motivated lively interdisciplinary discussions. We thank E.F. Redish and M.D. Girardeau for their excellent job, as chairmen of the second and third symposia, in promoting those discussions. In addition, sixteen related papers were contributed in a session on Few-Body Systems, on April 22nd, chaired by T.-K. Lim.

Special thanks are due the Office of Naval Research, which provided financial support for some of the speakers, and to its officers Drs. Bobbie R. Junker and Michael F. Shlesinger, for their help.

The editor appreciates the diligence of the authors in these proceedings, who provided the manuscripts within reasonable time and were patient in satisfying the requirements of the AIP editorial office. He thanks Frank S. Levin and Teck-Kah Lim, the present Chairman and Secretary-Treasurer of the topical group, respectively, for their help in the organization of the symposia. He is also grateful to Joanne Bratcher, Staff Assistant, Robin Bastanzi, Word Processing Operator, and the Physics and Chemistry Departments of the University of Florida for their help with budgetary matters and the preparation of these proceedings.

Gainesville, Florida, July 1987   David A. Micha

INTRODUCTION TO THE SYMPOSIA ON FEW-BODY SYSTEMS AND MULTIPARTICLE DYNAMICS

David A. Micha
University of Florida, Gainesville, Florida, 32611

Few-body systems (FBS) and multiparticle dynamics (MPD) are subjects of interest to many researchers, who have developed similar theoretical methods and experimental techniques in several areas of physics and chemistry. These developments are the result of parallel efforts made to analyze complex physical systems in terms of their simpler constituents.

Few-body systems in physics and chemistry are understood as consisting of a small number of well defined constituent structures which can be elementary, or composites with well defined properties over the duration of the phenomena of interest. Examples of few-body systems are light nuclei and atoms, small molecules, "fundamental particles" described in terms of their constituents, and even planetary systems involving both natural and artificial satellites.

Larger systems can frequently be described in terms of smaller constituents or in terms of collective excitations with well defined properties. Multiparticle dynamics refer to the methods and techniques that can be used to treat those larger systems in terms of their constituents or collective modes. Heavy nuclei, interacting fundamental particles, polyatomic molecules, clusters, and few-atom systems in condensed phases are examples of those larger systems.

Hence the topics of few-body systems and multiparticle dynamics cut across the disciplines of atomic, molecular, nuclear, intermediate- and high-energy particle physics, astrophysics, and mathematical physics, quantum chemistry, chemical dynamics, and statistical mechanics. They share many theoretical methods and experimental techniques and concepts. They are also required in many interdisciplinary studies such as those on fusion, design and construction of accelerators, and the properties of new materials studied with radiation or neutron scattering.

The following list includes some of the specific areas of active research on FBS and MPD:

- Computational methods for structure and properties of bound FBS
- Mathematical aspects of few-body scattering
- Collision dynamics of nuclear and molecular reactions
- Dynamical symmetries in FBS
- Localization and structural changes in FBS
- Resonances and transient states in FBS
- 3-body forces in 3-body systems
- Relativistic effects in FBS
- Long-range forces in FBS
- Chaotic motions in FBS
- FBS in electromagnetic fields
- Monte Carlo techniques for FBS
- Breakup and capture in few-body collisions
- Polarization in nuclear, atomic, and molecular scattering
- Quark models of nucleons in nuclei
- Annihilation and light emission in scattering
- Muonic atoms and molecules
- Few-hadron systems in bound and scattering states
- Clusters of nuclei, atoms, and molecules
- Composite-particle methods for nuclear and atomic systems
- Path integral methods in MPD

Wavepacket methods for FBS and MPD
Chaos and dissipation in MPD
Time-dependent selfconsistent field methods in MPD
2-body dynamics in condensed matter
Quantum Chromodynamics in nuclear FBS

Many of these areas have theoretical and experimental components. In addition, there are specific experimental techniques of importance in interdisciplinary work such as those of beam preparation, and particle detection and coincidence techniques.

Examples of research in these areas may be found in the proceedings of recent conferences[1-6] and in a recently launched journal[7]. However, because most conferences have concentrated on well established disciplines, such as nuclear or atomic physics, there is a need for additional interdisciplinary exchange and collaboration which these symposia and future ones hope to fulfill.

Two of the challenges to researchers on FBS and MPD are to communicate new results and recent progress to an interdisciplinary audience, and to show how theoretical methods and experimental techniques in several disciplines are interrelated. Symposia of the topical group should be beneficial in this regard.

Acknowledgments

The author thanks the members of the program committee of these symposia, and Frank Levin, for sharing with him their perspective on these topics.

References

1. *The Few-Body Problem,* edited by F. S. Levin, Nucl. Phys. **A353** (North-Holland, 1981).
2. *Proceedings of the Xth International Conference on Few-Body Problems in Physics*, edited by B. Zeitnitz, Nucl. Phys. **A416** (North-Holland, 1984).
3. *Few-Body Methods: Principles and Applications,* edited by T.-K. Lim, C. G. Bao, D. P. Hou, and H. S. Huber (World Scientific, 1986).
4. *Electronic and Atomic Collisions. Invited Papers to the XIVth ICPEAC,* edited by D. C. Lorents, W. E. Meyerhof, and J. R. Peterson (North-Holland, 1986).
5. *Proceedings of the XIth International Conference on Few-Body Systems in Particle and Nuclear Physics,* Tokyo, edited by T. Sasakawa, K. Nisimura, S. Oryu, and I. Ishikawa (North-Holland, 1987).
6. *Atomic Physics 10,* edited by H. Narumi and I. Shimamura (North-Holland, 1987).
7. *Few-Body Systems,* edited by H. Mitter and W. Plessas (Springer, 1986–).

THE TRINUCLEON BOUND-STATE SYSTEM

G. L. Payne
The University of Iowa, Iowa City, Iowa 52242

ABSTRACT

The trinucleon system can be used to test the validity of the traditional approach of nuclear physics which describes nuclei as a system of nonrelativistic nucleons interacting via two-body forces. It is now possible to solve bound-state problems to any desired accuracy. Thus, a disagreement between the numerical calculations and the experimental data can be attributed to a failure of the model and not to a lack of numerical accuracy. The use of the Faddeev equations to solve the three-body bound state is reviewed, and the various numerical techniques are discussed. The results of recent calculations by several different groups are presented, and the agreement is used to demonstrate the accuracy of the calculations. The evidence for three-body forces is reviewed, and the results of the recent calculations which include two-pion-exchange three-body forces are presented. The numerical calculations of the electromagnetic observables for the trinucleon are discussed, and comparisons are made with the experimental data. Topics covered include the rms charge radius, charge form factors, charge densities, Coulomb energy of ^{3}He, and magnetic moments.

INTRODUCTION

Much of the recent theoretical and experimental work in the three-nucleon problem has been directed at understanding certain low-energy properties of the trinucleon system. The detailed comparisons between theoretical calculations and the experimental data serve as a test of the traditional approach to nuclear physics. This approach is formulated in terms of nonrelativistic nuclear Hamiltonians which contain realistic two-nucleon intractions and possibly three-nucleon interactions. The fundamental question is whether this model which neglects subnuclear degrees of freedom, such as mesons or quarks, can provide an adequate model of the atomic nucleus. The trinucleon system plays an extremely important role in the investigation of the validity of the traditional model; for this system the solution of the model Hamiltonian may be numerically solved "exactly." That is, the numerical solution can be obtained with any desired accuracy; therefore, these calculations are a stringent test of the theoretical model. Any disagreement between the calculated and experimental values must be due to a failure of the model. For more complicated systems the present numerical methods cannot yield "exact" solutions, and one cannot determine whether any discrepancies between theory and experiment are due to the model or to the numerical approximations used.

In this paper the present status of the calculations for the bound state of the trinucleon system is reviewed. Accurate calculation of the energy eigenvalues and wave functions for the bound-state system has proved to be exceptionally difficult. Only a few years ago different calculational methods applied to the same model Hamiltonian gave different results for the bound-state energies. These differences were the result of the different numerical approximations used to solve the problem. The use of larger computers and the development of new numerical techniques led to more accurate results. However, it was only after several independent groups had solved the same model problem and obtained values for the binding energy which agreed to within 10 keV that we could claim to be able to solve the model problem "exactly." The results of various groups for the same model Hamiltonian will be presented below to illustrate the accuracy of the numerical calculations. Once the accuracy of these wave functions had been verified, they could then be used to calculate the expectation values for other physical observables of the bound-state system. It is the comparison of these calculations with the experimental data which provides the test of the model Hamiltonian.

In this paper we will be concerned only with the trinucleon bound-state results. For the scattering problem, the numerical problem is much more difficult, and the accuracy of the calculations has not been determined. While there have been many calculations for n-d and p-d scattering with realistic potentials, there has not been a detailed comparison between different groups solving the same problem. Until this comparison has been made, it will be difficult to judge our ability to solve the scattering problem. Hopefully, this goal will be achieved within the next few years.

In the next section, we review the Faddeev equations, and then we present some results for realistic interactions to illustrate the accuracy of the present calculations. In the following sections, we describe the calculations done using the wave functions obtained from the numerical solution of these equations. Also, we compare the experimental data with the theoretical predictions.

FADDEEV EQUATIONS

We assume that the trinucleon system can be described by the nonrelativistic Hamiltonian:

$$H = H_0 + V(\vec{x}_1) + V(\vec{x}_2) + V(\vec{x}_3) + W(\vec{x}_1, \vec{x}_2, \vec{x}_3)$$

$$= H_0 + V_1 + V_2 + V_3 + W \tag{1}$$

where H_0 is the free-particle kinetic energy operator, V_i is the two-body interaction between particles j and k, and $\vec{x}_i$ is the vector between particles j and k for i,j,k cyclic. The W is a possible three-body force, and for the models which we consider, it can be written in the form:

$$W = W_1 + W_2 + W_3 \tag{2}$$

where W_2 and W_3 are obtained from W_1 by the cyclic permutation of the arguments. Given the Hamiltonian, we wish to solve the Schrödinger equation

$$(H - E)|\Psi\rangle = 0 \tag{3}$$

For the bound-state system one can solve the Schrödinger equation directly using a method such as the Rayleigh-Ritz Variational Principle.[1,2] Also, one could write the Schrödinger equation as a differential equation and use a numerical procedure such as the hyperspherical expansion[3] to solve the equation. An alternate procedure[4] is to introduce the three Faddeev amplitudes and replace the Schrödinger equation by the three coupled Faddeev equations. For this case the total wave function is decomposed into three parts

$$|\Psi\rangle = \sum_{i=1}^{3} G_0(V_i + W_i)|\Psi\rangle = \sum_{i=1}^{3} |\Psi_i\rangle \tag{4}$$

where $|\Psi_i\rangle$ is the Faddeev amplitude, and

$$G_0(E) = (E - H_0)^{-1} \tag{5}$$

is the free-particle resolvent. Introducing the resolvent operator

$$G_i(E) = (E - H_i)^{-1} \tag{6}$$

where $H_i = H_0 + V_i + W_i$, Eq. (3) can be written in the form:

$$|\Psi_i\rangle = G_i(V_i + W_i)[|\Psi_j\rangle + |\Psi_k\rangle] \tag{7}$$

for $i = 1,2,3$. These three coupled equations comprise the Faddeev equations, and the solution of these equations is equivalent to the solution of the Schrödinger equation. To solve these equations, one must choose a particular representation, and different groups have chosen different representations. Possible choices are the momentum space representation,[5,6] the coordinate space representation,[7] and a combination of the two representations.[8] It is customary to use different coordinates for each of the Faddeev amplitudes. In coordinate space one normally uses the three sets of Jacobi coordinates

$$\vec{x}_i = \vec{r}_j - \vec{r}_k \tag{8a}$$

and

$$\vec{y}_i = \frac{1}{2}(\vec{r}_j + \vec{r}_k) - \vec{r}_i \tag{8b}$$

where $\vec{r}_i$ are the positions of the three nucleons, and i,j,k are taken cyclically. An example of these coordinates is shown in Fig. 1. Thus, in coordinate space the wave function is written in the form:

$$\Psi = \Psi(\vec{x}_1,\vec{y}_1) + \Psi(\vec{x}_2,\vec{y}_2) + \Psi(\vec{x}_3,\vec{y}_3) \tag{9}$$

In momentum space one uses the conjugate momentum variables $(\vec{p}_i,\vec{q}_i)$.

For the momentum space calculations it is customary to introduce the two-body t-matrices via

$$G_i V_i = G_0 t_i \tag{10}$$

and to write the bound-state Faddeev equations in the form:

$$|\Psi_i\rangle = G_0 t_i [|\Psi_j\rangle + |\Psi_k\rangle] \tag{11}$$

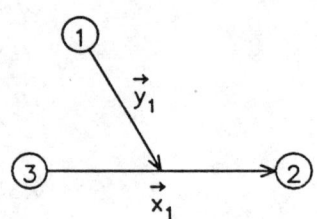

Fig. 1. Vectors $\vec{x}_1$ and $\vec{y}_1$ for the three-body system.

The three-body potential can then be included by summing the perturbation series.[9]

In both representations the three Faddeev amplitudes are projected onto a different complete set of channel wave functions which represent the orbital angular momentum and spin-isospin variables. These channel basis states are normally defined for channel α_i (which corresponds to either the coordinate or momentum space variables) as

$$|\alpha_i\rangle = |[(\ell_\alpha,s_\alpha)j_\alpha,(L_\alpha,S_\alpha)J_\alpha)]JM;(t_\alpha,T_\alpha)TM_T\rangle \tag{12}$$

where

ℓ_α = relative orbital angular momentum of particles j and k;
s_α = total spin angular momentum of particles j and k;
j_α = total angular momentum of particles j and k;
L_α = orbital angular momentum of particle i relative to particles j and k;
S_α = spin of particle i ($S_\alpha = 1/2$);
J_α = total angular momentum of particle i;
J = total angular momentum of the system ($J = 1/2$);
t_α = total isospin of particles j and k;
T_α = isospin of particle i ($T_\alpha = 1/2$);
T = total isospin of the system ($T = 1/2$).

In addition, potentials are replaced by potentials projected onto the channel functions. Thus, the three two-body potentials are written in the form:

$$V_i = \sum_{\alpha=1}^{N_c} \sum_{\alpha'=1}^{N_c} |\alpha_i'\rangle v_{\alpha'\alpha} \langle \alpha_i| \tag{13}$$

where N_c is the number of channels and $v_{\alpha'\alpha} = \langle \alpha_i'|V(\vec{x}_i)|\alpha_i\rangle$. A similar expansion is used for the W_i. The channel functions form a complete set; however, for the actual numerical calculations, the set must be truncated, and the early calculations used only a few channels (3-5). Consequently, the results were not very accurate. In order to obtain an accurate solution to the Faddeev equations, one must use enough channel functions to accurately describe both the Faddeev amplitudes and the potentials. In practice, the number of channels is increased until the solution has converged to the desired accuracy, where the convergence is determined by the value of the bound-state energy. The difficulty of the numerical calculations increases as the number of channels is increased. For the Schrödinger equation the binding energy increases monotonically as the number of channel states is increased. This is not the case for the Faddeev equations. The Faddeev operator is not a Hermitian operator; consequently, the equation cannot be derived from a variational principle. However, the Faddeev amplitudes calculated with a fixed value of N_c can be used to construct the total wave function, and this wave function can then be used to determine a variational upper bound for the binding energy whose magnitude increases monotonically with N_c.[10,11]

Typically, the number of channels is chosen by choosing the maximum value of the angular momentum j_α of the interacting pair (j,k). The initial calculations were performed using the channels with $j_\alpha \leq 1$ and positive parity; this case has five channels which corresponds to using a two-body potential which has only $^1S_0$ and $^3S_1-^3D_1$ interactions. Using a modern computer, calculations with $j_\alpha \leq 2$ and both positive and negative parities can be done; this case has 18 channels and yields binding energies within 100 keV of the converged result. In order to obtain a result for the binding energy with an error less than 10 keV, it is necessary to retain all channels with $j_\alpha \leq 4$, which corresponds to 34 channels. This is the largest calculation which has been done. Estimates of the contributions from the higher channels can be calculated using perturbation theory, and these calculations show that the higher channels contribute less than 10 keV additional binding energy.

The projection onto the N_c channel functions yields N_c coupled equations for the channel wave functions. In the momentum space calculations, the resulting equation is an integral equation which contains the unknown eigenvalue for the energy in a nonlinear form. To find the eigenvalue, Eq. (11) is rewritten in the form:

$$|\Psi_i\rangle = \lambda G_0 t_i [|\Psi_j\rangle + |\Psi_k\rangle] \qquad (14)$$

where the parameter λ has been added to the right-hand side of (14). This equation can be solved as an eigenvalue problem for λ with a fixed value of the energy. If the value assumed for the energy is the correct value, then λ will be unity. In practice, the value of the energy is varied until a value is found for which $\lambda = 1$. Normally, an iterative procedure is used to determine the eigenvalue λ.

In coordinate space the Faddeev equation is written as a differential equation. This is done by multiplying Eq. (7) by G_i^{-1}. The resulting equation is

$$[E - (H_0 + V_i + W_i)]|\Psi_i\rangle = (V_i + W_i)[|\Psi_j\rangle + |\Psi_k\rangle] \qquad (15)$$

where in coordinate space the kinetic energy operator is a differential operator. Introducing the N_c channel basis functions leads to N_c coupled differential equations for the channel wave functions. The energy E appears as a linear term in (15), and the equations can be rewritten as an eigenvlaue problem for the energy. However, for many channels the numerical calculations become prohibitive, and a procedure similar to the momentum space calculations is used.

NUMERICAL RESULTS

In Table I we give a brief summary of the results[11] for the Reid Soft Core[12] (RSC), Argonne[13] V_{14} (AV14), Super-Soft Core (C)[14] (SSC), Paris,[15] and Bonn[16] potential models. The results are given for various values of N_c, the number of channels for the truncated two-body potential.

TABLE I. Binding energies (in MeV) for ^{3}H calculated using various two-body potentials truncated to N_c channels. The experimental value is 8.48 MeV.

N_c	5	9	18	26	34
RCS	7.02	7.21	7.23	7.34	7.35
AV14	7.44	7.57	7.57	7.67	7.67
SSC	7.46	7.52	7.49	7.54	7.53
Paris[17]	7.30		7.38		
Bonn-r[18]					8.33
Bonn-q[19]	8.36	8.39	8.31		

All of the potentials except the Bonn underbind ^{3}H by approximately 1 MeV. From this table one can see that most of the binding energy comes from the first five channels, the projected potential with $j\alpha \leq 2$ and even parity. However, to obtain a result for the binding energy which is accurate to 10 keV, it is necessary to use 34 channels. To show the contribution to the binding energy from the various projected components of the two-body potential, we show in Tables II and III the matrix elements of V_j for $j\alpha \leq 6$ calculated with a 34-channel wave function. The terms with $j\alpha > 4$ represent the perturbation corrections to the binding energy. From these tables one can see that the higher states contribute approximately 10 keV additional binding energy.

In addition to the two-body forces which depend on the coordinates of two of the nucleons in the trinucleon system, there exist other forces which depend in a nontrivial manner on the coordinates of all three particles. The three-body forces are expected to be much weaker than the two-body interactions. The three-body force is not something that is made by nature; it results from the fact that in the traditional model of the trinucleon system some important

TABLE II. Potential energies (in MeV) for the RSC potential broken down according to j (the angular momentum of the interacting pair) and parity. The kinetic and total energies are also given.

j	0	1	2	3	4	5	6	Σ
$\langle V_j^+ \rangle$	-13.547	-43.859	-0.188	-0.116	-0.014	-0.009	-0.002	-57.735
$\langle V_j^- \rangle$	-0.176	0.227	-0.247	0.002	-0.006	0.000	0.000	-0.200
$\langle V_j \rangle$	-13.723	-43.632	-0.435	-0.114	-0.020	-0.009	-0.002	-57.935
$\langle T \rangle$								50.577
$\langle H \rangle$								-7.358

TABLE III. Potential energies (in MeV) for the AV14 potential broken down according to j (the angular momentum of the interacting pair) and parity. The kinetic and total energies are also given.

j	0	1	2	3	4	5	6	Σ
$\langle V_j^+ \rangle$	-13.590	-39.323	-0.134	-0.107	-0.008	-0.007	-0.001	-53.170
$\langle V_j^- \rangle$	-0.173	0.241	-0.242	0.004	-0.007	0.000	0.000	-0.177
$\langle V_j \rangle$	-13.763	-39.082	-0.376	-0.103	-0.015	-0.007	-0.001	-53.347
$\langle T \rangle$								45.670
$\langle H \rangle$								-7.677

degrees of freedom are suppressed. The neglect of these degrees of freedom can lead to the underbinding of the trinucleon system. Thus, there has been considerable interest recently in including a three-body force in the model Hamiltonian. In particular, the long-range effects of these degrees of freedom should be included; the short-range effects will be suppressed by the short-range repulsion of the two-body force. Consequently, most calculations have been concerned with pion and delta degrees of freedom.

There exist two different approaches to treating these sub-nuclear degrees of freedom. One approach is to treat explicitly the degrees of freedom associated with nonnucleonic components. This approach has been implemented by Hajduk and Sauer,[20] who include single Δ components in the wave function. Then the three-body forces mediated by a single Δ are implicit in the formalism. This model is much more complex because of the need to track the additional wave functions components, and the numerical solution of the model equations by Hajduk and Sauer is a calculational tour de force. This calculation finds that the introduction of the Δ-degrees of freedom leads to an additional 0.3 MeV binding energy. The second approach is to "freeze out" the nonnucleonic degrees of freedom and to include

them as separate three-body forces. This is the approach adopted by
the Tucson-Melbourne[21] and Brazil[22] groups. These groups have
focused on a two-pion exchange force between two nucleons as shown
in Fig. 2, the nonrelativistic reduction of this diagram yields the
two-pion exchange three-body potential. The addition of this
potential to the model Hamiltonian increases the binding energy by
approximately 1.5 MeV. The reason for the large difference in the
additional binding energy produced by the two approaches has not been
determined.

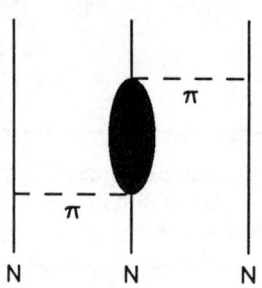

Fig. 2. Two-pion exchange three nucleon force.

To illustrate the effect of a three-body force, we give in Tables IV and V the results for the Tucson-Melbourne (TM) and the Brazil (BR) three-body forces used with the Reid Soft Core two-body potential. In addition, the variational bounds calculated using the wave function constructed from the Faddeev amplitudes are given. As noted above, the binding energy does not increase monotonically as the number of channels is increased; however, the variational bound does. These results show that 34 channels are required to obtain an accurate value
for the binding energy. Also, the addition of the three-body
potential has increased the binding energy by approximately 1.5 MeV.

TABLE IV. ^{3}H binding energies (MeV) for the RSC
plus the Tucson-Melbourne three-body force model and
the variational bounds.

N_c	5	9	18	34
$-E_F$	7.55	8.33	8.93	8.86
$-\langle H \rangle$	8.75	8.00	8.60	8.82

TABLE V. ^{3}H binding energies (MeV) for the RSC plus
the Brazil three-body force model and the variational bounds.

N_c	5	9	18	34
$-E_F$	7.66	8.77	8.70	8.89
$-\langle H \rangle$	8.28	8.60	8.72	8.90

Once we have obtained an accurate solution to the Faddeev
equations, we can use the resulting wave functions to calculate

additional properties of the bound-state system. Several examples are given in the next three sections.

COULOMB EFFECTS

Most investigations of the trinucleon bound state have assumed charge symmetry and charge independence. There are many mechanisms which can break this isospin symmetry. Primary among these is the Coulomb interaction between the two protons in ^{3}He. The mass difference between the mirror pairs ^{3}He and ^{3}H is a manifestation of this charge-symmetry breaking. After subtracting the mass difference between a free proton and a free neutron, a binding energy difference of 764 keV remains. It has been known for fifty years[23] that the bulk of this difference is due to the Coulomb interaction in ^{3}He; however, only recently has a qualitative understanding been achieved.[24-26] The effects of the Coulomb interaction can be studied by solving the Faddeev equations for the ^{3}He system both with and without the Coulomb potential between the protons. We have done this using several different realistic two-body and three-body nucleon-nucleon potentials.[27] Using the results of these calculations, we have generated a data set with many different binding energies. These data were then used to study the variation of the bound-state properties with the binding energy.

The charge form factors for the trinucleon systems, $F_{ch}(q^2)$, and the associated quantities derived from them, such as the charge density

$$\rho(r) = \frac{2}{\pi} \int_0^\infty \frac{\sin qr}{qr} F_{ch}(q^2) q^2 \, dq \qquad (16)$$

and the corresponding root-mean-square charge radius $\langle r^2 \rangle^{1/2}$, where

$$\langle r^2 \rangle = \int_0^\infty r^2 \, \rho(r) r^2 \, dr \qquad (17)$$

serve as tests for the quality of the wave function derived from the model Hamiltonian. The charge form factors are measured by electron elastic scattering at momentum transfer q, and the high-q values correspond to the short-range structure of the wave function. Thus, the magnitude of the secondary maxima and the position of the diffraction minima reflect the short-range correlations in the wave function. These features are also reflected in the behavior of $\rho(r)$ for small values of r. The charge radii are less sensitive functions of the short-range correlations, and we discuss these first.

In Fig. 3 we have plotted the root-mean-square charge radii for both ^{3}He and ^{3}H as a function of the binding energy for various potential models. While none of the model Hamiltonians reproduce the correct binding energy, the plot shows how the charge radii scale with the binding energy for realistic interactions. From this plot one can see that the charge radii exhibit a simple functional behavior, with only a small spread of values for a fixed value of the binding energy. The Coulomb energy shift is evident in the dashed curve, and the Coulomb barrier separates the two curves by approximately 0.04 fm. The Saclay results[28] are also shown on this figure (for E_B = 7.75 MeV, the binding energy of ^{3}He), and one can see that the Coulomb interaction is necessary to obtain agreement with the experimental results. If one assumes that the asymptotic part of the wave function dominates in the matrix element of r^2, then the rms radius should be inversely proportional to the square root of the binding energy. This is the case for the curves shown in Fig. 3. Thus, one would expect that since the Coulomb interaction decreases the binding energy, the charge radius should increase when a Coulomb potential is present. However, there is an additional effect due to the Coulomb barrier which acts to reduce the radius.

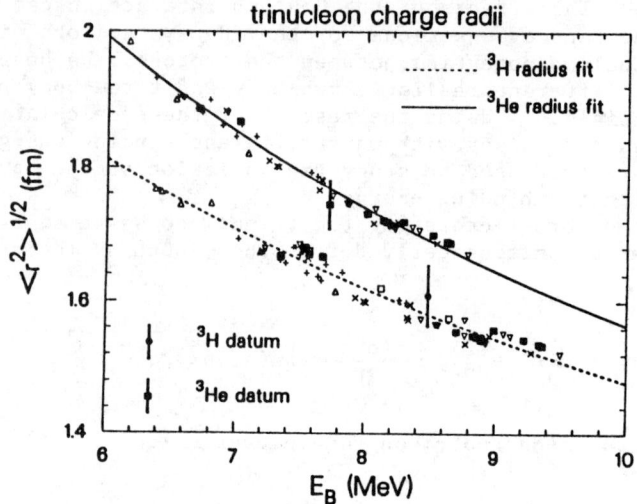

Fig. 3. Trinucleon rms charge radii versus binding energy for diverse potentials.

The numerical calculations for ^{3}He with and without the Coulomb can be used to test the accuracy of using first-order perturbation theory to calculate the Coulomb energy. This test shows that the perturbation approximation is valid, and we have plotted in Fig. 4 the Coulomb energies calculated with our model wave functions using a dipole proton form factor. The value of the Coulomb energy at the physical binding energy can be found from the fit to these model calculations, and the value is 652 keV. The estimated size of the second-order corrections is approximately 4 keV. Thus, the Coulomb interaction produces most of the binding energy difference. The remaining 100 keV must be due to other effects such as meson-exchange currents or relativistic effects.

A recurring problem for ^{3}He has been the charge form factor. The experimental behavior of the $F_{ch}(q^2)$ for large q values is

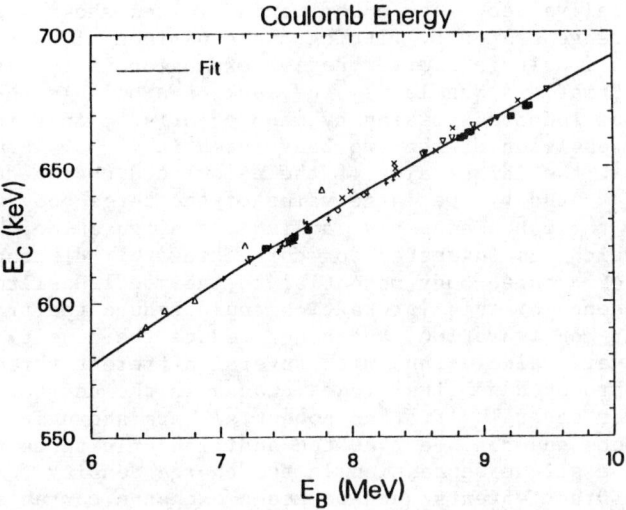

Fig. 4. ^{3}He Coulomb energy in first-order perturbation theory for the models in Fig. 3 as a function of the binding energy.

significantly different than the behavior predicted by the potential models for the trinucleon system. In configuration space this is reflected in the hole in the ^{3}He charge density which is not present in the model calculations. In Fig. 5 the point point-nucleon charge density calculated with the Reid Soft Core potential is compared with

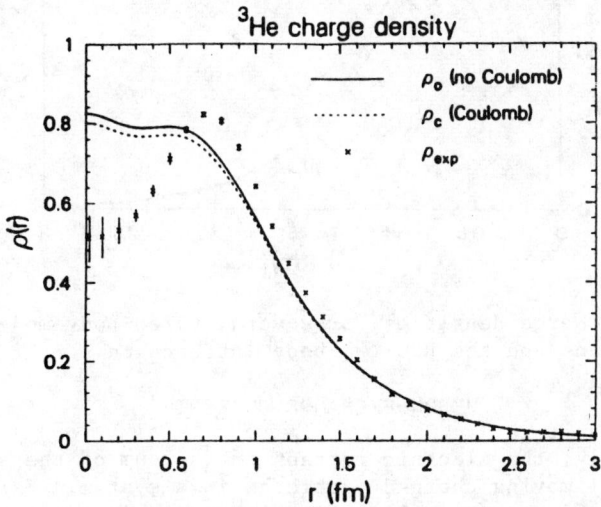

Fig. 5. The point-nucleon charge density for ^{3}He calculated using the RSC potential and the experimental results.

the experimental values. The experimental values shown with their
error bars were generated by fitting the experimental data for the
^{3}He form factors using a Fourier-Bessel expansion in the manner
described by Friar and Negele.[29] The lack of a hole in the theoretical density was found surprising by many theorists, in view of the
strong-range repulsion of the two-body potential. However, it has
been shown that the large value of the calculated charge density at
the origin is caused by the large value of the three-body wave function when the three nucleons have a linear configuration.[30] One
possibility which was suggested for correcting this discrepancy was
the addition of a three-body potential to the model Hamiltonian. The
angular dependence of this interaction could reduce the probability
for the linear configuration, and hence reduce the density near the
origin. However, calculations with several different three-body
interactions produced no significant change in the charge density.
The results for several different potentials are shown in Fig. 6.
From these plots one can see that the addition of a three-body force
only produces a slight depression in the charge density for small
values of r. Other effects such as meson-exchange currents are probably needed to reproduce the experimental data. These are discussed
in the next section.

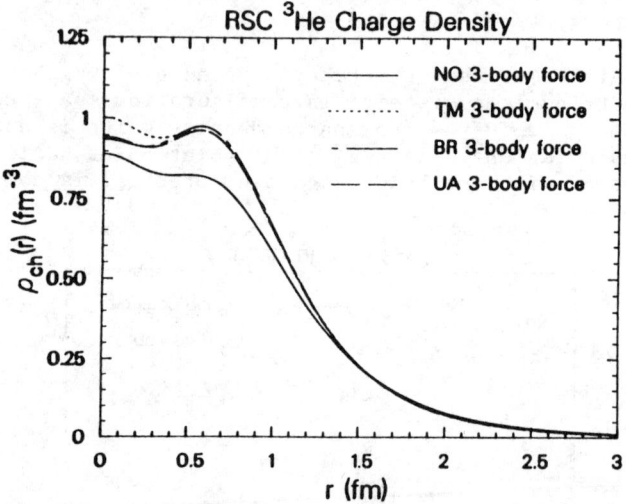

Fig. 6. Charge densities for several three-body model
Hamiltonians and the RSC two-body interaction.

MESON-EXCHANGE CURRENTS

Classically, the electric current is the sum of the separate
currents of all moving charged particles in a system. Failure to
keep track of all charges will obviously result in a lack of charge
or current conservation. Since the interaction between nucleons is
mediated by the exchange of mesons, it is expected that the nuclear

current is not simply a sum over the single-nucleon currents. The
distribution of charges and currents within a nucleus can be affected
substantially by the exchanged mesons; thus, in a nucleus we must
take into account the motion of virtual charged mesons as well as the
nucleons. Since both charge and angular momentum can be exchanged
between the nucleons via the mesons, one expects on semiclassical
grounds that the meson exchange currents will modify the static magnetic
moments of the ground states, as well as other properties of
the bound system. In this paper we consider only the magnetic moment
as an example of the effects of these currents. Calculations of
meson-exchange current corrections to the trinucleon magnetic moments
were first done in the early 1970's. One of the first papers was the
work of Harper, Kim, and Tubis[31] in 1972. Phenomenological wave
functions were used by Riska,[32] while Maise and Kim,[33] Strueve et
al.,[34] Torre and Goulard,[35] and Tomusiak et al.[36] all used wave
functions generated by Faddeev methods.

All of the papers mentioned above used perturbation theory for
the mesons, and they agreed that the dominant processes were those in
which a pion is exchanged between two nucleons. This result may seem
surprising at first, but it can be explained by the short-range repulsion
of the two-body nucleon-nucleon potential. Since this potential
is strongly repulsive for small separations of the nucleons, the
probability of finding nucleons close together is small. Therefore,
the matrix elements of short-range current operators are greatly suppressed,
and only the long-range operators should dominate. This is
illustrated in Fig. 7, which shows the two-body correlation function
$C_S(x)$, formed by integrating $|\Psi|^2$ over y and $\hat{x} \cdot \hat{y}$. One sees that
the maximum value of $C_S(x)$ falls between the one- and two-pion range
as indicated by the arrows on the diagram. The ρ-meson range corresponds
to very small values of $C_S(x)$, and consequently, the effects
of its current should be small.

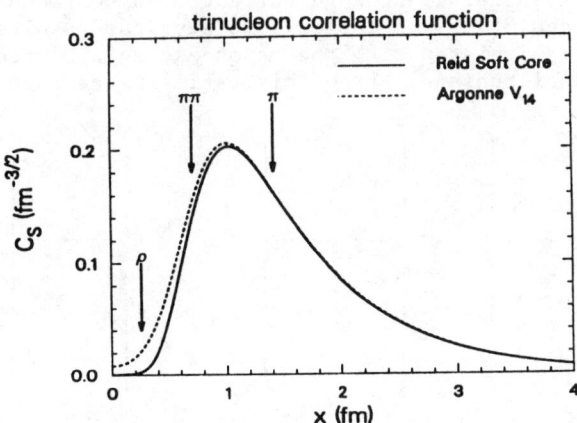

Fig. 7. Isoscalar trinucleon correlation function versus
the two-nucleon separation. The ranges for π, 2π, and ρ
exchange are marked by the arrows.

The need to include the meson-exchange currents is demonstrated in Table VI, where the results of the magnetic moments for ^{3}He and ^{3}H calculated in the impulse approximation are given for several different model interactions. In the impulse approximation, the currents arise only from the convection currents associated with the nucleons and the nucleon magnetization currents. The experimental values are also given in Table VI, and one can see that there is considerable disagreement between the model predictions and these values.

TABLE VI. Magnetic moments for the trinucleon system calculated using the impulse approximation and with the meson-exchange current (MEC) corrections added (in nuclear magnetons).

	Impulse		Impulse + MEC	
	$\mu(^3\text{He})$	$\mu(^3\text{H})$	$\mu(^3\text{He})$	$\mu(^3\text{H})$
RCS	-1.751	2.559	-2.051	2.859
AV14	-1.766	2.578	-2.063	2.875
RSC/TM	-1.753	2.555	-2.081	2.883
AV14/TM	-1.768	2.574	-2.093	2.899
Exp.			-2.128	2.979

The dominant meson-exchange corrections are due to the processes shown in Fig. 8. These diagrams show how a pion influences the electromagnetic interaction of a nucleus; the cross and wiggly lines denote an external electromagnetic interaction which produces a pion (photopion production) denoted by a dashed line that is then absorbed by a nucleon. The process in (a) is the pair term, and (b) is the true-exchange term. The process in (c) involves the virtual excitation of the Δ-isobar, and it is called the isobar term. The derivation of the corresponding exchange currents can be found in any of several references.[37,38] In Table VI we give the results for the magnetic moment calculated with the meson-exchange corrections for several model Hamiltonians. From this table one can see that the

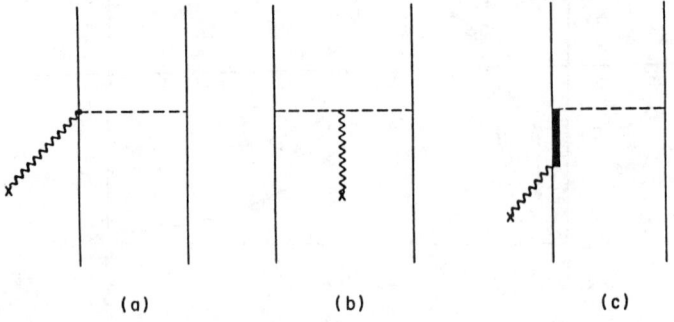

Fig. 8. (a) Seagull or pair diagram; (b) pion (true exchange) current diagram; (c) Δ-isobar diagram.

contributions of these currents are not negligible, and that with the addition of these terms the agreement with the experimental values is improved considerably.

ASYMPTOTIC NORMALIZATION

The trinucleon asymptotic normalization constants provide direct evidence for the existence of tensor forces, and physically they echo the internal dynamics of the wave function through the overall normalization. The asymptotic normalization constants are defined such that their value is unity when the effective interaction in the asymptotic channel is zero range.[39] Using the Jacobi coordinates, shown in Fig. 1, the asymptotic ^{3}H wave function is written as:

$$\Psi(\vec{x}_1,\vec{y}_1) \xrightarrow[y_1 \to \infty]{} C_S N \frac{e^{-\beta y_1}}{y_1} \{[Y_0(\hat{y}_1) \times \chi^{1/2}(1)]^{[1/2]} \times \phi^{[1]}(\vec{x}_1)\}^{[1/2]} \frac{\eta}{\sqrt{2}}$$

$$+ C_D N \frac{e^{-\beta y_1}}{y_1} \left(1 + \frac{3}{\beta y_1} + \frac{3}{\beta^2 y_1^2}\right)$$

$$\times \{[Y_2(\hat{y}_1) \times \chi^{1/2}(1)]^{[3/2]} \times \phi^{[1]}(\vec{x}_1)\}^{[1/2]} \frac{\eta}{\sqrt{2}} \quad (18)$$

where β is given in terms of the three-body binding energy, B_3, and deuteron binding energy, B_2, as $\beta^2 = 4M[B_3 - B_2]/(3\hbar^2)$. $N = \sqrt{2\beta}$ is the zero-range normalization, $\chi^{1/2}$ is the spin-1/2 function, $\phi^{[1]}$ is the deuteron wave function, and η is the isospin-1/2 function for three particles in which particles 2 and 3 (the deuteron) are coupled to isospin 0. For ^{3}He, one replaces $e^{-\beta y_1}$ by the Whittaker function $W_{-\kappa,1/2}(2\beta y_1)$, replaces $e^{-\beta y_1}[1 + 3/(\beta y_1) + 3/(\beta^2 y_1^2)]$ by $W_{-\kappa,5/2}(2\beta y_1)$, and replaces N by

$$N_W = \left[\frac{\frac{1}{2}\Gamma(3+\kappa)\Gamma(2+\kappa)}{{}_3F_2(\kappa,2,1+\kappa;3+\kappa,2+\kappa;1)}\right]^{1/2} \quad (19)$$

where $\kappa = 2Me^2/(3\hbar^2\beta)$.

In Fig. 9 we plot the S-wave asymptotic normalization constant, C_S, calculated with the wave functions for our various model Hamiltonians. From this plot one can see that the value of C_S has a simple scaling relationship with the binding energy. A similar plot for the D-wave asymptotic normalization constant, C_D, is shown in Fig. 10. Finally, Fig. 11 shows the ratio $\eta = C_S/C_D$ calculated for the various model Hamiltonians, together with the experimental value. Once again, we find that for realistic potentials, the traditional model will reproduce the experimental value if the model has the correct binding energy. All of the two-body potentials used for this plot have the one-pion tail, and thus they have a strong tensor force. This is necessary in order to fit the D-state asymptotic normalization.

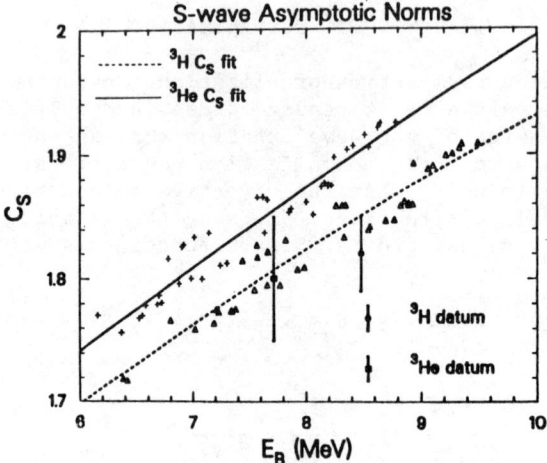

Fig. 9. The S-wave asymptotic normalization versus the binding energy for various potential models.

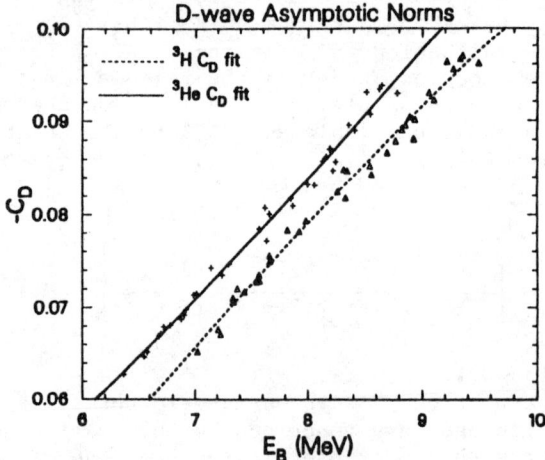

Fig. 10. The D-wave asymptotic normalization versus the binding energy for various potential models.

CONCLUSIONS

We have now reached the point where the Faddeev equations can be solved "exactly" for the bound-state three-body problem, either in momentum space or in coordinate space. These solutions have become

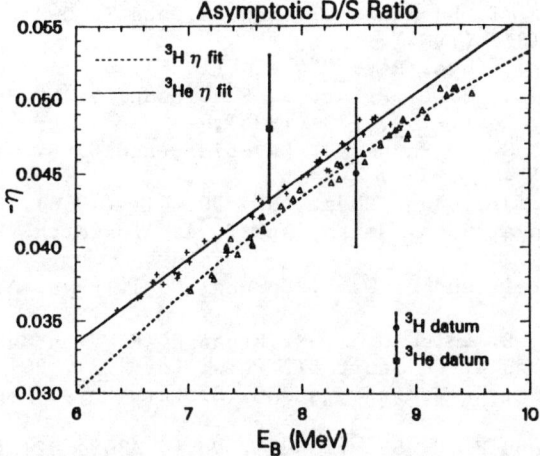

Fig. 11. The D/S ratio of the asymptotic normalization constants versus the binding energy for various potential models. The experimental results are also shown.

an important tool which can be used to test our understanding of the nuclear system. Using these "exact" wave functions, one can reproduce many of the experimental properties of the trinucleon bound-state system. However, there exist data which depend upon the short-range correlations that are not reproduced by the traditional models. Including the effects of meson-exchange currents can remove many of the discrepancies, but the question of whether one must include other subnuclear degrees of freedom such as quarks is still not answered. Only through more detailed calculations can we determine the need for including these degrees of freedom in our model calculations.

REFERENCES

1. J. Carlson, V. R. Pandharipande, and R. B. Wiringa, Nucl. Phys. A401, 59 (1983).
2. C. Ciofi delgi Atti and S. Smith, Phys. Rev. C32, 1090 (1984).
3. J. L. Ballot and M. de la Ripelle Fabre, Ann. Phys. (NY) 127, 62 (1980).
4. W. Glöckle, The Quantum Mechanical Few-Body Problem (Springer-Verlag, New York, 1983), and references therein.
5. E. P. Harper, Y. E., Kim, and A. Tubis, Phys. Rev. Lett. 28, 1533 (1972); W. Glöckle, G. Hasberg, and A. R. Neghabian, Z. Phys. A300, 217 (1982).
6. C. Hadjuk and P. U. Sauer, Nucl. Phys. A322, 329 (1979).
7. S. P. Merkuriev, C. Gignoux, and A. Laverne, Ann. Phys. (NY) 99, 30 (1976); G. L. Payne, J. L. Friar, B. F. Gibson, and I. R. Afnan, Phys. Rev. C22, 823 (1980).

8. S. Ishikawa, T. Sasakawa, T. Sawada, and T. Ueda, Phys. Rev. Lett. 53, 1877 (1984).
9. A. Bömelburg, Phys. Rev. C34, 14 (1986).
10. R. B. Wiringa, J. L. Friar, B. F. Gibson, G. L. Payne, and C. R. Chen, Phys. Lett. 14B, 273 (1984).
11. C. R. Chen, G. L. Payne, J. L. Friar, and B. F. Gibson, Phys. Rev. C33, 1740 (1986).
12. R. V. Reid, Jr., Ann. Phys. (NY) 50, 411 (1968).
13. R. B. Wiringa, R. A. Smith, and T. A. Ainsworth, Phys. Rev. C29, 1207 (1984).
14. R. de Tourreil and D. W. L. Sprung, Nucl. Phys. A201, 193 (1973).
15. M. LaCombe, B. Loiseau, J. M. Richard, R. Vinh Mauh, J. Cote, P. Pires, and R. de Tourreil, Phys. Rev. C21, 861 (1980).
16. R. Machleidt, K. Holinde, and C. Elster, Phys. Rep., to be published.
17. C. Hadjuk and P. U. Sauer, Nucl. Phys. A369, 321 (1981).
18. T. Sasakawa, Nucl. Phys. A463, 327c (1987).
19. R. A. Brandenburg, G. S. Chulick, R. Machleidt, A. Picklesimer, and R. M. Thaler, Los Alamos Report LA-UR-86-3700.
20. C. Hadjuk and P. U. Sauer, Nucl. Phys. A405, 581 (1983).
21. S. A. Coon, M. D. Scadron, P. C. McNamee, B. R. Barrett, D. W. E. Blatt, and B. B. J. McKeller, Nucl. Phys. A317, 242 (1979).
22. H. T. Coelho, T. K. Das, and M. R. Robilotta, Phys. Rev. C28, 1812 (1983).
23. H. Bethe and R. F. Bacher, Rev. Mod. Phys. 8, 82 (1936).
24. M. Fabre de la Ripelle, Fizika 4, 1 (1972).
25. R. A. Brandenburg, S. A. Coon, and P. U. Sauer, Nucl. Phys. A294, 305 (1978).
26. J. L. Friar and B. F. Gibson, Phys. Rev. C18, 908 (1978).
27. J. L. Friar, B. F. Gibson, and G. L. Payne, Phys. Rev. C, to be published.
28. F.-P. Juster et al., Phys. Rev. Lett. 55, 2261 (1985).
29. J. L. Friar and J. W. Negele, Adv. Nucl. Phys. 8, 219 (1975).
30. J. L. Friar, B. F. Gibson, E. L. Tomusiak, and G. L. Payne, Phys. Rev. C24, 665 (1981).
31. E. P. Harper, Y. E. Kim, A. Tubis, and M. Rho, Phys. Lett. 40B, 533 (1972).
32. D. O. Riska, Nucl. Phys. A350, 227 (1980).
33. M. A. Maize and Y. E. Kim, Nucl. Phys. A420, 365 (1984).
34. W. Strueve, Ch. Hadjuk, and P. U. Sauer, Nucl. Phys. A405, 620 (1983).
35. J. Torre and B. Goulard, Phys. Rev. C28, 529 (1983).
36. E. L. Tomusiak, M. Kimura, J. L. Friar, B. R. Gibson, and G. L. Payne, Phys. Rev. C32, 2075 (1985).
37. M. Chemtob and M. Rho, Nucl. Phys. A163, 1 (1971).
38. J. L. Friar, Ann. Phys. (NY) 104, 380 (1977).
39. J. L. Friar, B. F. Gibson, D. R. Lehman, and G. L. Payne, Phys. Rev. C25, 1616 (1982).

MONTE CARLO TECHNIQUES FOR GROUND-STATE PROPERTIES

K.E. Schmidt
Department of Chemistry and Courant Institute of Mathematical Sciences,
New York University, New York, NY. 10003

ABSTRACT

Monte Carlo calculations of the ground-state properties of various few-body systems are described. The variational Monte Carlo method is used to calculate upper bounds to the ground-state energy. The resulting variational wave functions can be used to calculate other properties such as the density profile and potential and kinetic energies. Typically, the variational trial functions are also used as the initial and guiding functions in the Green's function Monte Carlo (GFMC) method, which can solve for the ground-state of the Schroedinger equations with only statistical errors. The GFMC method is described and its intimate relationship with path integral techniques shown. The fermion problem is discussed. Results for variational and GFMC calculations in ^{5}He as well as various atomic systems are discussed.

INTRODUCTION

I will briefly describe the variational and Green's function Monte Carlo methods[1,2], and give some results of their application to few-body systems. Monte Carlo methods have been applied to numerous few-body problems including atoms and molecules[3], nuclei[4], and quark models of hadrons[5]. I will describe some results for light nuclei and for some atoms and molecules.

The Hamiltonian used is for the nonrelativistic Schroedinger equation,

$$H = -\sum_i \frac{\hbar^2}{2m_i} \nabla_i^2 + V, \qquad (1)$$

where V is the potential which can have spin-isospin etc. dependence for a realistic nuclear model. For atomic problems V is the coulomb interaction of the electrons and nuclei.

THE VARIATIONAL METHOD

The variational method uses the well known property,

$$E_0 \leq \frac{<\Psi_T|H|\Psi_T>}{<\Psi_T|\Psi_T>} \equiv <H>, \qquad (2)$$

where E_0 is the exact ground-state energy and Ψ_T is any trial function of the appropriate symmetry. For purely central potentials, the spin-isospin variables can be easily summed over leaving only a spatial dependence,

© American Institute of Physics 1987

$$\langle H\rangle = \frac{\int dR\Psi_T^*(R)H\Psi_T(R)}{\int dR\Psi_T^2(R)}$$

$$= \int dR\left[\frac{\Psi_T^*(R)H\Psi_T(R)}{\Psi_T^2(R)}\right]\frac{\Psi_T^2(R)}{\int dR\Psi_T^2(R)}. \qquad (3)$$

Here R stands for the 3A coordinates of the A particles, and for fermions $\Psi_T(R)$ is antisymmetric under interchange of coordinates of like spin-isospin particles.

Eq. 3 is in the form

$$\langle H\rangle = \int dR F(R)P(R), \qquad (4)$$

where

$$P(R) = \frac{\Psi_T^2(R)}{\int dR\Psi_T^2(R)}. \qquad (5)$$

and

$$F(R) = \frac{\Psi_T^* H\Psi_T(R)}{\Psi_T^2(R)}. \qquad (6)$$

P(R) can be interpreted as a normalized probability distribution and $\langle H\rangle = \bar{F}$ the average of F(R) under this probability distribution. The Metropolis et. al. method[6] is then normally used to calculate $\langle H\rangle$. The error bar is given as usual by $\frac{\sigma}{\sqrt{N}}$ where N is the number of independent samples, and

$$\sigma^2 = \overline{F^2} - \bar{F}^2 \qquad (7)$$

is the variance of F(R). For $\Psi_T = \Psi_0$ the exact ground-state wave function, the variance is zero since F(R) is independent of R. For good trial functions the variance of the energy is small. Once a good trial function is known, other expectation values can be easily calculated by substituting the appropriate operator for H in Eq. 3. Typically, the variance is larger for these quantities, but not unreasonably so.

The calculations for spin-isospin dependent potentials are more complicated since good trial functions must include the effects of the dependence of the potential. A general form for the trial function is $\Psi_T(R,S)$ where S indicates the spin-isospin states of the A particles. The expectation value of H is then given by

$$\langle H\rangle = \frac{\sum_{S,S'}\int dR\Psi_T^*(R,S)H_{S,S'}\Psi_T(R,S')}{\sum_S \int dR\Psi_T^*(R,S)\Psi_T(R,S)} \qquad (8)$$

Typically, the spin-isospin sums are done explicitly on the computer, and the R integral calculated as before using the Monte Carlo method. Eq. 8 is rewritten as

$$<H> = \int dR \left[\frac{\sum_{S,S'} \Psi_T^*(R,S) H_{S,S'} \Psi_T(R,S')}{\sum_S \Psi_T^*(R,S) \Psi_T(R,S)} \right] P(R) \quad (9)$$

with

$$P(R) = \frac{\sum_S \Psi_T^*(R,S) \Psi_T(R,S)}{\int dR \sum_S \Psi_T^*(R,S) \Psi_T(R,S)}, \quad (10)$$

and the integrals calculated as in the purely central case.

I will show some results of the application of the variational method to α–n scattering or ^{5}He[7]. The results are given with the Urbana v14 potential[8] and the effect of the Urbana model V three nucleon interaction is shown[9]. A major reason why ^{5}He is interesting is our desire to calculate the properties of weakly bound p-shell nuclei like ^{6}Li and ^{7}Li. Accurate calculation of these nuclei will only be possible if our model can accurately predict the properties of ^{5}He.

The scattering phase shifts are extracted from a variational calculation by solving for the ground-state with modified boundary conditions that the wave function go to zero if the α–n separation $\geq R_n$.[7,10] If the α–n interaction is zero at this separation, the phase shift in the angular momentum l channel is

$$\tan \delta_l = \frac{j_l(kR_n)}{n_l(kR_n)}, \quad (11)$$

where j_l and n_l are spherical bessel functions and

$$k = [2m_R E_{sep}/\hbar^2]^{\frac{1}{2}}. \quad (12)$$

Here m_R is the reduced mass of the α–n system, and

$$E_{sep} = E_5 - E_\alpha - E_n, \quad (13)$$

is the separation energy.

The trial wave functions for ^{4}He and ^{5}He are given by

$$\Psi_T = S \prod_{i<j} F_{ij} \Phi, \quad (14)$$

where S is a symmetrizing operator which operates on the correlation operators F_{ij}, which in general do not commute, and Φ is a model wave function of one-body states coupled to the correct angular momentum.

$$F_{ij} = f_{ij}^c [1 + u_3 \sum_p u_{ij} O_{ij}^p] \quad (15)$$

where u_3 is a small three-body correlation, f_{ij}^c is a central Jastrow correlation that keeps the particles apart, and u_{ij}^p are operator correlations. The subscripts ij indicate as usual that they are functions of the interparticle distance. The O_{ij} are 1, $\sigma_i \cdot \sigma_j$, and the tensor operator as well as $\tau_i \cdot \tau_j$ times these operators. Details of the calculation of the F_{ij} are given in reference 7.

A correlated sampling technique is employed to calculate the separation energy. Particularly when the α and n are well separated, the variance of the α particle energy dominates the variance of both the α particle and ^{5}He. Correlated sampling of the integrals allows some of this variance to be canceled.

Suppressing spin sums, the E_α and E_5 are written as

$$E_\alpha = \frac{\int dr_1 dr_2 dr_3 dr_4 dr_5 \Psi_\alpha^* H_\alpha \Psi_\alpha G(r_{5\alpha})}{\int dr_1 dr_2 dr_3 dr_4 dr_5 \Psi_\alpha^* \Psi_\alpha G(r_{5\alpha})}, \tag{16}$$

and

$$E_5 = \frac{\int dr_1 dr_2 dr_3 dr_4 dr_5 \Psi_5^* H_5 \Psi_5}{\int dr_1 dr_2 dr_3 dr_4 dr_5 \Psi_5^* \Psi_5}. \tag{17}$$

Defining

$$P(R) = \frac{\Psi_5^* \Psi_5}{\int dr_1 dr_2 dr_3 dr_4 dr_5 \Psi_5^* \Psi_5}, \tag{18}$$

gives

$$E_\alpha = \frac{\int dr_1 dr_2 dr_3 dr_4 dr_5 \left[\frac{\Psi_\alpha^* H_\alpha \Psi_\alpha G(r_{5\alpha})}{P(R)} \right] P(R)}{\int dr_1 dr_2 dr_3 dr_4 dr_5 \left[\frac{\Psi_\alpha^* \Psi_\alpha G(r_{5\alpha})}{P(R)} \right] P(R)}, \tag{19}$$

and

$$E_5 = \int dr_1 dr_2 dr_3 dr_4 dr_5 \left[\frac{\Psi_5^* H_5 \Psi_5}{\Psi_5^* \Psi_5} \right] P(R). \tag{20}$$

By sampling from P(R) and using the same set of configurations for the calculations of E_α and E_5, the variance of the difference in the energies is reduced.

Results for the calculated energy n the j=3/2 and j=1/2 states are shown in Table I for a spacing of 7.5 fm. As can be seen, the variance is lower for the energy difference than for the α particle alone. Without the correlated subtraction it would be more than twice as large as shown. The breakdown in the energy is among the various potential terms is also shown.

Table I. The energy and various components for ^{5}He and ^{4}He Calculated at R_n=7.5fm using the variational Monte Carlo method.

state	<H>	two–body	2–πthree–body	three–body
^{4}He	−29.6±0.3	−136.0±2.0	−10.4±0.4	3.9±0.2
^{5}He(J=3/2)	4.2±0.2	−14.0±1.2	−0.9±0.2	0.5±0.1
^{5}He(J=1/2)	6.5±0.2	−6.0±1.0	−0.2±0.1	0.2±0.1

Table II shows the calculated separation energy and resulting phase shifts for various separations as well as a comparison with experimental results.[11] As can be seen, the j=3/2 values are 1-2 Mev removed from experiment. This discrepancy will have to be removed before accurate 6 and 7 body nuclei calculations can be performed.

Table II. The ernergy and calculated phase shift with comparison with experiment for various values of R_n.

J	R(fm)	E_{MC}(MeV)	E_{EXPT}(MeV)	δ_{MC}(deg)	δ_{EXPT}(deg)
1/2	7.5	6.5±0.2	5.5	42	57
1/2	10.0	3.9±0.1	3.7	33	40
1/2	12.5	2.9±0.1	2.9	18	30
3/2	7.5	4.2±0.2	2.2	80	125
3/2	10.0	2.9±0.1	1.8	63	128
3/2	12.5	2.1±0.1	1.2	45	100

THE GREEN'S FUNCTION MONTE CARLO METHOD

The Green's function Monte Carlo (GFMC) method allows us to improve upon the variational method just given. It can integrate the Schroedinger equation numerically with only statistical errors. It corresponds to converting the time dependent Schroedinger equation

$$H\Phi = -\frac{\partial \Phi}{i \partial t}, \qquad (21)$$

to imaginary time $\tau=it$,

$$H\Psi = -\frac{\partial \Psi}{\partial \tau}. \tag{22}$$

Eq 22 is a diffusion equation in 3A dimensions; it is easy to simulate diffusion on a computer. The solution for $\tau \to \infty$ is formally

$$\Psi(\tau) = e^{-H\tau}\Psi(0). \tag{23}$$

Expanding $\Psi(0)$ in terms of the eigenstates of H gives

$$\Psi(0) = \sum_n a_n |n\rangle, \tag{24}$$

and substituting into Eq 23,

$$\Psi(\tau) = e^{-E_0\tau}[a_0|0\rangle + e^{-(E_1-E_0)}|1\rangle + \cdots] \tag{25}$$

which converges to the lowest state with nonzero a_n. The GFMC method also has a time integrated form with a Green's function given by

$$\frac{1}{H+E_C} = \int_0^\infty d\tau \exp(-[H+E_C]\tau), \tag{26}$$

where E_C is a constant added to make the spectrum of $H+E_C$ positive. The GFMC equation is then

$$|\psi^{(n)}\rangle = \frac{E_T+E_C}{H+E_C}|\psi^{(n-1)}\rangle \tag{27}$$

where E_T is a trial energy. Eq. 27 also iterates to $|0\rangle$ for large n.

To calculate Eq. 27, using the Monte Carlo method, $\frac{E_T+E_C}{H+E_C}$ needs to be sampled. Details of the sampling methods can be found in references 13 and 14. The basic method is to write the identity

$$\frac{1}{H+E_C} = \frac{1}{H_U+E_C} + \frac{1}{H+E_C}(H-H_U)\frac{1}{H_U+E_C}. \tag{28}$$

H_U is a Hamiltonian (or more precisely, a set of Hamiltonians) with a known Green's function $\frac{1}{H_U+E_C}$. Typically, we choose it to be the Hamiltonian for a constant potential inside a domain around the current position such that $H_U > H$. This makes all terms in Eq. 28 positive. They can then be interpreted as probabilities and the infinite sum implied by Eq. 28 can be sampled using the Monte Carlo method. I should note that for E_C large, the first term dominates Eq. 28. Taking this limit and dropping the higher order terms corresponds to the short time approximation that is often used.

In R-space, Eq. 27 becomes

$$\psi^{(n+1)}(R)=(E_T+E_C)\int dR' G(R,R')\psi^{(n)}(R'), \qquad (29)$$

with

$$G(R,R')=\langle R|\frac{1}{H+E_C}|R'\rangle, \qquad (30)$$

and the integrals are done using the Monte Carlo method.

The GFMC method is simply related to a path integral as can be seen by calculating the expectation value of an operator O,

$$\frac{\langle\psi^{(n)}|O|\psi^{(n)}\rangle}{\langle\psi^{(n)}|\psi^{(n)}\rangle}=\frac{\int\prod_{k=1}^{2n+2}dR_k\psi^{(0)}(R_1)\prod_{i=1}^{n}G(i,i+1)O\prod_{j=n+1}^{2n+2}G(j,j+1)\psi^{(0)}(R_{2n+2})}{\int\prod_{k=1}^{2n+2}dR_k\psi^{(0)}(R_1)\prod_{j=1}^{2n+2}G(j,j+1)\psi^{(0)}(R_{2n+2})} \qquad (31)$$

where R_i specify the path and $G(i,j)=G(R_i,R_j)$. The iterative scheme used by the standard GFMC method, Eq. 29, corresponds to moving step by step along the path rather than sampling and storing the whole path. The iteration of Eq. 28 corresponds to correcting an element of the path by inserting intermediate points in the path. The whole GFMC method can be formulated in terms of an exact path integral method, however this does not seem to be particularly useful for ground-state properties. It may be useful at temperatures greater than zero.

Before giving some results of the GFMC method, I must mention the problems associated with calculating fermion properties. These are discussed in detail in reference 2. Monte Carlo methods interpret integrals and sums as probability densities times a function to average. The probability density necessarily must be positive definite. The function to average should be as nearly constant as possible for low variance. Unfortunately, in fermi problems as well as for any excited state of the Hamiltonian, the wave function must necessarily change sign and therefore cannot be interpreted as a probability. If the sign is included in the function to average, a large variance results. Note that there is no analytical problem with fermions since Eq. 25 clearly shows that given a purely antisymmetric starting wave function, the GFMC method will iterate to the correct fermion ground state. The problem is that in a naive application of the Monte Carlo method, the variance grows exponentially with the iteration number. It is sometimes possible to get useful numbers from such a transient estimation scheme[2] before the variance grows to large. The more usual case is to run out of computing power before the result has converged.

This growth can be understood from the diffusion analogy. Sample a set of walker positions from the absolute value of an antisymmetric wave function. As the walkers diffuse the positive walkers will diffuse into the neighborhood of the negative walkers. In principle, their contributions should cancel. They do cancel in the averages, but in the variance they add. The signal dies with respect to the noise as the equations are iterated. For a few particles it is possible to identify walkers that cancel and drop them from the calculation of both the averages and the variance.[2,12] However, as the number of particles increases, the density of walkers is small in configuration space and no significant cancellation occurs.

A simple expedient is to employ an approximate solution. One simple approximation is the fixed node approximation. This corresponds to solving the Schroedinger equation with the constraint that the nodes of the wave function are the same as those of a trial function. Since the nodes for the ground state of a one-dimensional system are given by the positions where the fermions overlap, the fixed node approximation is exact in one dimension. In higher dimensions, there are additional nodes. The fixed node approximation gives an upper bound to the ground-state energy corresponding to the wave function with the lowest energy given the nodes of the trial function.

In Tables III-V give some representative results of the GFMC method for some atoms and molecules.[3] Table III gives the results for the Be atom. This is a severe test of the fixed node approximation because of the near degeneracy of the 2s-2p orbitals. The trial functions used are self consistent field orbitals times a Jastrow factor for the single ζ[15], and RSW[16] results. Clearly, the nodal structure of these wave functions is inadequate. Results with the inclusion of additional determinants which mix both the 2s and 2p orbitals are shown as GPA[23] and MCSCF[24]. These improved wave functions which contain only a few determinants give 100 percent of the correlation energy within error bars.

Table III. The electronic energy and correlation energy of Be computed by the GFMC method. Energy in hartree.

Ψ_{trial}	E	σ	E_{corr}	% $_{corr}$
single-ζ	−14.640	0.006	0.067	71±9
RSW	−14.652	0.003	0.079	84±4
GPA	−14.665	0.002	0.092	98±2
MCSCF	−14.667	0.002	0.094	100±2
Expt	−14.6674	...	0.0944	100

The energies of the first two excited states of Be are shown in Table IV. The ground state was represented by the MCSCF form above. For the excited states a two determinant form with the correct symmetry was constructed from single ζ orbitals.

Table IV. The excitation energy of the Be atom computed by the GFMC method. The standard deviation (σ) refers to the largest value for either state of the transition. Energy in eV.

Transition	E_{GFMC}	σ	E_{Expt}
$^1S \rightarrow {}^3P$	2.90	0.08	2.73
$^1S \rightarrow {}^1P$	5.42	0.08	5.28
$^3P \rightarrow {}^3P$	2.52	0.08	2.55

Some preliminary results on the water molecule are shown in Table V. The trial wave function was taken from reference 19. They are compared with CEPA calculations from reference 20, and experimental values.[21]

Table V. The electronic energy and correlation energy of water (R_{OH}=1.81 bohr, theta=104.45) computed by the GFMC method. SCF-I and SCF-II refer to the minimal Slater and double-zeta for oxygen 2p functions basis of reference 19. Energy in hartree.

ψ_{trial}	E	σ	E_{corr}	$\%_{corr}$
SCF-I	−76.238	0.040	0.170	46±24
SCF-II	−76.402	0.015	0.335	91±4
CEPA	−76.3683	...	0.301	81
Expt.	−76.4376	...	0.370	100

CONCLUSION

The variational and GFMC methods been described, and some representative results given for few-body systems. Both methods have the virtue of giving upper bounds to the energy with only statistical errors. The GFMC method can give exact results for spatially symmetric systems. The major task ahead is to find methods to extract useful quantities from these calculations with low variance.

ACKNOWLEDGEMENTS

This work was supported by the United States Department of Energy under contract DE-AC02-79ER10353. Some of the results shown were obtained in collaboration with M.H. Kalos, J.W. Moskowitz, and J. Carlson whom I wish to thank.

REFERENCES

1. D.M. Ceperley and M.H. Kalos, in *Monte Carlo Methods in Statistical Physics,* ed. by K. Binder, Topics in Current Physics, 7, (Springer, Berlin, Heidelberg, New York, 1979).
2. K.E. Schmidt and M.H. Kalos, in *Applications of the Monte Carlo Methods in Statistical Physics,* ed. by K. Binder, Topics in Current Physics, 36, (Springer, Berlin, Heidelberg, New York, 1984).
3. J.W. Moskowitz, K.E. Schmidt, M.A. Lee, M.H. Kalos, J. Chem. Phys. 76, 1064 (1982); J. W. Moskowitz and K.E. Schmidt, J. Chem. Phys. 85, 2868 (1986); J.B. Anderson, J. Chem. Phys. 63, 1499 (1975); 65, 4121 (1976); 73, 3897 (1980); P.J. Reynolds, D.M. Ceperley, B.J. Alder, and W.A. Lester, Jr., *ibid.* 77, 5593 (1982).
4. J. Lomnitz-Adler, V.R. Pandharipande, and R.A. Smith, Nucl. Phys. A361, 399 (1981).
5. T.J. Goldman, K.E. Schmidt and G.J. Stephenson, Jr., AIP Conference Proceedings 150underline, (1986).
6. N. Metropolis, A.W. Rosenbluth, H. Rosenbluth, A. Teller, and E. Teller, J. Chem. Phys. 21, 1087 (1953).
7. J. Carlson, K.E. Schmidt, and M.H. Kalos, Condensed Matter Theories 1, Ed. by F.B. Malik (Plenum, New York, 1986); J. Carlson, K.E. Schmidt, and M.H. Kalos, to be published in Phys. Rev. C.
8. I.E. Lagaris and V.R. Pandharipande, Nucl. Phys. A355, 331 (1981).
9. J. Carlson, V.R. Pandharipande, and R.B. Wiringa, Nucl. Phys. A401, 59 (1983).
10. J. Carlson, V.R. Pandharipande, and R.B. Wiringa, Nucl. Phys, A424, 427 (1984).
11. R. A. Arndt and L.D. Roper, Nucl. Phys. A209, 447 (1973).
12. D. Arnow, M.H. Kalos, M.A. Lee, and K.E. Schmidt, J. Chem. Phys. 77, 5562 (1982).
13. M.H. Kalos, D. Levesque, and L. Verlet, Phys. Rev. A9, 2178 (1974).
14. K.E. Schmidt, in Lectures Notes in Physics 273, 363 (1987).
15. E. Clementi and D. L. Raimondi, J. Chem. Phys. 38, 2686 (1963).
16. C. C. J. Roothaan, L. M. Sachs, and A. W. Weiss, Rev. Mod. Phys. 32, 186 (1960).
17. R. McWeeny and B. T. Sutcliffe, Proc. R. Soc. London Ser. A273, 103 (1963).

18. E. Clementi, "Atomic Energy Tables" Supplement to IBM J. of Res. Develop. 9, 2 (1965).
19. S. Aung, R. M. Pitzer, and S. I. Chan, J. Chem. Phys. 49, 2071 (1968).
20. W. Meyer and P. Rosmus, J. chem. Phys. 63 , 2356 (1975).
21. B. J. Rosenberg and I. Shavitt, J. Chem. Phys. 63, 2162 (1975).

QUARKS AND NUCLEON-NUCLEON CORRELATIONS

Gerald A. Miller
University of Washington, Seattle, WA. 98195

INTRODUCTION

From the title you can see that I am taking a different direction than the previous speakers. Profs. Payne and Schmidt solve the Schroedinger equation. The forces between nucleons are assumed to be known.

Working with quarks is different. Guesswork is required. This is true even though Quantum Chromodynamics (QCD) is THE theory of the strong interaction. In this theory quarks and gluons are somehow responsible for nuclear binding. The problem is, we really don't know how to use quarks to compute nuclear properties. We simply don't know the forces. Thus quark-nuclear physics is very different from what you have heard here so far.

Furthermore, it is very well-known that nuclei are made up mainly of neutrons and protons. There are many examples of nucleonic successes: the shell model and reactions such as (d,p) in which a neutron is put into the nucleus are just two.

We should remember that the neutron-proton picture of the nucleus describes mainly single nucleon processes. In these the nucleon may be regarded as isolated. In that case, the quark composition of the proton is not relevant.

However, the nucleus does have many nucleons in continuous motion. Sometimes two nucleons overlap. In that case the quark structure might be relevant. Our approach[1,2,3] is to use quarks to describe close encounters of two nucleons. Thus we use quarks to describe the nucleon-nucleon correlation function.

The ultimate goal of such work is to employ the nucleus as a tool to learn about or verify theories of quark confinement. This is a possible hope because the nucleus provides a wide variety of examples and processes not available with single hadrons.

OUTLINE

Here is the outline of the rest of the talk.

1. The basic method of treating quarks will be described. The nucleus is to be described as a collection of nucleons and six-quark bags. A six-quark bag is six confined quarks moving about some common center. The aim here is to discuss a phenomenological treatment of quark degrees of freedom suitable for applications to many different problems. In this, we use quarks to describe only the <u>non-nucleonic</u> aspects of the wave function. Prof. Faessler will present a quark treatment of the entire wave function.

2. The next step is to discuss some applications to ^{3}He and ^{3}H, discusssed already in this session. Quarks influence computations of the binding energy difference between the two nuclei, charge densities and also magnetic moments.

We shall argue that the quarks provide contributions that are helpful in explaining experimental data. However, it is difficult to find an effect that requires a quark explanation. I therefore broaden the discussion to include:

3. Certain pion-nucleus reactions in which overlapping nucleons are needed.

In each of the examples the procedure is to use quarks to describe nucleon-nucleon wave functions at small separations.

BUILDING THE NUCLEUS FROM NUCLEONS AND SIX-QUARK BAGS

The first question is "When and how should one use quarks?" One answer is: always. Simply solve QCD. However, this would be very difficult because much work would be devoted to building the nucleon degrees of freedom that are known to exist.

I want to find a shortcut that will allow a focus on the quark aspects of the physics. This is to use the information that the conventional nucleonic picture is a good description of the long range aspects. Consider, for example, two nucleons bound in a nucleus. At large separations one employs the conventional nucleon-nucleon wave function, Fig. 1. Here r is the distance between the two nucleons. Other coordinates and labels are suppressed for simplicity. The quantity ψ_{NN} is generally computed asssuming point-like nucleons and nucleon-nucleon forces as described in the earlier talks.

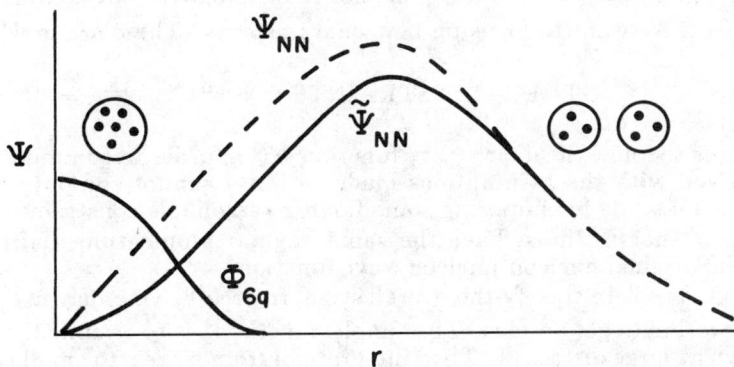

Fig. 1. Schematic picture of quark modifications to the NN wave function.

How might the composite nature of the nucleon modify ψ_{NN}? I present a schematic answer here. Suppose now that the nucleons are sizeable objects of three quarks. At large separations the nucleons do not overlap, so their finite extent does not matter. Next let the nucleons move toward each other. At some point the nucleons start to overlap. Then effects such as the quark-quark Pauli principle and gluon exchanges between nucleons can occur. If the volume of the overlapping region is small, one may expect the system mainly consists of two nucleons, but with a modified wave function, $\tilde{\psi}_{NN}(r)$.

Given the scenario I am describing there must be another component or channel in the full wave function Ψ. This new channel should be orthogonal to $\tilde{\psi}_{NN}$. Call this new piece the six-quark wavefunction. Φ^{6q}. Then one has

$$\Psi = \tilde{\psi}_{NN} + \Phi^{6q} . \tag{1}$$

One way to insure the orthogonality is to require that Φ^{6q} consist of a product of unusual "baryons" each carrying color. This is a useful separation, because operators that do not depend on color have no matrix elements that connect $\tilde{\psi}_{NN}$ and Φ^{6q}. Eq. (1) is simply an extension of standard coupled channel techniques.

Confinement requires that it takes an infinite amount of energy to separate two "baryons" carrying color. Then one may reasonably expect that Φ^{6q} is concentrated near the origin, as indicated in Fig. 1. The picture of Ψ consisting of an ordinary component and one like Φ^{6q} comes out of detailed resonating group calculations such as to be shown by Prof. Faessler. See also Ref. 4.

Thus, the idea is to use nucleons at large separations and quarks at small values of r.

How can we further describe Φ^{6q}? We take it to consist of six quarks in a single spherical bag centered at $r = 0$. The wave function is assumed to be a spatially symmetric state. For example, in a six-quark state with no orbital angular momentum each quark is in the lowest (posiitve) energy state of the confining bag or potential. This spatially symmtric state is called the [6] configuration, in the language of Young tableaux. The use of the [6] is an easy way to implement (approximately but accurately) the orthogonality of $\tilde{\psi}_{NN}$ and Φ^{6q}. Of course, we expect that Φ^{6q} represents a small part of the wave function. The [42] configuration stressed by Prof. Faessler is included (in our treatment) in the term $\tilde{\psi}_{NN}$. We want to focus on new quark aspects. These are in Φ^{6q}.

OTHER CONSTRAINTS ON Φ^{6q}

One needs to know the quark wave function Φ^{6q} in order to compute matrix elements. Even with the assumptions made so far, I cannot compute it from theory. Thus I'll settle for imposing some further reasonable constraints.

It is clear that Φ^{6q} must have the same angular momentum, parity, and isospin as the original nucleon-nucleon wave function.

The next step is to specify the overall strength of Φ^{6q}. One may use probability conservation to get an idea. The simplest method[4] is to assume that $\tilde{\psi}_{NN}$ is well known at large distances. Then integrate in from large r to small r, using the presumably well-known nucleon-nucleon interaction. At some separation r_0 one no longer believes the nucleon-nucleon (NN) wave function. For distances $r < r_0$ we use quarks instead of nucleons. But the original NN wave function has some short distance probabilty. We simply give this probabilty (P_{6q}) to the quarks.

The major question we face is: What is the value of P_{6q}? Obtaining definitive experimental verification that it is not zero would be very helpful.

It is evident that P_{6q} depends strongly on the chosen value of r_0. One can obtain reasonable upper and lower limits from the regions of validity of the meson exchange and quark theories of nuclear forces. The quark theory does not work at very large distances. On the other hand, meson exchange theories resort to phenomenology to describe the short distance behavior. Thus I believe that the range

$$0.7 \text{ fm} < r_0 < 1.2 \text{ fm} \tag{2}$$

is reasonable.

Here I will take $r_0 = 1.0$ fm. This value, about equal to the radius of a nucleon, has a nice geometric interpretation. It is the separation at which the edge of one nucleon is at the center of the other. For such separations the volume of overlap is large. If 1 fm is used one finds that about 5% of the probablity is carried by Φ^{6q}.

^{3}He - ^{3}H BINDING ENERGY DIFFERENCE

It is well known that the ^{3}H nucleus is more bound than the ^{3}He nucleus. This is not surprising. ^{3}H has two neutrons and one proton, while ^{3}He has two protons and one neutron. The Coulomb repulsion between the two protons reduces the binding energy. After removing the Coulomb energy and other electromagnetic effects one finds that about 80 KeV of attraction is still to be explained[5].

We try to find the missing attraction as a difference between the neutron-neutron and proton-proton interactions. There is no long range contribution coming from one pion exchange. So I shall look for an explanation in terms of quarks and gluons.

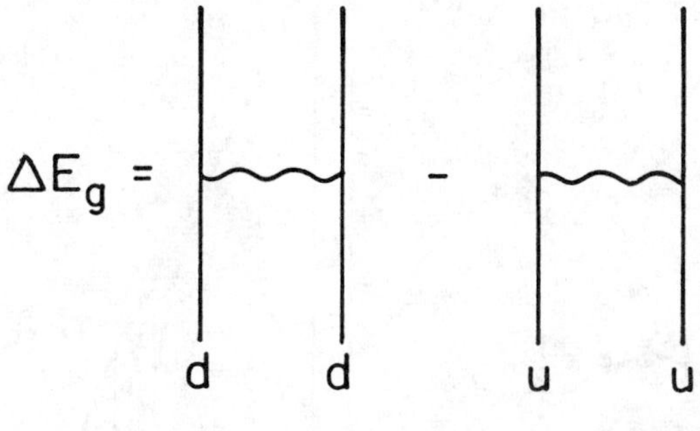

Fig. 2. Gluonic contribution to charge dependent quark-quark forces.

Consider one-gluon exchange between quarks, Fig. 2. QCD is analogous to QED. Thus there is an interaction between the (color) magnetic dipole moments of the quarks. The dipole moments depend on the masses of the quarks. The mass of the down quark (d) is bigger than that of the up quark (u). Thus the interaction between two down quarks is different than between two up quarks. One finds an energy difference, ΔE_g:

$$\Delta E_g = -(m_d - m_u)C < 0 . \qquad (3)$$

If ΔE_g were the only contribution, the neutron would be less massive than the proton. This is not the case. However, the effect of ΔE_g is compensated by the influence of the quark mass difference.

The question that concerns us here is the comparison between three and six quark systems. The neutron proton mass difference involves one dd pair vs. one uu pair. Thus $m_n - m_p$ has one factor of ΔE_g.

The six quark bag formed by two neutrons consists of 4d and 2u quarks. The one made by two protons has 4u and 2d quarks. Thus 5 (6 - 1) factors of ΔE_g enter into the mass difference between the 6 quark bags. This is considerably larger than the factor of 2 that you might expect to get from using two nucleons. Thus there is an increase of binding. Numerical evaluation of this effect is a calculated increase of binding by about the needed 80 KeV[6]. The strength of the effect is determined by P_{6q} and by C which is controlled by the well-known delta-nucleon mass difference.

Thus gluon exchanges between quarks can solve the long-standing problem of explaining the binding energy difference.

^{3}He AND ^{3}H CHARGE DENSITIES

In a fundamental treatment of photon-nucleus reactions the photon would be absorbed by a quark. In our model, the quark is inside a nucleon or a six quark bag. Both the 3 and 6 quark bags move through the nucleus. For photons of low momentum the important distance in the problem is the radius of the nucleus. At higher momentum transfers the smaller sizes of the nucleon or six-quark bag are relevant.

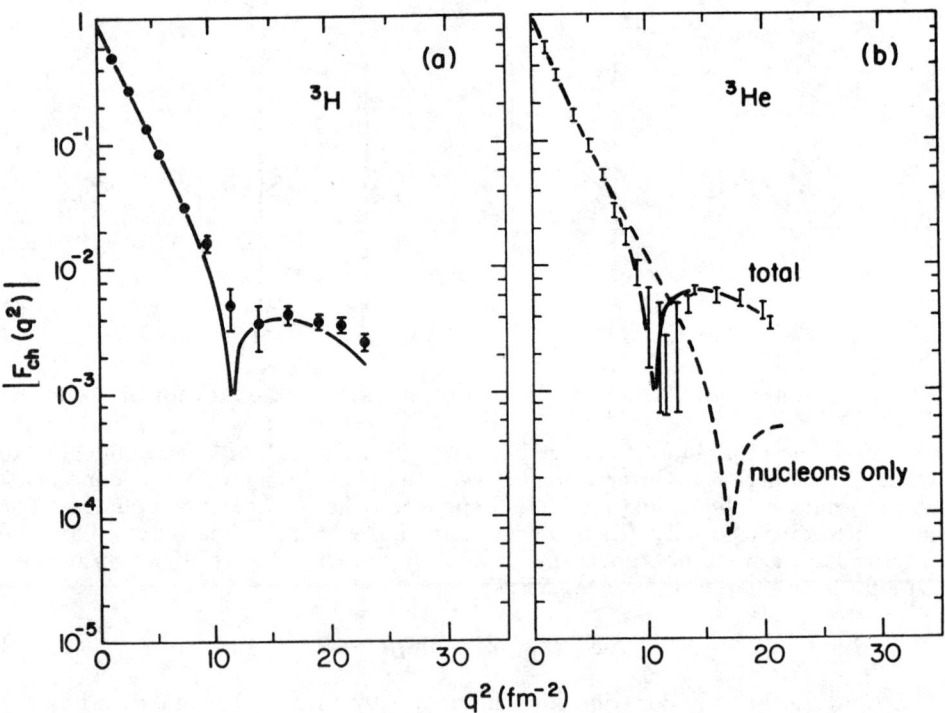

Fig. 3. (a) ^{3}H and (b) ^{3}He charge form factors. Calculation - Ref. 10. Data - Ref. 11.

In many models the radius of the six quark bag is larger than that of the nucleon[3,7]. For example, in the bag model the influence of quarks hitting the inner wall of the confining bag creates an outward pressure. This pressure increases if there are more quarks inside the bag. Thus the radius of a 6 quark bag is larger than the nucleon. The net result is a decrease in the charge density at distances near the nuclear center.

Detailed calculations of such effects have been made by Kisslinger & Hoodbhoy[8] and by others[9]. The recent results of Kisslinger, Ma & Hoodbhoy[10] are shown in Fig. 3. The charge form factors (Fourier transforms of the charge densities) measured in elastic electron-nucleus scattering are compared with the data of Ref. 11. The effects of using only nucleons (dashed) curve are considerably improved by including the 6 quark bags (solid curve). The results for both nuclei are in excellent agreement with the data. It has been quite difficult, but not impossible, to explain both sets of data with more conventional theories. This is because the pion exchange current (an isovector term) gives different results for the two nuclei.

A further test is to calculate the nuclear distribution of magnetism. I am told that calculations are in progress.

Once again, we see that quarks provide effects helpful in describing the data.

SUMMARY FOR ^{3}He AND ^{3}H

It seems that using quarks helps to explain nuclear properties. Calculations of the binding energy difference, charge densities, and also magnetic moments[12] can be improved by including specific quark effects.

But the effects we compute with quarks are small. This means that many alternate explanations of any given phenomenon are possible.

In order to obtain more definitive information we can try to find reactions where the effects of 6 quark bags are large. Thus I would like to discuss some pion-nucleus reactions which require overlapping nucleons.

ELASTIC PION-NUCLEUS DOUBLE CHARGE EXCHANGE REACTIONS

In double charge exchange (DCX) reactions a positively charged pion interacts with a nucleus, say ^{14}C. The final state consists of a negatively charged pion and the residual nucleus (^{14}O) in its ground state. (This means that the final nucleus is a member of the ground state isospin multiplet of the target. This kind of DCX process is called "elastic".) The key point is that the pion loses two units of charge. Thus, two nucleons (at least) must participate. The basic process is $\pi^+ + nn \rightarrow \pi^- + pp$.

Since two nucleons are involved, the short distance contributions must be included. Still, one may ask why the small separation region is important. Consider a simple process without quarks, Fig. 4. Here we see two separate single charge exchange reactions. Note that the π^0 is a virtual particle. Thus one must integrate over its momentum (Q). In doing the integrals, it turns out that Q can sometimes be quite high. Large values of Q correspond to small separations r, so including quarks might be useful.

The 6-quark mechanism of double charge exchange is shown in Fig. 5. A π^+ lands on a d quark converting it to a u quark. A d quark can emit the π^- and become a u quark. The pion-quark interaction, $H_{\pi q}$, is constrained to reproduce the pion-nucleon coupling constant. Furthermore the form of the interaction is constrained by various symmetry principles. $H_{\pi q}$ takes the same

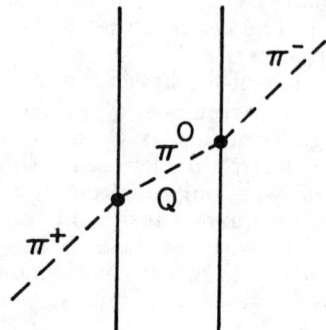

Fig. 4. Sequential mechanism for DCX. Q is the momentum of the virtual pion.

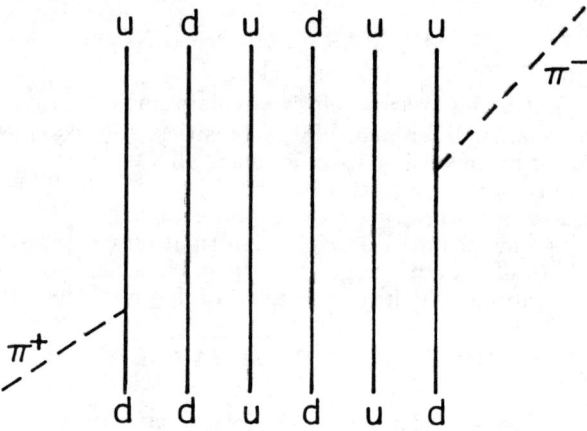

Fig. 5. Six-quark bag mechanism for DCS.

form in many different bag models that respect chiral symmetry[2,7]. (There are other diagrams[13,15], but this is the largest.

The crucial elements of the matrix element, M, are shown below. (For more details see Refs. 13-15.)

$$M \sim \sum_n \frac{<6q|H_{\pi q}|n><n|H_{\pi q}|6q>}{E_\pi - E_n} P_{6q} \qquad (4)$$

Each of the six quark states have a completely symmetric spatial wave function with quarks in the lowest positive energy (l=0) state. The initial and final 6-quark states have the same quantum numbers as the $^1S_0$ nucleon-nucleon states. The pion-quark interaction is an axial-vector operator. This means that there are only two intermediate states with the specified spatial wavefunction. One has $S = 1$ and $T = 0$. The other has $S = 1$ and $T = 2$. The energies[16] are 290 and 580 MeV respectively. The relevant six-quark probability is about 6%[13,14].

The energy E_π is the total energy of the pion, including the mass. We shall show a result for a small pion kinetic energy (50 MeV) for which the energy denominator of Eq. (4) is large.

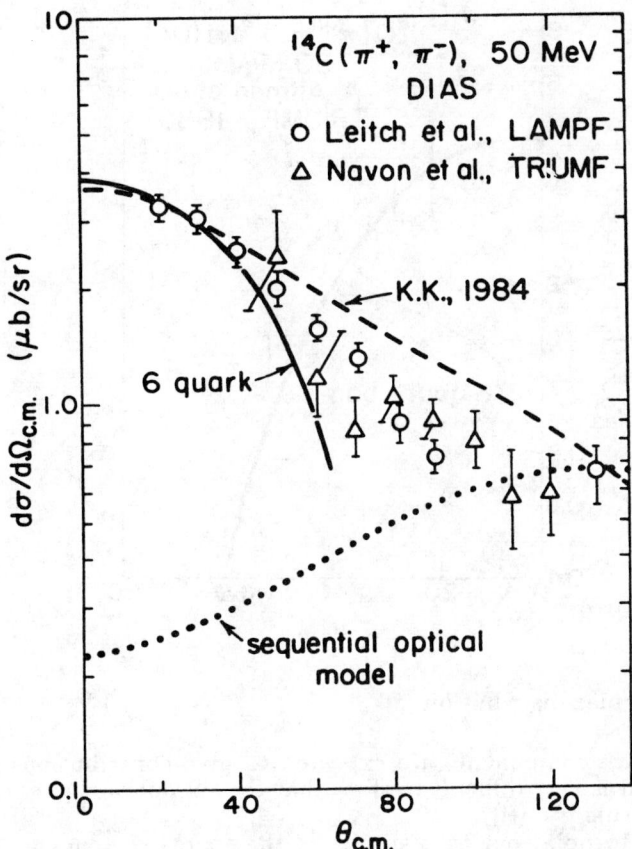

Fig. 6. DCX angular distribution ^{14}C.

The 50 MeV data[17,18] for the elastic (π^-, π^-) reaction on a ^{14}C target is shown in Fig. 6. The first calculation to be done[18] is shown as the dotted curve. It is a simple version of the diagram of Fig. 4, in which the sum on intermediate nuclear states is restricted to the first excited state of ^{14}N. The 6 quark bag calculation[13] is shown as the solid line. The curve stops at about 60° because non-quark terms were expected to contribute at larger angles. The dashed line is the calculation of Karapiperis and Kobayashi (KK)[19]. This is a more sophisticated nucleonic calculation (no quarks at all!) in which more excited states are included. The 6 quark bag effects are expected to be significant. However, the success of the nucleonic calculation indicates that the data do not require quarks.

The six quark bag calculation works reasonably well in describing the ^{18}O data of Altman et al.[20] Fig. 7. So does the calculation of KK.

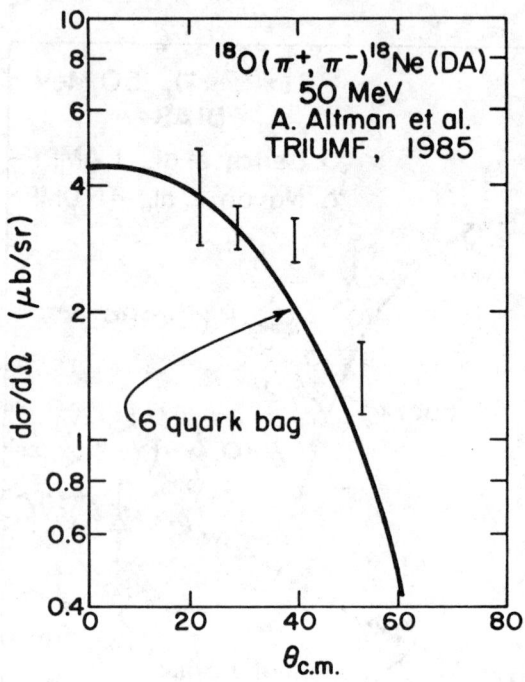

Fig. 7. DCX angular distribution ^{18}O.

The six-quark components are expected to give contributions of relevant sizes. However, it is very difficult to show that the six-quark cluster explanation is unique (or to rule it out!).

I now wish to point out that studies of the energy dependence of forward (π^+, π^-) cross sections may provide more definitive evidence.

Recall that (Fig. 5, Eq. 4) one expects a resonant enhancement of the DCX cross section if E_π is close to the energy of the six-quark intermediate state. The energies of such states are not known very well, but it is reasonable to expect that some "bumps" might appear.

A sample calculation[14] is shown in Fig. 8. The predicted enhancement in the forward cross section is quite striking. It is entirely due to the 580 MeV state. The effects of the pion-nucleon (3,3) resonance significantly reduces the computed DCX cross section for total pion energies of about 300 MeV. The predictions at higher energies are not influenced as much and some enhancement is expected to remain.

The appearance of such a peak would require the existence of nuclear six-quark bags. I hope that the relevant experiments may take place soon.

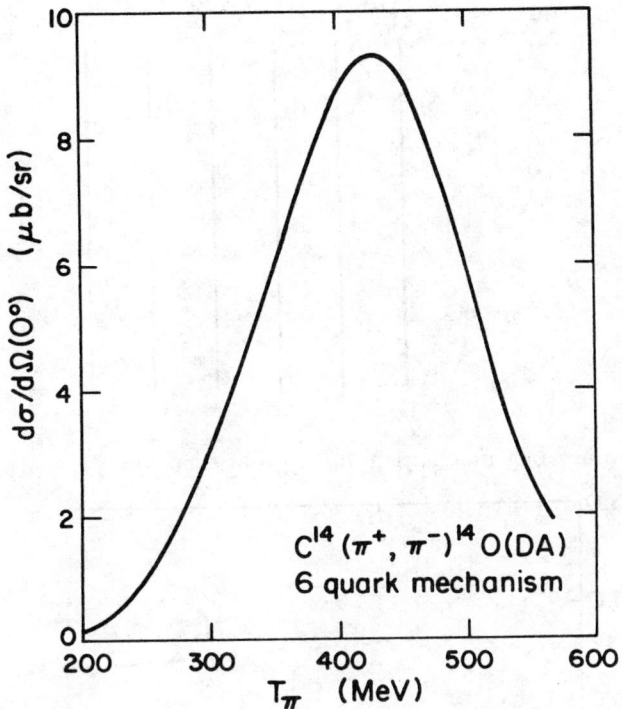

Fig. 8. Energy dependence of the forward DCX cross section.

PION-NUCLEUS ABSORPTION

I want to talk about the process of $\pi^- + (pp) \to np$. The cross section can be obtained[21] from the reaction $\pi^- + {}^3He \to np + n$.

Pion absorption is a high momentum transfer process. For example, the absorption of 65 MeV (kinetic energy) pions leads to a momentum transfer (to the nucleons) of about 600 MeV/c. This corrresponds to short distances (about 1/3 fm). It may then be reasonable to use quarks.

The absorption on a pp pair is particularly interesting. Processes involving an intermediate $\Delta - N$ state usually dominate the reaction mechanism. However such contributions are strongly suppressed by various conservation laws[21]. One may be free to search for other mechanisms.

Our quark mechanism for pion absorption is shown in Fig. 9. The dynamics is the same as for the DCX reactions. Once again the major uncertainties involve the probability amplitudes for the 6-quark bag components. The initial 6q wavefunction is again the one with the quantum numbers of the $^1S_0$ nucleon-nucleon state. There are three final states of relevance. These have the quantum numbers of $^3S_1$, $^3D_1$ and $^3P_0$ states. Some detatils are to be found in Ref. 3.

The results of our calculation of the total cross section are shown in Fig. 10. The three solid curves are the results using the 6-quark mechanism. There is strong dependence on the value of r_0. Our calculation uses quarks only. The most recent conventional (no quarks) calculation is that of Ref. 22. It

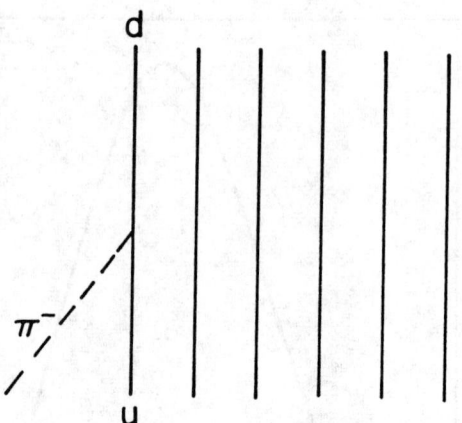

Fig. 9. Six-quark bag mechanism for pion absorption.

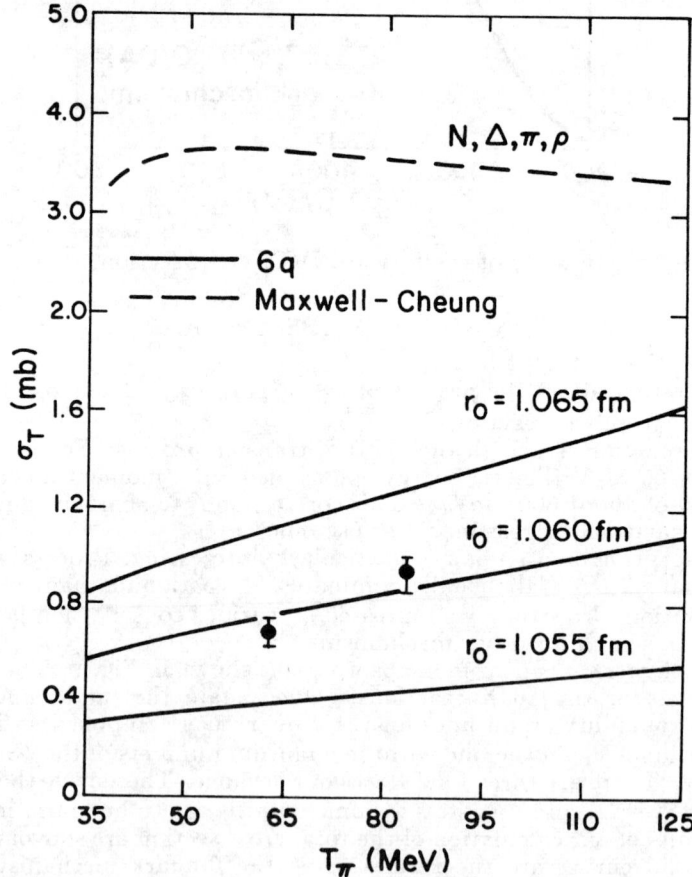

Fig. 10. Total cross section for $\pi^- + pp \to np$ as a function pion kinetic energy.

overestimates the cross sections by considerable amounts.

However, we are just at the beginning of trying to understand this reaction. Better calculations of the angular distributions in which quark and nucleonic effects are both included are in progress[23]. Data for spin observables are also required[24].

In any case, we see that quarks are expected to provide contributions of relevant sizes.

SUMMARY

Quark effects seem to have a significant influence on a variety of nuclear properties. A striking and unique quark signature is still missing. Future (π^+, π^-) and pion absorption experiments may provide the necessary definitive evidence.

REFERENCES

1. L. S. Kisslinger, Phys. Lett. 112B, 307 (1982); E. M. Henley, L. S. Kisslinger, and G. A. Miller, Phys. Rev. C 28, 1277 (1983).
2. G. A. Miller, *Int. Rev. of Nucl. Phys.*, Vol. 1 ed. W. Weise (World Scientific Publications, Singapore, 1984), p. 189.
3. G. A. Miller, *Workshop on Nuclear Chromodynamics*, ed. S. Brodsky and E. Moniz (World Scientific Publishing, Singapore, 1986), p. 343.
4. M. Oka. and K. Yazaki, Nucl. Phys. A402, 477 (1983); Y. Yamauchi and M. Wakamatsu, Nucl. Phys. A457, 621 (1986).
5. R. A. Brandenburg *et al.*, Nucl. Phys. A294, 305 (1978).
6. V. Koch and G. A. Miller, Phys. Rev. C 31, 602 (1985); C 32, 1106E (1985).
7. See A. W. Thomas, Adv. Nucl. Phys 13, 1 (1983) and many references therein.
8. P. Hoodbhoy and L. S. Kisslinger, Phys. Lett. 146B, 163 (1984).
9. M. Namiki, K. Okano and N. Oshimo, Phys. Rev. C 25, 2157 (1982); M. A. Maize and Y. Kim, Phys. Rev. C 31, 1923 (1985); J. P. Vary, S. A. Coon and H. J. Pirner, in *Few Body Problems in Physics*, ed. B. Zeitnitz (North Holland, Amsterdam, 1984), p. 683.
10. L. S. Kisslinger, W-h. Ma and P. Hoodbhoy, Nucl. Phys. A, 1986.
11. F.-P. Juster *et al.*, Phys. Rev. Lett. 55, 2261 (1985); J. S. Mc Carthy *et al.*, Phys. Rev. Lett. 25, 885 (1970); R. G. Arnold *et al.*, Phys. Rev. Lett. 40, 1429 (1978).
12. G. Karl, G. A. Miller and J. Rafelski, Phys. Lett. 143B, 326 (1984).
13. G. A. Miller, Phys. Rev. Lett. 53, 2008 (1984); See also G. A. Miller in *Proceedings of the LAMPF Double Charge Exchange Workshop*, LA-10550-C, ed. H. Baer and M. Leitch, p. 193.
14. G. A. Miller, Phys. Rev. C 35, 3777 (1987).
15. H. W. Baer and G. A. Miller, Comm. Nucl. Part. Phys. 15, 269 (1986).
16. P. J. Mulders and A. W. Thomas, J. Phys G 9, 1159 (1983).
17. I. Navon *et al.*, Phys. Rev. Lett. 52, 105 (1984).
18. M. J. Leitch *et al.*, Phys. Rev. Lett. 54, 1482 (1985).
19. T. Karapiperis and M. Kobayashi, Phys. Rev. Lett. 54, 1230 (1985), to be published Ann. Phys. (N.Y.).
20. A. Altman *et al.*, Phys. Rev. Lett. 55, 1273 (1985).

21. M. A. Moinester *et al.*, Phys. Rev. Lett. 52, 1203 (1984); K. A. Aniol *et al.*, Phys. Rev. C 33, 1714 (1986).
22. O. V. Maxwell and C.Y. Cheung, Nucl. Phys. A454, 606 (1986).
23. G. A. Miller and A. Gal, in progress.
24. E. Piasetsky, D. Ashery, M. Moinester, G. A.Miller and A. Gal, Phys Rev Lett. 57, 2135 (1986.)

EIGENFUNCTIONS AND PERIODIC ORBITS IN CLASSICALLY CHAOTIC SYSTEMS: ORDER IN CHAOS?

Eric J. Heller, Patrick W. O'Connor, and John N. Gehlen

Departments of Chemistry and Physics
University of Washington
Seattle, WA 98195

ABSTRACT

We study the quantum mechanics of a Hamiltonian system which is classically chaotic: the stadium billiard. We have examined many of the eigenstates of the stadium, up to about the 10,000th. Complex periodic orbits play an active role in shaping the eigenstates.

INTRODUCTION

The nature of the quantum eigenstates of classically chaotic Hamiltonian systems is the subject of much current research[1-3]. Some of the reasons for the keen interest are (1) desire to understand the quantum manifestations of the classical nonlinear dynamics and chaos which has been the object of so much work and progress in the last decade, (2) a desire to extend the well developed semiclassical methods of quasiperiodic dynamics into the chaotic domain, and (3) a need to extend the considerable body of results on the quantum eigenvalue spectra of classically chaotic systems[4] to the eigenfunctions, which contain more information than the eigenvalues.

For quasiperiodic classical dynamics, the semiclassical eigenfunctions can be written

$$\Psi_E(\vec{x}) = \sum_n a_{n,E}(\vec{x}) e^{iS_n(\vec{x})/\hbar + i\varphi_n} \quad (1)$$

where $S_n(\vec{x})$ is the classical action, φ_n is a phase correction which depends on the caustics that the orbit touches (the Maslov phase) and $a_{n,E}(\vec{x})$ is a real amplitude. The quasiperiodic classical trajectory may arrive at the position $\vec{x}$ in several distinct ways, distinguished by different momentum $\vec{p}_n = \vec{\nabla} S_n(\vec{x})$ (but $p = |\vec{p}_n| = \hbar k$ is fixed by the local kinetic energy and is independent of n) and classical probability $|a_{n,E}(\vec{x})|^2 = \det[\partial \vec{p}_n / \partial \vec{J}]$, where $\vec{J}$ are the classical action variables. Equation (1) is a sum over these distinct ways of reaching $\vec{x}$, and is simply the familiar statement, common to so many wave equations, that in the short wavelength limit the "rays" are the gradients of the associated wave fronts. In the short wavelength (high energy, small $\hbar$) limit, we may locally take the waves to be plane waves with a wave length determined (in the Schrodinger case) by the local kinetic energy, or (in the optics case) by the local refractive index.

Quasiperiodic motion implies that the sum over n in Eq. (1) contains a finite number of terms; i.e., a trajectory or ray accesses a given point $\vec{x}$ heading in finitely many directions.

Chaotic ray motion implies that each point $\vec{x}$ is accessed by infinitely many distinct rays, and that ray directions are random. There is no clean theory leading to a direct extension of Eq. (1) to the classically chaotic case, but M. Berry conjectured[4a] that Eq. (1) still applies, at least qualitatively, with the added reasonable assumption that each time a ray returns to a given region, its history since the last visit generates a random phase. That is, $S_n(\vec{x})$ is not in phase with $S_{n+1}(\vec{x})$, etc. Locally, the eigenfunction is a superposition of an unlimited number of plane waves of fixed wavevector magnitude but random amplitude, phase, and direction:

$$\Psi^{rand.}(\vec{x}) = \sum_n a_n e^{i\vec{k}_n \cdot \vec{x}}, \qquad (2)$$

where a_n includes a random phase factor. By the central limit theorem, $\Psi^{rand.}(\vec{x})$ is then Gaussian random. Berry showed that it has the coordinate space correlation function (for the two dimensional case)

$$\begin{aligned} C(\vec{x}, \vec{x}+\vec{\delta}) &= \int \Psi^{rand.*}(\vec{x}) \Psi^{rand.}(\vec{x}+\vec{\delta})\, d\vec{x} \\ &= (const.) \cdot J_0(k\delta) \end{aligned} \qquad (3)$$

where J_0 is the Bessel function of zero order and where k and δ are the magnitudes of $\vec{k}_n$ and $\vec{\delta}$.

RANDOM PLANE WAVE SUPERPOSITIONS

It seems desirable to know what a random eigenfunction actually looks like. This provides a standard against which to judge the eigenfunctions of classically chaotic systems. Although technically the characterization of $\langle \vec{x} | \Psi^{rand} \rangle$ as "Gaussian random with Bessel function spatial correlation" is a fairly complete statement about the ideally random eigenfunction, it does not give us a good idea of the appearance of such a function. Comparisons of $\langle \vec{x} | \Psi^{rand} \rangle$ with Gaussian random speckle patterns (as in scattered laser light) have often been made. These analogies are not completely correct.

It is easy to explicitly generate a random superposition of plane waves with equal wavevector magnitude, and plot the result in coordinate space. This has no doubt been done many times before, but perhaps regions only a few wavelengths across were plotted. When large regions are plotted, say more than 30 wavelengths across, some surprising structures appear.

The existence of these structures can be discovered by investigating another correlation function involving two coherent states instead of two positions. Consider first the overlap between a coherent state $|\alpha\rangle$ and a particular random wave superposition $|\Psi^{rand.}\rangle$. We choose the coherent state to have the same average wavevector magnitude as the fixed k used in the plane wave superposition by requiring that $|\vec{p}|^2 = \hbar^2 k^2$, where $\vec{p}$ is $\langle \alpha | \hat{\vec{p}} | \alpha \rangle$. In coordinate space, the coherent state (Gaussian wave packet) reads $\langle \vec{x} | \alpha \rangle \equiv \gamma_{\vec{x}_\alpha \vec{p}_\alpha}(\vec{x}) = \exp[-A(\vec{x}-\vec{x}_\alpha)^2/2\hbar + i\vec{p}_\alpha \cdot (\vec{x} -$

$\vec{x}_\alpha)/\hbar$]. It has average position $\vec{x}_\alpha$ and momentum $\vec{p}_\alpha$. This Gaussian is a flexible test function which balances position uncertainty and momentum uncertainty in a known way: $\Delta p \sim \sqrt{\hbar A/2}, \Delta x \sim \sqrt{\hbar/(2A)}$. The Gaussian includes the previous spatial correlation as the limit $A \to \infty$, but this limit corresponds to complete uncertainty of the magnitude and direction of the momentum.

The statistics of the amplitude $\langle \alpha | \Psi^{rand.} \rangle$ is Gaussian random if we examine a large number of positions $\vec{x}_\alpha = \langle \alpha | \hat{\vec{x}} | \alpha \rangle$ and directions $\hat{\vec{p}}_\alpha$ of the coherent state. Suppose that by chance we had picked an $|\alpha\rangle$ with a very large overlap with $|\Psi^{rand.}\rangle$. Now, we evolve the coherent state for a short time using the free particle propagator. It moves in the direction of $\vec{p}_\alpha$ a distance $|\vec{p}_\alpha| t/m$, and spreads a little, but the magnitude of its overlap with $|\Psi^{rand.}\rangle$ is unchanged, since $|\langle \Psi^{rand.}|\exp(-iHt/\hbar)|\alpha\rangle| = |\langle \Psi^{rand.}|\exp(-iEt/\hbar)|\alpha\rangle| = |\langle \Psi^{rand.}|\alpha\rangle|$. Thus, at the position $\vec{x}_\alpha + \vec{p}t/m$, a coherent state with the same momentum also has anomalously large overlap with $|\Psi^{rand.}\rangle$. If we move far enough away, however, significant spreading of the wave packet (the evolving coherent state) in the time necessary to get there will degrade the argument. This reasoning suggests that the regions of large amplitude of $\langle \vec{x} | \Psi^{rand.} \rangle$ occur in short segments, rather than as isolated spikes. Since the momentum is aimed along the axis of these segments, the associated nodal lines will run perpendicular to them. The special feature which causes this effect and which is not present in the usual speckle phenomenon is the use of a fixed wavevector magnitude in the plane waves composing $|\Psi^{rand.}\rangle$. If mixed magnitudes are used, the replacement $\langle \Psi^{rand.}|\exp(-iHt/\hbar) = \langle \Psi^{rand.}|\exp(-iEt/\hbar)$ is no longer correct and the argument leading to the suggestion of segments no longer applies. However, semiclassical considerations force us to use fixed magnitude at each $\vec{x}$.

The linear segments are obvious in Fig. (1a), which shows a contour plot of $|\langle \vec{x} | \Psi^{rand.} \rangle|^2$ over a region 60 wavelengths on a side. The function $|\Psi^{rand.}\rangle$ is a superposition of 10,000 cosine waves, each with a random orientation $\vec{k}$, a random phase shift, and a Gaussian random amplitude.[†] The linear segments take the form of a random network of ridges. This is quite different than a typical speckle pattern, shown in Fig. (1b). A range of wavevector magnitudes is used to produce the speckle pattern.

It may be puzzling that the ridges are not so apparent in a laser speckle pattern. After all, the laser has a fixed wavevector magnitude. In typical laser speckle geometries, one must remember that the retina of the observer represents a two-dimensional slice through a three-dimensional wavefront. The bottom line is that the effective wavevector does vary, so the ridges are suppressed.

---

† It is not necessary to use nearly so many plane waves, or even to use random magnitudes, to get substantially the same wavefunction plots.

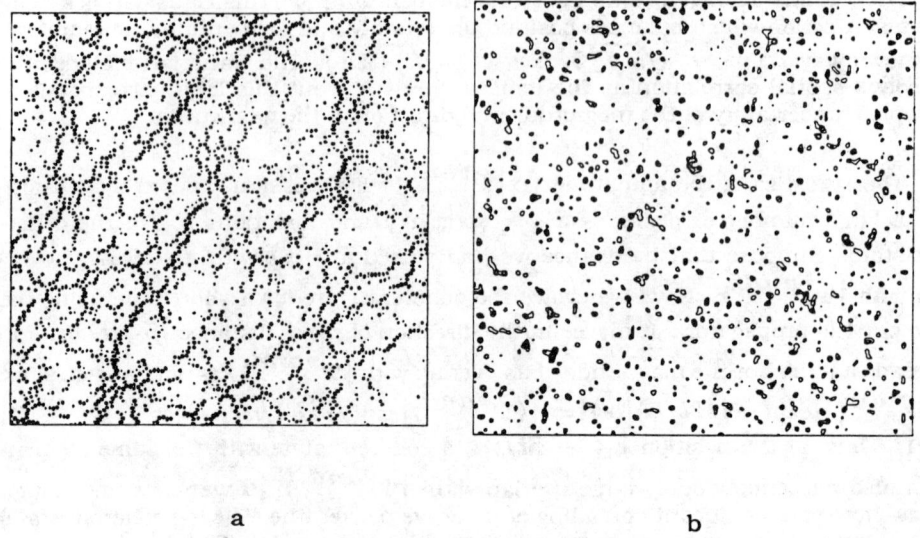

Fig. (1) a. A superposition of 10,000 plane waves of random direction, amplitude, and phase shift. The wavevector magnitude was the same for all the plane waves used. The plot is 60 wavelengths on a side. b. A speckle pattern, produced in the same way except that various wavevector magnitudes were used.

COHERENT STATE CORRELATION FUNCTIONS

We now proceed to quantify the effects seen in the coordinate space plot Fig. (1a), and discussed qualitatively above. First, we clean up a dusty corner of the problem: the arguments given here and in the literature have so far been in terms of superpositions of *complex* plane waves, but for systems with time reversal symmetry, one can always take the eigenfunctions to be real. For every plane wave $a_n \exp(i\vec{k}_n \cdot \vec{x})$ we must have another, $a_n^* \exp(-i\vec{k}_n \cdot \vec{x})$. This will affect some of the properties of the random wavefunctions, but the plots of $|\Psi^{rand.}(\vec{x})|^2$ are not affected. The quasi-linear structures remain true for both the real and the complex random superpositions.

The eigenfunction is to be modeled with random superpositions of phase-shifted standing cosine waves, rather than traveling plane waves. Abbreviating the plane wave with wavevector magnitude k traveling at an angle θ with respect to the x-axis as $|\theta\rangle$, the cosine waves take on the form $|\Psi^{rand.}\rangle = \sum_j [a_j |\theta_j\rangle + a_j^* |\theta_j + \pi\rangle]$. The correlation function we choose to investigate is

$$C(\alpha,\beta) = \left\langle (P_\alpha^\Psi - P)(P_\beta^\Psi - P) \right\rangle, \tag{4}$$

where $P_\alpha^\Psi = |\langle\alpha|\Psi^{rand.}\rangle|^2$, $P = \langle|\langle\alpha|\Psi^{rand.}\rangle|^2\rangle = \langle|\langle\beta|\Psi^{rand.}\rangle|^2\rangle$. We have chosen to examine the cross correlation of *probabilities* rather than *amplitudes* because this directly probes the correlation of large overlap of two different coherent states with a random cosine wave superposition, eliminating the "dephasing" effects which can occur when amplitudes are considered. (In the present context we are not interested in the dephasing).

Inserting the expression for $|\Psi^{rand.}\rangle$ into Eq.(4), discarding terms random in phase, and replacing sums by integrals where appropriate, we get

$$C(\alpha,\beta) = \int d\theta \int d\theta' \, \langle\alpha|\theta\rangle\langle\theta|\beta\rangle\langle\beta|\theta'\rangle\langle\theta'|\alpha\rangle$$

$$+ \int d\theta \int d\theta' \, \langle\alpha|\theta\rangle\langle\theta'|\beta\rangle\langle\beta\theta + \pi\rangle\langle\theta' + \pi|\alpha\rangle \tag{5}$$

We examine a piece of this equation, $c_{\alpha,\beta} = \int d\theta \, \langle\alpha|\theta\rangle\langle\theta|\beta\rangle$. This integral is related to Bessel functions $J_n(x)$: these may be written

$$J_n(x) = \frac{i^{-n}}{2\pi} \int_0^{2\pi} e^{ix\,\cos(\theta+\theta')+in(\theta+\theta')} \, d\theta,$$

where we have added a phase θ' to the integrand. The integral is however independent of θ', *even for complex θ'*; we arrive at

$$c_{\alpha,\beta} = e^{-2h^2 k^2/A} J_0(k\sqrt{\vec{z}\cdot\vec{z}}) \tag{6a}$$

where

$$\vec{z} = \vec{x}_\alpha - \vec{x}_\beta - i(\vec{p}_\alpha + \vec{p}_\beta)/A. \tag{6b}$$

After a little more algebra we obtain

$$C(\alpha,\beta) = e^{-4h^2 k^2/A} \left\{ |J_0(k\sqrt{\vec{z}\cdot\vec{z}})|^2 + |J_0(k\sqrt{\vec{z}'\cdot\vec{z}'})|^2 \right\} \tag{7}$$

where $\vec{z}$ is as before and $\vec{z}' = \vec{x}_\alpha - \vec{x}_\beta - i(\vec{p}_\alpha - \vec{p}_\beta)/A$.

In the limit $A \to \infty$, the Gaussians become coordinate space delta functions, and the correlation function $C(\alpha,\beta)$ reduces to $\simeq |J_0(k\cdot r)|^2$, where $r = |x_\alpha - x_\beta|$, in agreement with the previous coordinate space results.

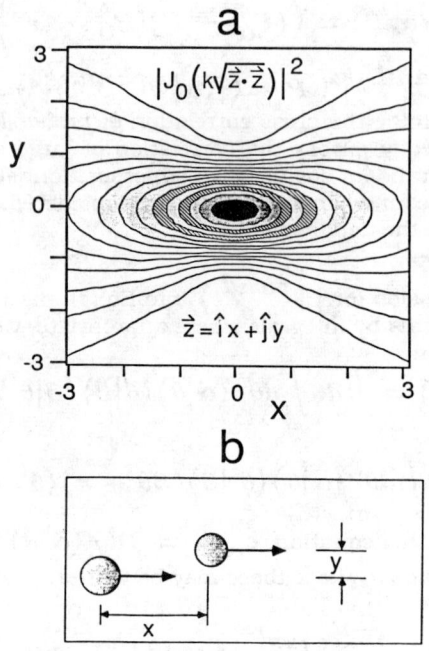

In Fig. (2a), we plot $C(\alpha,\beta)$ for $k = 2\pi$, $A = 1$ for the geometry shown in Fig. (2b). The tendency for the overlap to be large in the direction of motion (x in this case) is evident. The manifestation of this in coordinate space plots of $|\Psi^{rand.}\rangle$ is the quasi-linear structures seen in Fig. (1a).

The semi-circular structures seen in Fig. (1a), which surround regions of low amplitude, can be understood as follows. Imagine placing a Gaussian coherent state $|\alpha\rangle$ in one of these low amplitude regions, with some direction $\vec{k}$. The magnitude of the overlap, which is small, must remain the same as $|\alpha\rangle$ is propagated into the surrounding region, which has regions of much larger amplitude $|\langle\vec{x}|\Psi^{rand.}\rangle|$. Only if the nodal structure is quite different can the magnitude remain small. The propagating packet has its nodal lines perpendicular to $\vec{k}$, and note that the nodes of the semi-circular regions surrounding the low amplitude domains are approximately radial from the center of the domain. This contrasting nodal structure guarantees that the overlap $|\langle\alpha(t)|\Psi^{rand.}\rangle|$ will remain small.

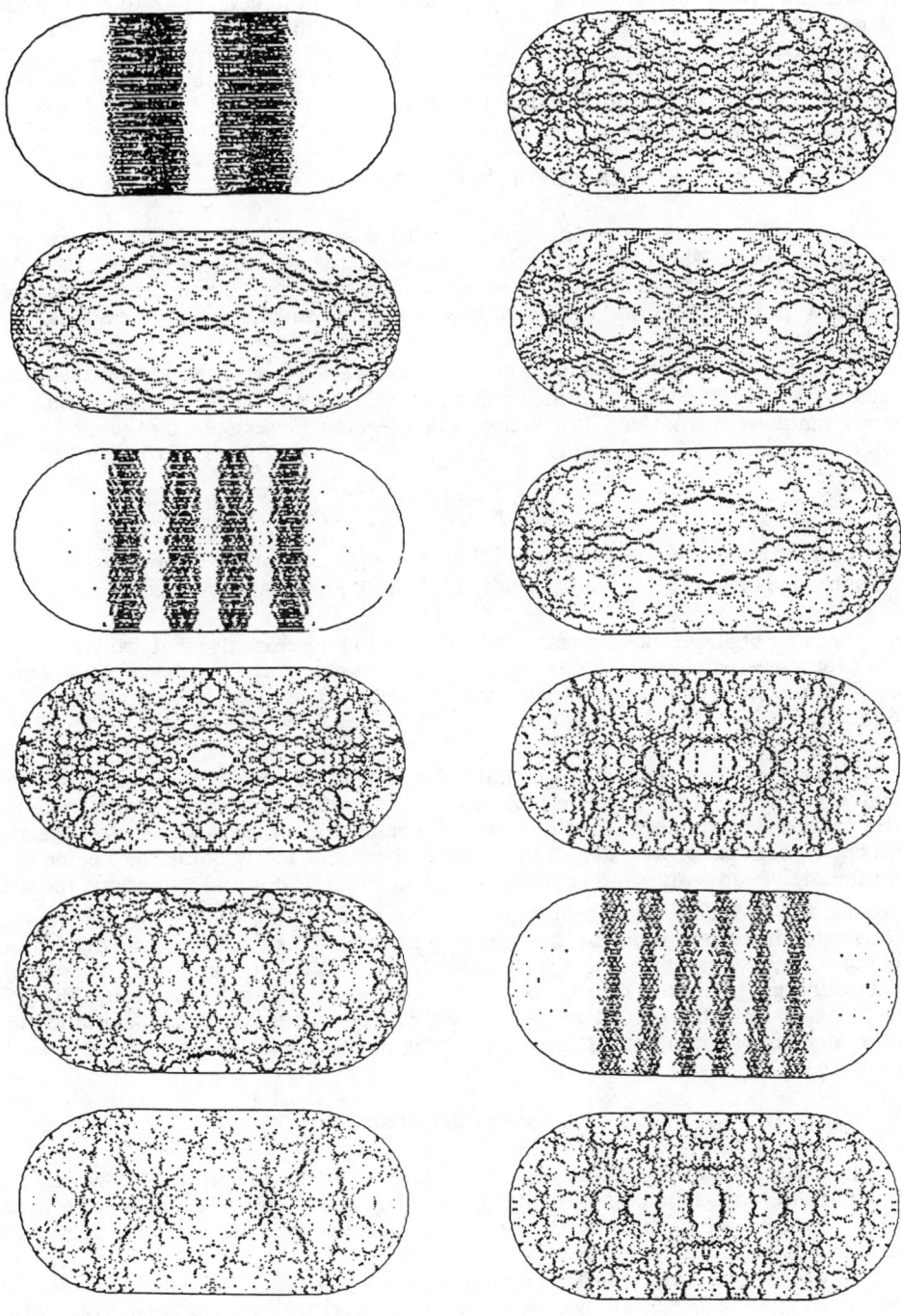

showed that those orbits with $\omega/\lambda > 1$ are expected to contribute to the scarring of some of the states. This is a stringent requirement, and only a few orbits with very few bounces satisfy it. The theory makes no statement about orbits with $\omega/\lambda < 1$. However, even in the relatively low energy plots which are not shown here (see Refs. (1)), there was already evidence that the scars were stronger than would have been expected from this theory.

SCARS AND LONG PERIODIC ORBITS

One of the theoretical predictions is that the scars should become narrower as energy increases. This has been born out in the eigenstates, up to about the 10,000th, that we have investigated so far. Therefore, as energy gets higher, the opportunity arises to "resolve" finer and finer details of the more convoluted periodic orbits.

The longer an orbit takes to return, the smaller is ω/λ. Only the shortest orbits have $\omega/\lambda \geq 1$ Thus the existing theory suggests that the longer periodic orbits may never manifest themselves, but it does not preclude the existence of scars around such complicated periodic orbits.

cousins

In most cases, the periodic orbits that appear to govern the eigenstates are more complex cousins of the simplest orbits. † Figure 5 shows six cousins.

Because of all the close cousins, it is difficult to know exactly what orbit is responsible for a given state, or if indeed there is a one-to-one correspondence. But coming up with a cousin or a member of a class of orbits related to a given state is not too difficult.

In Figure 6, we show an eigenstate and associated multibounce periodic orbit, with $\omega/\lambda \ll 1$, This periodic orbit is not symmetrical; it needed to be reflected on the y axis to be symmetrized. Orbits can be symmetrical, or they can require reflection on one or both axes. In order for orbits with about this many bounces to come back on themselves to within the accuracy of the drawing, the initial conditions must be specified to about 16 significant digits, or about 10^{-16} cm in absolute distance on the page. This is less than the diameter of a nucleus! (The trajectory was determined using a 60 bit floating point algorithm with steps taken to minimize errors in the determination of bounce directions. 15-20 bounces is about the maximum that can be handled with this precision. The algorithm was written by D. McLean and runs on a Macintosh computer. Copies of the program are available on request from the authors.)

pseudocaustics

In a few of the eigenstates, the scars show obvious curvature. How can this be, since the rays lie along straight paths? The answer can be seen in some of the periodic orbits, which have structures which are analogous to ordinary caustics in

---

† By a cousin, we mean loosely an orbit that has roughly the same shape as another orbit but which takes more or fewer bounces, and may temporarily visit a region not visited by a cousin.

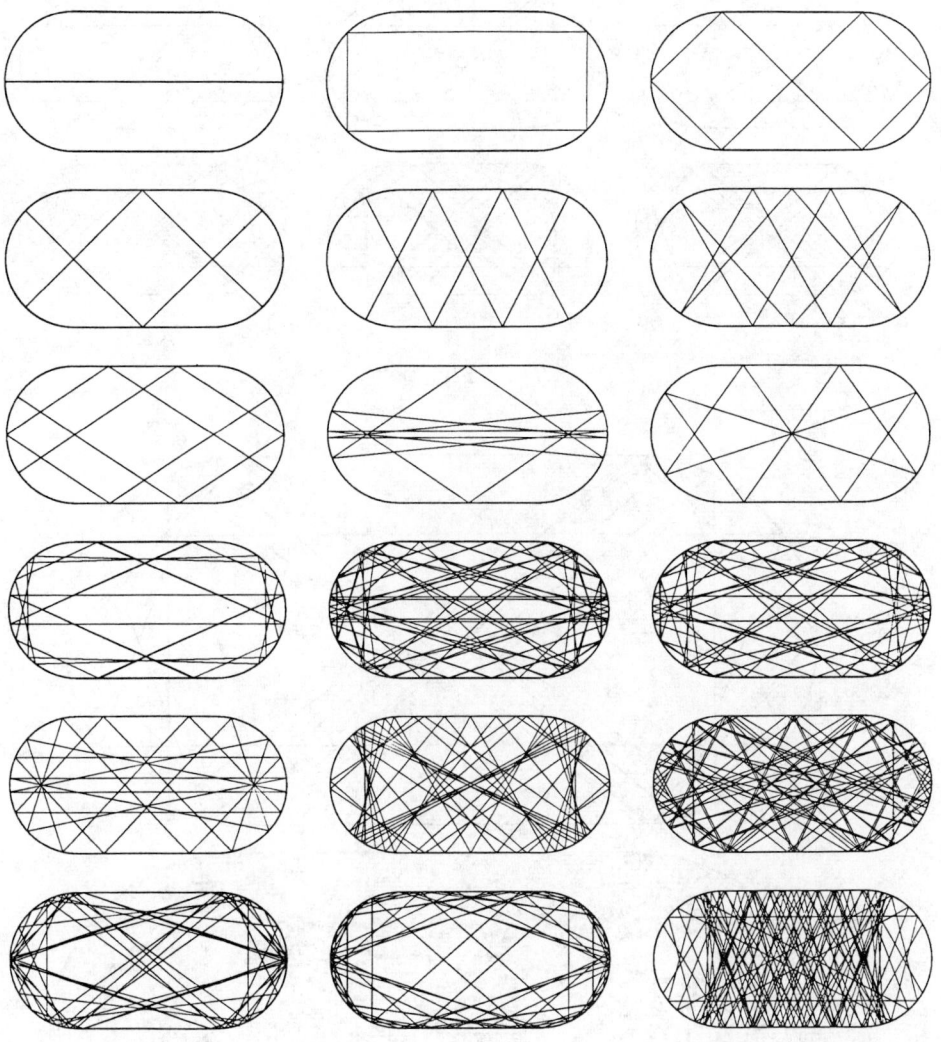

Fig. 4

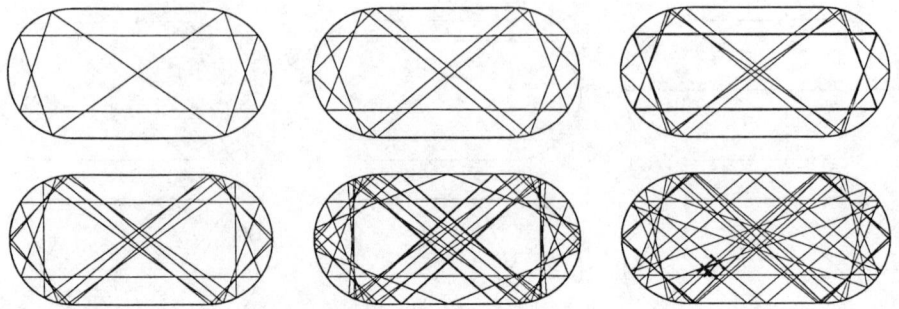

Fig. 5

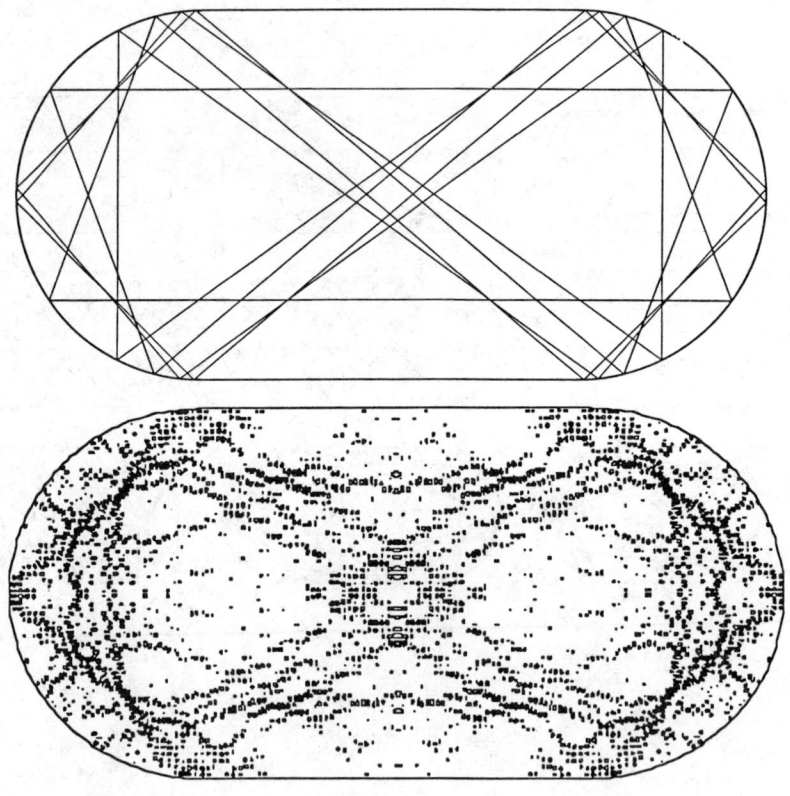

Fig. 6

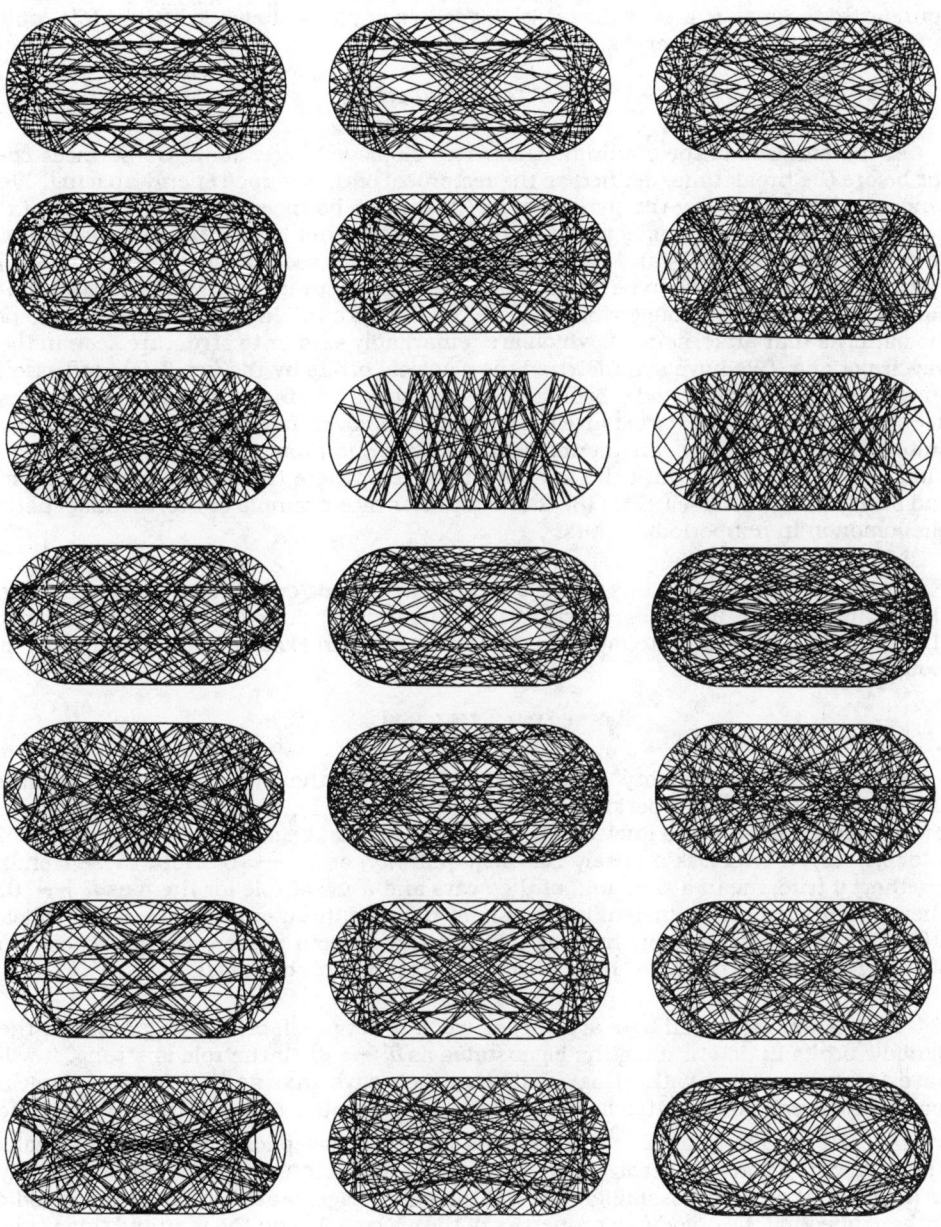

Fig. 7

quasiperiodic motion. There are technically no caustics in the chaotic domain, but if a periodic orbit is long enough it may develop caustic-like curves which are tangent boundaries to parts of a periodic orbit. A good example is shown in the middle orbit in the second to last row of Fig. (4).

longer "ordered" orbits

At the energies of the stadium eigenstates displayed here, about 30 bounces occur before the break time, defined as the reciprocal of the average energy spacing. We cannot accurately follow the dynamics for this many bounces, but one can qualitatively argue that an orbit affected by (initially) small errors is near to the dynamics of an orbit launched closeby and run exactly. The patterns seen in a typical (technically) inaccurate orbit would then be similar to those in an accurate one. With this assumption, it pays to look at longer (30 or more) bounce orbits. In Fig.(7) we see some of the patterns that arise, some of which are remarkably similar to structure seen in the wavefunctions. (We have symmetrized the classical orbits by the two 2-fold reflection lines). These are not periodic orbits; since the orbits are inaccurate anyway it does not seem useful to look for closure of the orbit. Moreover, one can argue that for any finite number of bounces, an aperiodic orbit is closely approximated by a periodic one, which looks just like it until the final few bounces, where the periodic orbits differs and finally closes on itself. Fig. (8) shows a rather nice example of the pseudocaustic phenomonon in nonperiodic orbits.

If one observes such orbits on a computer, the tracing out of the pseudocaustic is almost always done in stages. First, some of the tangent lines are laid down, but not all. The trajectory returns to lay down more only after an excursion to other type(s) of motion.

CONCLUSIONS

Based on the scar theory[1], our original guess was that the scars would localize around only the simplest periodic orbits (with $\omega/\lambda > 1$), of which there are only a few. Moreover, they would grow narrower (but no deeper, because the theory suggests a "depth" ω/λ, which is a purely classical quantity) as $\hbar \longrightarrow 0$. Both these trends together, if true, mean a "healing" of the scars and a *weak* role for them as $\hbar \longrightarrow 0$. They would become an insignificant part of the eigenfunction as $\hbar \longrightarrow 0$, which otherwise would presumably become just like the random plane wave superposition of Fig. (1a), with no rhyme or reason to the orientation of the ridges.

The results presented here are suggestive, but not proof, of a stronger role for the periodic orbits in determining the eigenstates as $\hbar \longrightarrow 0$. If the role is strong, it will have to be reconciled with classical limit, which says that as $\hbar \longrightarrow 0$, a typical localized wavepacket must follow a typical, i.e. chaotic, classical path for a time which goes to infinity as $\hbar^{-2}$. (After this time the wavepacket necessarily breaks up.) Since any wavepacket is made up of many eigenstates, in fact infinitely many as $\hbar \longrightarrow 0$, there is a possibility that the individual eigenstates could be nonergodic without affecting the ergodicity properties of the wavepacket on the required timescale.

In comparing the random plane wave superpositions with the eigenstates of the stadium, one notices that locally many of the stadium eigenfunctions have much in common with the random states. Indeed some numerical checks suggest that many

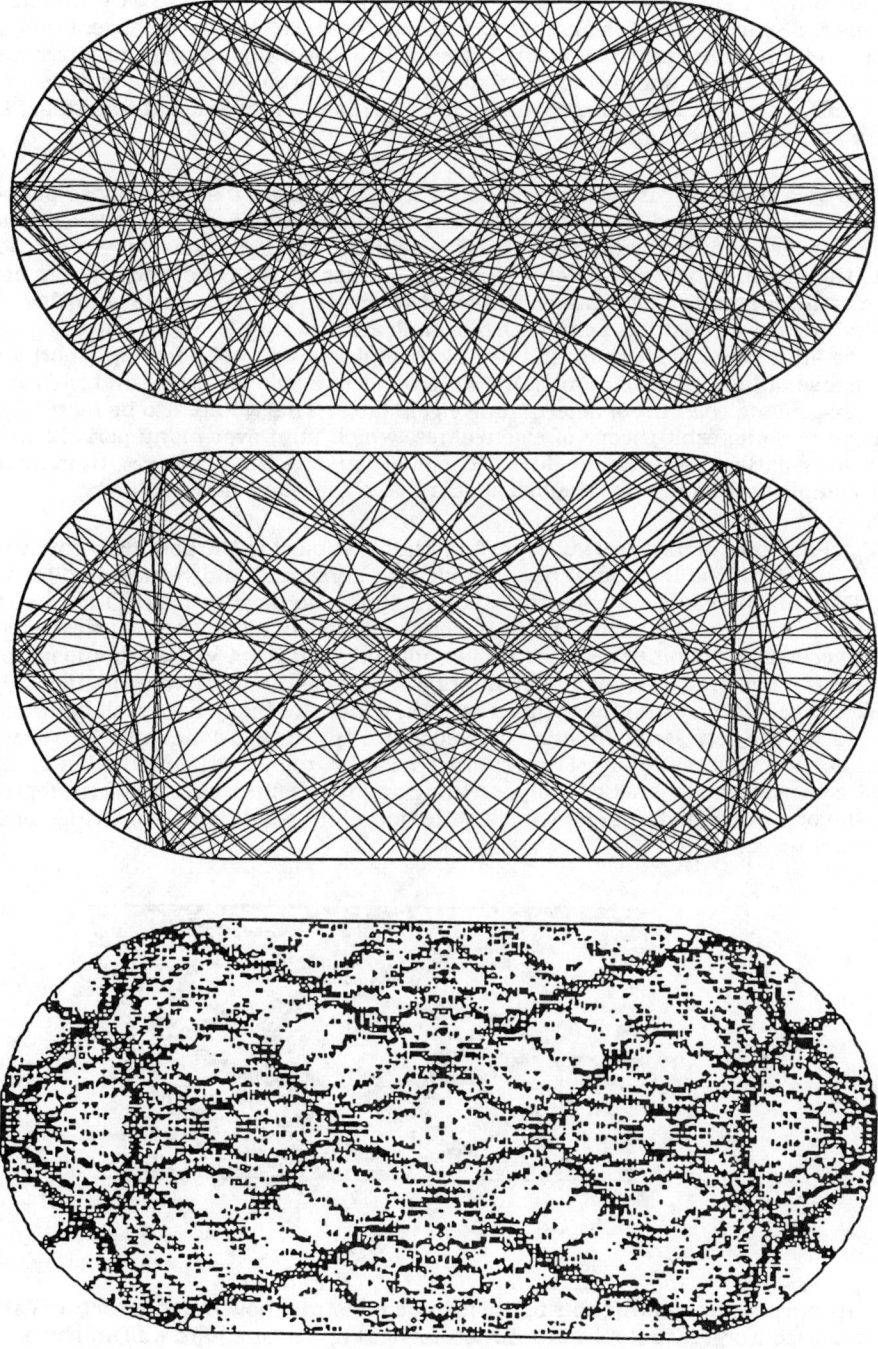

Fig. 8

or most (but not all) of the eigenstates of the stadium satisfy Eq. (3), for example. So perhaps the scarring effect is another example of "order in chaos": the eigenfunctions of the stadium have made use of the ridges which naturally occur in random plane wave superpositions, organizing them along periodic orbits in such a way as to do little damage to most measures of randomness. This is a loose idea at this stage, and we are attempting to quantify it.

The idea does suggest something about how the stadium states choose periodic orbits to localize around. First, we note that the ridges are about a wavelength wide. Second, if the stadium states are to look locally like random plane wave superpositions, the distance between neighboring ridges must be typical of the random case. Therefore, the stadium eigenstates will want to choose periodic orbits with the right density of rays in position space. Orbits that are too simple would leave big gaps with too little intensity. Orbits that are too complex would not show up as orbits at all because of lack of resolution. (This does suggest a role for orbits and their more complex cousins *collectively* determining eigenstates. This would also be more in line with the periodic orbit theory of eigenvalues, which sum over many periodic orbits to get information about eigenvalues.) The suggestion is that as $\hbar \longrightarrow 0$, more and more complex orbits become the loci of scars, and the scars never go away.

Fig. (9) shows plane wave superposition that is subject to the boundary conditions that it vanish on the horizontal boundaries, and horizontal and vertical midlines. (It is odd-odd in the same sense of the stadium states we have shown). Otherwise the function is random, or as random as the constraints will allow. It is 30 wavelengths in the vertical direction. There are no periodic orbits (except vertical bouncing ball) because the box ends are open. The boundaries cause the local wavefunction ridges to reflect specularly, as do the symmetry lines, which are really hard wall boundaries too. Indeed, the correlation function derived above applies to a coherent state reflected off a flat hard wall boundary. (This can bee seen by applying the method of images, familiar from electromagnetic theory, creating an image coherent state which together with the original coherent state satisfies the boundary condition). So, a ridge near a wall *must* reflect off it.

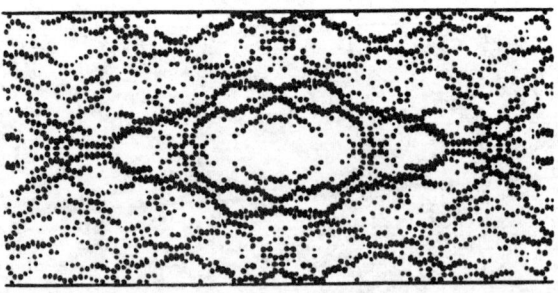

Fig. 9

Our purpose in showing this figure is to demonstrate how much structure can be generated from organizing the naturally occuring ridges near a hard wall and by having two symmetry reflection axes, but no periodic orbits. The resulting wavefunction is not unlike some of the stadium states. This lends credence to the idea that the stadium eigenfunctions are "organized chaos" and, at the same time, cautions us not to read too much into the stadium plots.

Much more numerical and theoretical work will be needed before the stadium eigenfunctions are fully understood.

REFERENCES

1. E.J. Heller, *Phys. Rev. Lett.* **53**, (1984) 1515; E.J. Heller, in Quantum Chaos and Statistical Nuclear Physics, ed. T.H. Seligman and H. Nishioka, Springer, (1986) 162.

2. (a) S.W. McDonald and A.N. Kaufman, *Phys. Rev. Lett.* **42**, (1979) 1189; (b) S.W. McDonald, Ph. D. Thesis, Lawrence Berkeley Laboratory Report LBL 14837 "Wave Dynamics of Regular and Chaotic Rays".

3. (a). M. Tabor, *Physica* **6D** 195(1983); (b) M. Shapiro and G. Goelman, *Phys. Rev. Lett.* **29**, (1984) 1714; (c).G. Hose and H.S. Taylor, *Phys. Rev. Lett.* **51**, (1983) 947; (d) B. V.Chirikov "An Example of Chaotic Eigenstates in a Complex Atom", preprint; (e) M. V. Berry and M. Robnik J. Phys. **A 19** (1986) 1365; (f) T. H. Seligman, J. J. M. Verbaarschot, and H. A. Weidenmuller Phys. Lett.**167B**, (1986) 365 ; (g) M. Shapiro and R. D. Taylor and P.Brumer, *Chem. Phys.Lett.* **106**,(1984) 325; (h) D.L. Shepelyansky, *Phys. Rev. Lett.* **56**, (1986) 677; (i) B. Eckhardt, "Exact Eigenfuctions for A Quantized Map", preprint.

4. (a) M.V. Berry, "Semiclassical Mechanics of Regular and Irregular Motion" In Chaotic Behaviour of Deterministic Systems, edited by G. Looss, R. Helleman, and R. Stora, (North-Holland, New York,1983),p. 171 (b) S. Fishman , D.R. Grempel, and R.E. Prange, *Phys. Rev. Lett.* **49** 509(1982); (c) R. Balian and C. Bloch, Annals of Phys.**69**,76(1972); (d) G. Casati, I. Guarneri, and F. Valz-Gris, *Phys. Rev. A* **3** (1984) 1586; (e) M. Robnik, J. Phys. **A17**, (1984) 1049; (f) M. V. Berry and M. Robnik, J. Phys. **A17**, (1984) 2413; (g) P. Pechukas, *Phys. Rev. Lett.* **51**, (1983) 943; (h) M.C. Gutzwiller, *Physica* **5D**, (1982) 183.

THE QUARK MODEL AND THE NUCLEON-BARYON INTERACTION

Amand Faessler
Institut für Theoretische Physik
Universität Tübingen
Auf der Morgenstelle 14
D-7400 Tübingen
West Germany

ABSTRACT

The ^{3}S, ^{1}S and P wave phase shifts of the nucleon-nucleon interaction are calculated in the six quark model using the resonating group method. For large distances the model is supplemented by π, σ, ρ and ω-meson exchange. The role of the orbital $[42]_r$ symmetry for the short range repulsion is studied. It is shown that at short distances the orbital $[42]_r$ symmetry plays an important role which is even enlarged by the colour magnetic interaction. The $[42]_r$ symmetry enforces the short range repulsion by a node which it requests at short distances. The mechanism is complicated by the fact, that the orbital $[6]_r$ symmetry is also admixed. We show that for meson exchanges which mediate the long range behaviour we can now use the SU_3 flavour ratios of the meson-nucleon

---

* Supported by the Deutsche Forschungsgemeinschaft

coupling constants even for the ω-nucleon coupling. For the ω-meson one had to use in the OBEP's an ω-N coupling constant twice to three times as large as predicted by SU_3 flavour to describe the short range repulsion. We also calculated the nucleon-hyperon interaction and describe the NΛ-scattering in agreement with the data. The model is also applied to study the EMC effect.

INTRODUCTION

The nucleon-nucleon (N-N) scattering yields phase shifts[1], which are characteristic for a strong short range repulsion. This short range repulsion has been described in the past by the exchange of the ω-meson between the nucleons[2]. But to obtain the observed repulsion one needed to increase the ω meson-nucleon coupling constant $g^2_{\omega NN}/4\pi$ to twice or even three times the value predicted by SU_3-flavour in relation to the ρ meson-nucleon coupling ($g^2_{\omega NN}/4\pi = 9 \cdot g^2_{\rho NN}/4\pi = 4.5$). All the other meson-nucleon coupling constants follow roughly the SU_3-flavour relations. The discrepancy for the ω meson-nucleon coupling reflects the fact that ω exchange is not the mechanism responsible primarily for the short range repulsion.

With the advent of the quark model and QCD one suggested that the intrinsic quark structure of the two interacting nucleons can explain this short range repulsion[3,4]. These early trials have serious short comings: (i) They used[3,4,5] the Born-Oppenheimer approximation, which would be only justified, if the effective mass of the quarks is small compared to that of the nucleon. This is not the case in the constituent quark model. Even in the current quark model the energy eigenvalue of the quark is about one third of the nucleon mass. (ii) Another serious short coming of these calculations is the neglect of the orbital $[42]_r$ symmetry for the six quarks at distance zero between the two nucleons. The importance of the $[42]_r$ symmetry has first been pointed out by Neudatchin and coworkers[6,7]. The ansatz for the two nucleon wave function as three quarks in an oscillator potential at $\vec{r}/2$ and three quarks in another oscillator potential at $-\vec{r}/2$ suppresses the $[42]_r$ symmetry like[8]

$$F(r) = [1 - f^2 - f^4 + f^6]^{1/2}$$

$$f(r) = \exp(-r^2/4b^2) \qquad (1)$$

$$b^2 = \frac{\hbar}{m\omega}$$

at small distances r. Harvey[5] included the $[42]_r$ symmetry even at small distances by allowing the excitation of two quarks in p states in the Born-Oppenheimer approximation. He obtained a large effect. The repulsive core[3,4] disappeared. This opened up again the question for the nature of the short range repulsion.

Oka and Yazaki[9] and Faessler and coworkers[8,10] showed that a non-adiabatic treatment with the help of the resonating group method including the $[42]_r$ symmetry yields hard core phase shifts for the ^1S and ^3S interaction between two nucleons. This short range repulsion is also strongly influenced by the colour magnetic interaction.

In detail we summarize shortly in chapter two the model and discuss a possible mechanism for the short range repulsion. In chapter three we give the results and in chapter four we discuss the nucleon-hyperon interaction. In an additional chapter we apply the model to the EMC effect.

INTERACTION OF TWO NUCLEONS IN THE QUARK MODEL

At short distances between two nucleons we describe the nucleon-nucleon (NN) interaction by the exchange of gluons and quarks. The gluon exchange is determined by the quark gluon vertex.

$$\mathcal{L}_{int}^{QCD} = i(g/2)\bar{q}_i \gamma^\mu \lambda_{ik}^{(a)} q_k G_\mu^{(a)} \qquad (2)$$

Here $\lambda^{(a)}$ are the eight Gell-Mann colour SU_3 matrices. q_i are the Dirac spinors of the quarks with colour i. Eq. (2) includes a sum over repeated indices $\mu = 0,1,2,3$, $a = 1,2,\ldots 8$ and i,k = red, blue, yellow. The one gluon exchange between two quarks in the non-relativistic reduction is given by[11]:

$$V_{(i,j)}^{OGEP} = \frac{\alpha_s}{4} \lambda_i^{(a)} \cdot \lambda_j^{(a)} [\frac{1}{r_{ij}} - \frac{\pi}{m_q^2} \delta(\vec{r}_{ij})(1 + \frac{2}{3}\vec{\sigma}_i \cdot \vec{\sigma}_j)] + \ldots \qquad (3)$$

m_q is the constituent quark mass and $\alpha_s = g^2/4\pi$ the strong fine structure constant. The dots indicate terms of tensor and two-body spin-orbit nature of the quark-quark interaction. They do not play a role if we restrict us to ^1S and ^3S interaction between nucleons. But they have to be included for the higher partial waves. The quark-quark interaction (3) is the leading term only for large momentum transfer and therefore short distances between nucleons. But even there one should not take (3) literally. It probably gives only the rough dependence. Its quality is improved by fitting the parameters

$m_u = m_d, m_s$ and α_s to the nucleon and the Δ mass.

The total six quark Hamiltonian (for different quark masses see eq. (14))

$$\hat{H}_6 = \sum_{i=1}^{6} [m_i + \frac{p_i^2}{2m_q}] \qquad (4)$$

$$+ \sum_{i<j=1}^{6} [V^{OGEP}(i,j) - \lambda_i^{(a)} \cdot \lambda_j^{(a)} \, a \, r_{ij}]$$

must include also a colour confinement term. The parameter a [MeV fm^{-1}] is adjusted to the charge root mean square radius of the proton including the pion cloud[12].

For large distances one can not exchange colour objects like quarks and gluons. Thus we go back there to meson exchange. But we have to guarantee asymptotic feedom. This is done by allowing the coupling of the mesons to the quarks only near the surface of the nucleon.

$$g_{qq\mu}(r) = c_\mu \, r^2 \qquad (5)$$

The free parameters c_μ will be adjusted for each meson μ to the meson-nucleon coupling constants[13] at zero momentum transfer $g^2_{\pi NN}(q=0)/4\pi = 14.1$, $g^2_{\sigma NN}(q=0)/4\pi = 5.65$ and $g^2_{\rho NN}(q=0)/4\pi = 0.5$. For the ω-meson coupling we shall see that the flavour SU_3 value $g^2_{\omega NN} = 9 \cdot g^2_{\rho NN}$ yields a good agreement for the NN phase shifts. This is opposed to the OBEP's[2,13] where the ω-nucleon coupling constant has to be blown up by a factor two to three to describe the short range repulsion.

In a first step we include only the exchange of the two lightest mesons (m_π = 140/MeV; m_σ = 520 MeV). The ansatz for the resonating group wave function is

$$\psi_{6q} = A\{|\overline{NN}\rangle \chi_N(r) \qquad (6)$$

$$+ |\overline{\Delta\Delta}\rangle \chi_\Delta(r) + |\overline{CC}\rangle \chi_C(r)\}$$

The Kohn-Hulthen variational principle

$$\langle \delta\hat{\psi}_{6q}|\hat{H}_6 - E|\psi_{6q}\rangle = 0 \qquad (7)$$

yields for ^1S and ^3S channels three coupled integral equations for the relative wave functions $\chi_N(r)$, $\chi_\Delta(r)$ and $\chi_C(r)$. If one describes the six quarks wave function by three quarks in an oscillator at $-\vec{r}/2$ and by three quarks in an oscillator at $\vec{r}/2$, the antisymmetrized wave function has the form[8]:

$$A\{|NN,r\rangle\} = \tfrac{1}{3} D(r) [6]_r \{33\}_{ST}[222]_C\rangle \qquad (8)$$

$$+ \tfrac{1}{3} F(r)[[42]_r\{33\}_{ST}[222]_C\rangle - [42]_r\{51\}_{ST}[222]_C\rangle]$$

with:

$$D(r) = [1 + 9f^2 + 9f^4 + f^6]^{1/2}$$

$$F(r) = [1 - f^2 - f^4 + f^6]^{1/2}$$

$$f = \exp(-r^2/4b)^2$$

Instead of (8) we are using a basis which allows that the variational principle chooses freely the amplitudes of the different orbital symmetries characterized by Young tableaux.

$$\begin{pmatrix} |\overline{NN}\rangle \\ |\overline{\Delta\Delta}\rangle \\ |\overline{CC}\rangle \end{pmatrix} = A \begin{pmatrix} |(\{33\}_{ST}[222]_C) \; [\tilde{6}]\rangle \\ |(\{33\}_{ST}[222]_C) \; [\tilde{42}]\rangle \\ |(\{51\}_{ST}[222]_C) \; [\tilde{42}]\rangle \end{pmatrix} \qquad (9)$$

$$A = \begin{pmatrix} \frac{1}{3} & \frac{2}{3} & \frac{2}{3} \\ -\sqrt{\frac{4}{45}} & -\sqrt{\frac{16}{45}} & -\sqrt{\frac{25}{45}} \\ \sqrt{\frac{4}{5}} & -\sqrt{\frac{1}{5}} & 0 \end{pmatrix}$$

The above basis states contain no spatial dependence. ST and colour C are only coupled to the symmetry conjugate to the orbital one. The states $|\overline{NN}\rangle$, $|\overline{\Delta\Delta}\rangle$ and $|\overline{CC}\rangle$ contain also the internal spatial variables for each nucleon. A solution of the resonating group problem (7) yields for each symmetry its own radial dependence. The ansatz (9) includes quark excitations in higher orbital states. If one solves the relative wave function by a coupled system of Hill-Wheeler-Griffin equations

$$\sum_{\beta=N,\Delta,C} \int dr' [\langle \alpha,r|\hat{H}_6 + V_\pi + V_\sigma |\beta,r'\rangle \qquad (10)$$

$$-E\langle \alpha,r|N|\beta,r'\rangle] \cdot \chi_\beta(r') = 0$$

it is useful to look to the wave function in the following form:

$$\tilde{\chi}_\alpha(r) = \sum_\beta \int <\alpha,r|N|\beta,r'>^{1/2} \chi_\beta(r')dr' \qquad (11)$$

According to (9) it can be decomposed into the symmetry basis. Fig. 1 shows such relative wave functions in the ^3S channel. One sees a node in the [42] orbital symmetries. The nodes are reflected in the norm-kernels in oscillator functions, which are diagonal in the symmetry basis.

$$<n,l=0|N_{[6]_r}|n,l=0> = 1 + 9 \left(\frac{1}{3}\right)^{2n} \qquad (12)$$

$$<n,l=0|N_{[42]_r}|n,l=0> = 1 - \left(\frac{1}{3}\right)^{2n}$$

One sees that the 0s state is Pauli forbidden for $[42]_r$. This yields the node in the orbital [42] symmetric relative wave functions. If we would have only $[42]_r$ the node would guarantee a hard core phase shift. (One should stress, that the relative wave function is not unique. For the $[42]_r$ symmetry one can add any admixture of 0s oscillator functions without changing any observable. This can remove the node in the relative wave function. It depends only on the way how the non-orthogonal basis states |NN>, |ΔΔ> and |CC> are orthogonalized[24]. The non-orthogonality is here included by the overlap kernel $<\alpha,r|N|\beta,r'>$. In discussing the relative wave function in the NN channel it is natural to orthogonalize the |ΔΔ> and |CC> channels relative to |NN>. But this orthogonalization and the possible removal of the Pauli forbidden 0s admixture in the $[42]_r$ symmetry does not change any observable).

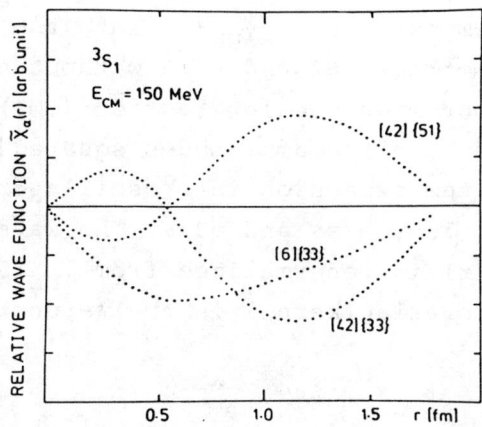

Fig. 1. Relative ^{3}S wave functions in the symmetry basis at E_{CM}=150 MeV as a function of the distance r between the two nucleons. (m_q = 336 MeV, α_s = 1.3, a_c = 41 MeV fm^{-1}).

The colour magnetic part of the quark-quark wave function is essential in enlarging the [42] orbital symmetry in the relative NN wave function.

At the end of this chapter we are hopefully convinced that the node of the $[42]_r$ symmetry is responsible for the hard core phase shift. But the $[42]_r$ symmetry has only its important position due to the colour magnetic part of the quark-quark force.

Table I. Decomposition of $\tilde{\chi}_{NN}(r)$ into the spatial symmetries [6] and [42] without and with the colour magnetic interaction (CMI). The numbers given are the amplitudes squared $|\langle ns|\tilde{\chi}_{NN}[f]\rangle|^2$ for the expansion into oscillator functions $|0s\rangle$, $|1s\rangle$, $|2s\rangle$ and $|3s\rangle$. The wave function $\tilde{\chi}_{NN}(r)$ is renormalized from $\chi_{NN}(5)$, so that the overlap kernel is the δ-function.

	[f]	$|0s\rangle$	$|1s\rangle$	$|2s\rangle$	$|3s\rangle$
without	[6]	0.450	0.045	0.005	0
CMI	[42]	0	0.160	0.039	0.001
with	[6]	0.132	0.062	0.026	0.011
CMI	[42]	0	0.219	0.188	0.088

This result is also supported by table I. It shows the decomposition of the relative wave function between two nucleons $\chi_{NN}(r)$ into the spatial symmetries [6] and [42] and their expansion into oscillator wave functions $|n,\ell=0\rangle$ calculated in the NN channel only. The results with the colour magnetic force (with CMI) shows the $[42]_O |1s\rangle$ as the strongest admixture. While without CMI this role is played by $[6]_O |0s\rangle$.

RESULTS

The solution of the coupled system of resonating group equations (10) yields the relative wave functions $\chi_N(r)$, $\chi_\Delta(r)$ and $\chi_C(r)$ for the channels with two asymptotic nucleons, two asymptotic Δ's and the six quark hidden colour state, respectively. The asymptotic form of the relative wave function in the NN-channel gives the phase shift for the nucleon-nucleon scattering. This calculation is performed including quark and gluon exchange $\hat{H}_6$ and the exchange of π- and σ-mesons. The parameters are adjusted as described in chapter 2. The results are given for the ^1S and ^3S in figures 2 and 3, respectively.

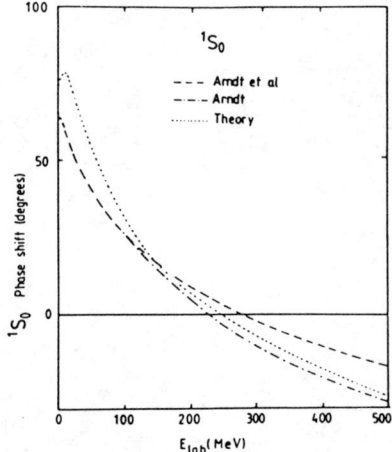

Fig. 2 ^1S nucleon-nucleon phase shifts as a function of the laboratory energy. The dotted curve[12] gives the results of the theory presented. The dashed and the dashed-dotted line are two different sets of experimental phase shifts[15]. (m_q = 336 MeV, α_s = 1.3 and a_c = 41 MeV fm^{-1}).

In the results presented in figures 2 and 3 we included only the exhange of the two lightest mesons, the π- and σ-meson. It is interesting to see what happens if we include also in addition the ρ- and the ω-mesons. This question is especially exciting since in the one meson exchange potentials the ω-meson is solely responsible for the short range repulsion. Out of these reasons one adjusts in the OBEP's the ω-nucleon coupling constant to a value which is by a factor two to three times larger than the flavour SU_3 value derived from the ρ-nucleon coupling. If we now have the correct nature of the short range repulsion, we should get a satisfactory fit to the phase shifts, if we use the flavour SU_3 value $g^2_{\omega NN}/4\pi = 4.5$. Figure 6 shows the ^3S phase shifts for different ω-NN coupling constants in radians as a function of the nucleon bombarding energy. The factor g_0 is defined as the deviation from the flavour SU_3 value.

The figure shows that for a reasonably good nuclear radius one gets a good agreement for the phaseshifts for a value of the parameter $g_0 = 1$ in agreement with the flavour SU_3 symmetry. These results support strongly our conviction that we found indeed the correct nature of the short range nucleon-nucleon repulsion.

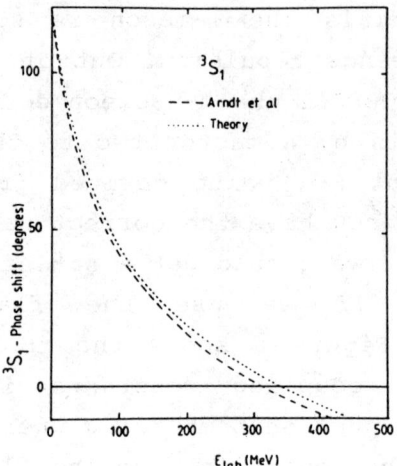

Fig. 3   ³S nucleon-nucleon phase shifts as a function of the laboratory energy of the nucleon projectile. The dotted curve is the present calculation[12] and the dashed curve represents the experimental phase shifts[15]. (For parameters see caption of Fig. 2).

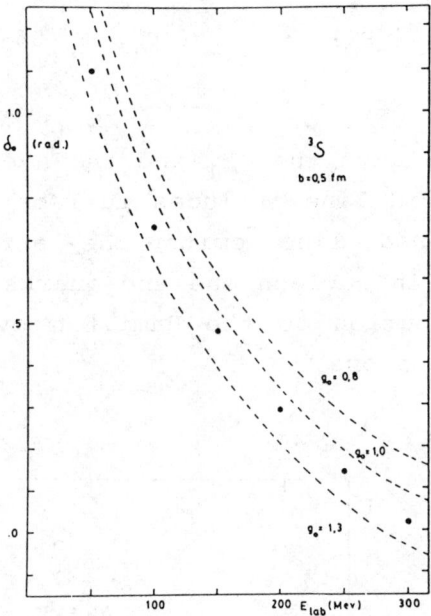

Fig. 4 ^{3}S nucleon-nucleon phase shifts in radians as a function of the laboratory energy for the oscillator length b = 0.5 fm. The ω-nucleon coupling constant is chosen as $g^2_{\omega NN}/4\pi = 4.5\, g_0$. The data are taken from ref. 15.

Recently we have extended this model to include also the finite size of the pions. This has the advantage that one can take into account the pion cloud in determining the energy of the proton and still keep the nucleon stable at a finite radius[16]. For a point pion the pion self energy exerts on the nucleon such a pressure that it is compressed to the radius zero. In addition we calculated also the $^1P_1$ and the averaged ^3P_J partial waves.

$$\overline{\delta(^3P_J)} = \frac{1}{9}[\delta(^3P_0) + 3\delta(^3P_1) + 5\delta(^2P_2)] \qquad (13)$$

Figures 5 and 6 give the $^1P_1$ and the averaged ^3P_J phase shifts. The solid line includes full antisymmetrization while the dashed line omits the antisymmetrization between quarks in nucleon one and quarks in nucleon two for the contribution to the Hamilton overlap from the exchange of the pions.

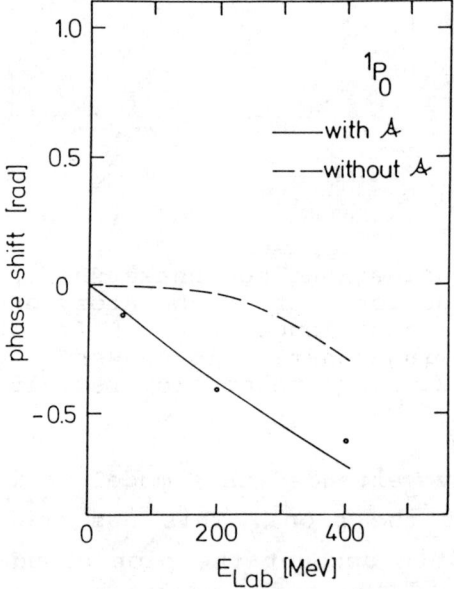

Fig. 5 $^1P_1$ phase shift in radians as a function of the laboratory nucleon bombarding energy. The solid line includes the full antisymmetrization of the quarks in nucleon one and two while the dashed line neglects for the pion exchange contribution this antisymmetrization. The dots indicate the experimental values.

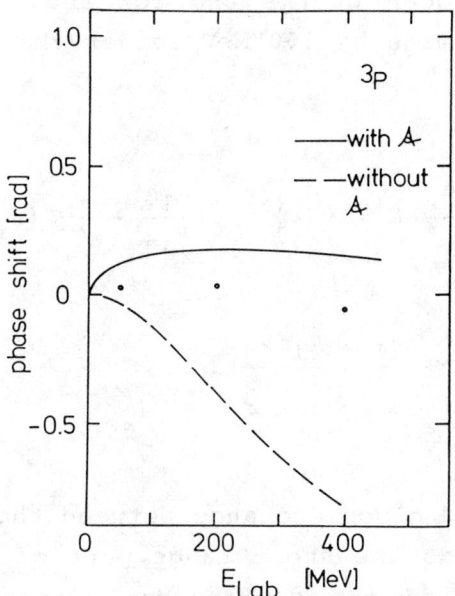

Fig. 6 ^3P_J averaged according to equation (13) phase shifts in radians as a function of the laboratory bombarding energy of the nucleon. The solid line includes the full antisymmetrization between quarks in nucleon one and quarks in nucleon two. The dashed line neglects this antisymmetrization for the contribution of the pion exchange.

THE LAMBDA-NUCLEON INTERACTION

The lambda-nucleon (Λ-N) can be calculated in the same way as the nucleon-nucleon interaction. The one-gluon exchange (3) must now include the different masses for the up and down quarks and for the strange quark. We choose this mass by 150 MeV larger than the up and down quark masses[18].

$$V^{OGEP}_{(i,j)} = \frac{\alpha_s}{4} \lambda_i^{(a)} \cdot \lambda_j^{(a)} [\frac{1}{r_{ij}} - \frac{\pi}{2} \delta(r_{ij}) \tag{14}$$

$$\cdot (\frac{1}{m_i^2} + \frac{1}{m_j^2} + \frac{4}{3} \frac{\vec{\sigma}_i \cdot \vec{\sigma}_j}{m_i m_j})]$$

The quark and gluon exchange between the hyperon and the nucleon yields the short range part of the interaction. One has to add the confinement potential (4) and for large distances meson exchange. The meson exchange potential is a generalization of the pion potential of ref. 19 to include all pseudoscalar mesons and also mixing of the octet η_8 with the singlet η_1. A mixing angle of $\Theta = -23°$ is assumed.

$$V^{PSM}_{(i,j)} = \frac{1}{3} \frac{g_{qqM}}{4\pi} \frac{\mu^2}{4M_A M_B} \frac{\Lambda^2}{\Lambda^2 - \mu^2} e^{-\mu^2 b^2/3} \tag{15}$$

$$\cdot \mu [\frac{e^{-\mu \cdot r_{ij}}}{\mu r_{ij}} - \frac{\Lambda^3}{\mu^3} \cdot \frac{e^{-\Lambda r_{ij}}}{\Lambda r_{ij}}] \vec{\sigma}_i \cdot \vec{\sigma}_j \cdot O^F_{ij}$$

with: $g_{qq8} = \frac{3}{5} g_{NN\pi} = \frac{3}{5} 14.2$

$g_{qq1} = g_{NN\eta_1} = \frac{\sqrt{5}}{6} g_{NN\pi}$

Here μ is the mass of the pseudoscalar meson in question, Λ a size parameter describing the finite size of the meson, M_A and M_B are the masses of the baryons A, B between which the mesons are exchanged and O_{ij}^F is an operator in flavour space according to table II.

Table II. Meson parameters

	μ (MeV)	Λ (MeV)	O_{ij}^F
π	138	832	$\sum_{K=1}^{3} \lambda_K^F(i) \cdot \lambda_K^F(j)$
K	495	832	$\sum_{K=4}^{7} \lambda_K^F(i) \cdot \lambda_K^F(j)$
η	549	4000	$\cos\theta \cdot \lambda_8^F(i) \cdot \lambda_8^F(j) - \sin\theta$
η'	958	4000	$\sin\theta \cdot \lambda_8^F(i) \cdot \lambda_8^F(j) + \cos\theta$

There are two independent coupling constants g_{qq8} and g_{qq1} describing the coupling of the octet or singlet mesons with the quarks. Because the coupling constants are assumed to be independent of flavour, they can be determined by looking to the long range part of (15) and comparing it with the one meson exchange potential for the NN interaction. In this way one finds the relations given in eq. (15).

In addition to these potentials acting on the quark level we add a phenomenological σ meson potential on the baryon level with a microscopic calculated form factor[20] representing the contributions of two pion exchange.

$$V_\sigma(r) = -\frac{g_\sigma^2}{4\pi} \frac{1}{2m^2 R^2 r} \begin{cases} 1-e^{-mr} -e^{-2mR} \sinh(mr) \\ \quad \text{for } r<2R \\ [\cosh(2mR)-1]e^{-mr} \\ \quad \text{for } r<2R \end{cases} \quad (16)$$

Here m=520 MeV is the mass of the σ-meson and R=0.72 fm is a size parameter fitted to the form factor.

In the kinetic energy we use an average quark mass. These approximations to use a mean mass in the kinetic energy is unavoidable if one uses the same symmetry space as for the nucleon-nucleon interaction. The parameters of the short range part of the interaction are fixed by requiring that the mass differences $M_\Lambda - M_N$ = 177 MeV, $M_\Sigma - M_N$ = 253 MeV are reproduced and that the stability condition $dM_N/db = 0$ is satisfied. The parameters obtained by this procedure are shown in table III for two values of the oscillator length b without and with the meson cloud included. One sees that the meson cloud contributes to about one half of the mass splittings.

Table III. Parameter sets for OGE

b (fm)	0.50	0.55	0.50	0.55
$m_u = m_d$ (MeV)	313	313	313	313
m_s (MeV)	535.2	535.2	425.4	443.2
α_s	0.82	1.09	0.26	0.48
a_c (MeV/fm^2)	85.2	47.0	58.3	33.2
$(M_\Lambda - M_N)_{Meson}$ (MeV)	0	0	74.7	61.9
$(M_\Sigma - M_N)_{Meson}$ (MeV)	0	0	135.3	114.2
$g_\sigma^2/4\pi$	0	0	2.3	2.6

Fig. 9 shows the total elastic cross section for Λp-scattering as a function of the incident Λ-momentum in the laboratory frame. The theoretical results are compared with the experimental data given in the references 21 to 23.

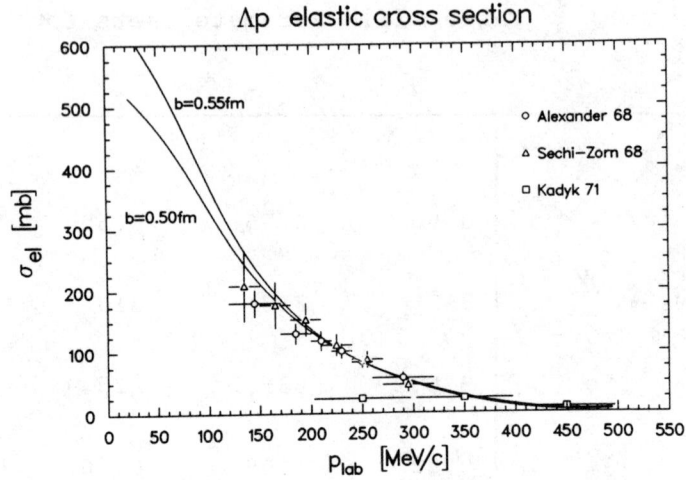

Fig. 7 Total elastic cross section for Λp-scattering as a function of the incident Λ-momentum in the laboratory frame. Two theoretical curves are compared with the experimental data from refs. 21-23. The two theoretical curves are distinguished by the two sets of parameters adjusted for the two different oscillator lengths b=0.5 fm and b=0.55 fm.

The comparison between theory and experiment for the Λ-proton scattering shows a surprisingly good agreement between theory and experiment.

THE SIGMA-NUCLEON INTERACTION

For the Σ-nucleon interaction we use the same parameters as for the Λ-nucleon interaction. Thus this part of the calculation is totaly parameter free. Figures 8 and 9 show the results for the T=1/2 $^3S_1$ and $^1S_0$ and the T=3/2 $^3S_1$ and $^1S_0$ phase shifts, respectively. The results are given for the parameters of the last two colummns in table III. They include for the Λ-N and the Σ-N splitting also the contributions of the meson clouds. The results are given in table 3 and figures 8 and 9 for two different oscillator lengths b=0.50 fm and b=0.55 fm corresponding to two different root mean square radii of the quark content of the baryon b=$\langle r^2 \rangle^{1/2}$.

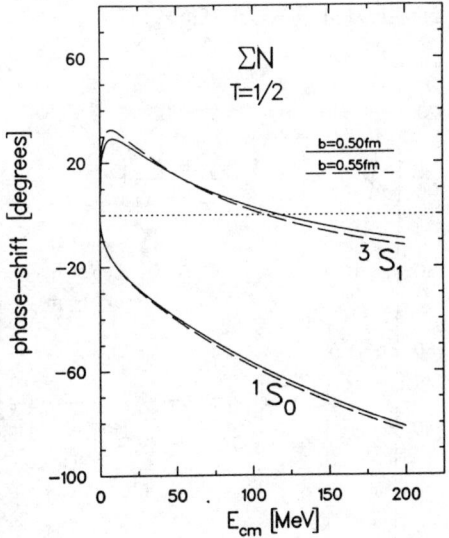

Fig. 8   Phase shifts of the Σ-nucleon scattering for the T=1/2 $^3S_1$ and $^1S_0$ partial waves. The results are given for two different oscillator lengths b=0.50 fm and b=0.55 fm corrresponding to two different root mean square radii of the quark content of the baryons.

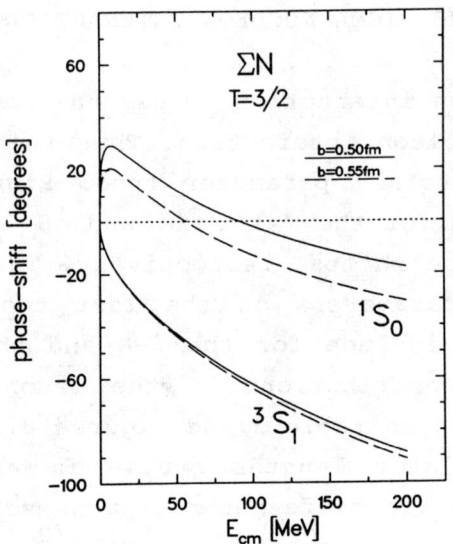

Fig. 9 Σ-nucleon T=3/2 $^3S_1$ and $^1S_1$ phase shifts for b=0.50 fm and b=0.55 fm.

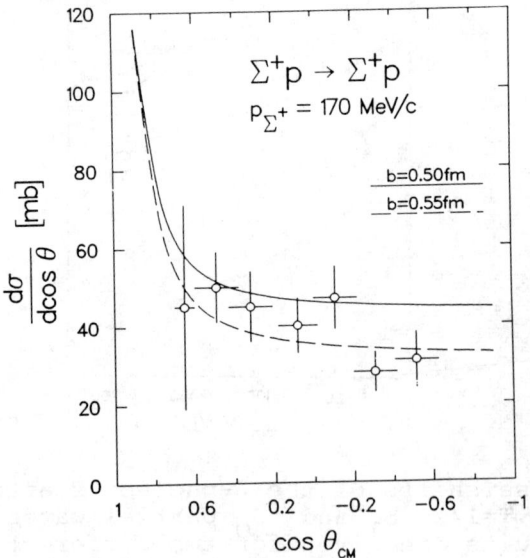

Fig. 10 Differential Σ^+p cross section at P_{Σ^+} = 170 MeV/C.

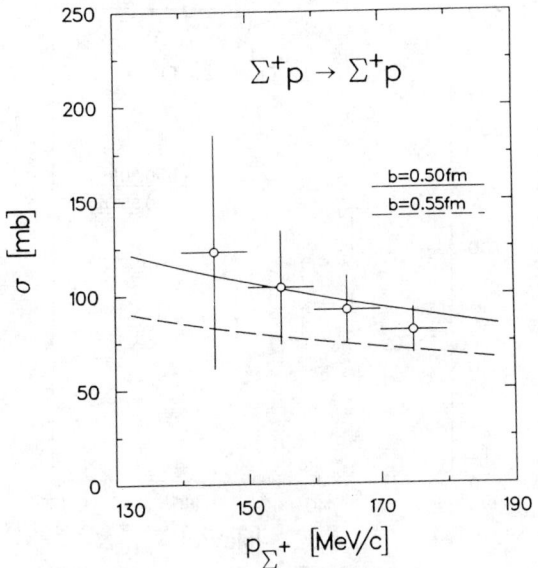

Fig. 11 Total $\Sigma^+ p$ cross section as a function of P_{Σ^+} in MeV/c.

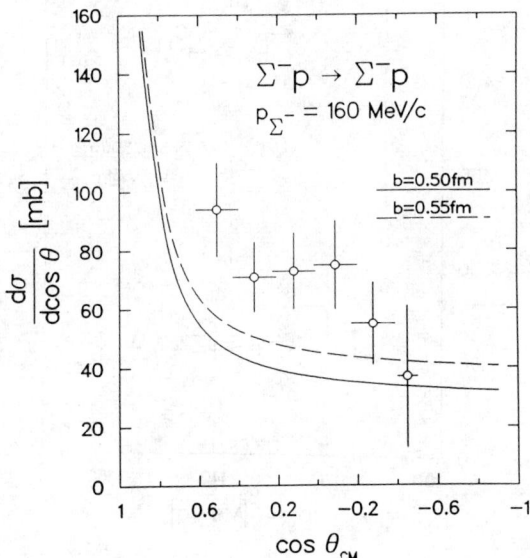

Fig. 12 Differential $\Sigma^- p$ cross section at P_{Σ^-} = 160 MeV/C.

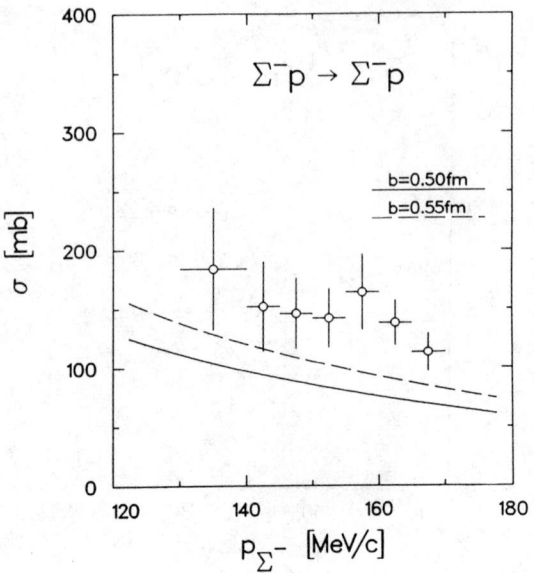

Fig. 13   Total Σ^-p cross section as a function of P_{Σ^-} in MeV/C.

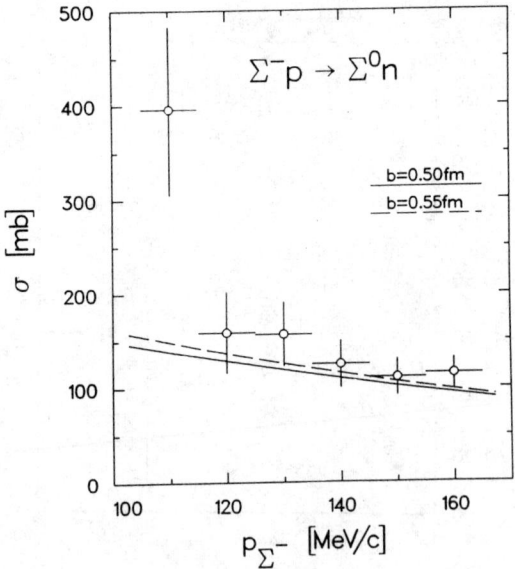

Fig. 14   Charge exchange $\Sigma^-p \rightarrow \Sigma^0p$ in cross section as a function of P_{Σ^-}.

The figures 10 to 14 show a comparison of the measured cross sections at low energies with the theory. One finds a surprising good agreement in view of the fact that these data a all calculated without fitting a single parameter to them.

THE EMC EFFECT

We have also extended[25] the method described above to the investigation of the EMC effect[26,27]. The quark momentum distribution in nucleons inside nuclei are calculated by solving for the nucleon-nucleon scattering in the nucleus a generalized Bethe-Goldstone equation with quark degrees of freedom in a nonrelativistic quark model, including Pauli blocking due to other nucleons and six-quark bags with different radii[25]. The structure functions[27] of a nucleon in ^{56}Fe and ^{107}Ag are calculated including also the binding effects[28] due to the average binding energy $|\varepsilon|$ of a nucleon, which leads to a rescaling of x.

$$x^A = x^{free} \frac{m_N}{m_N - |\varepsilon|} = \frac{Q^2}{2P \cdot q} \qquad (17)$$

Figures 15 and 16 give the ratios of the structure functions F_2 for Fe and Ag over the one of the deuteron as a function of the scaling variable x.

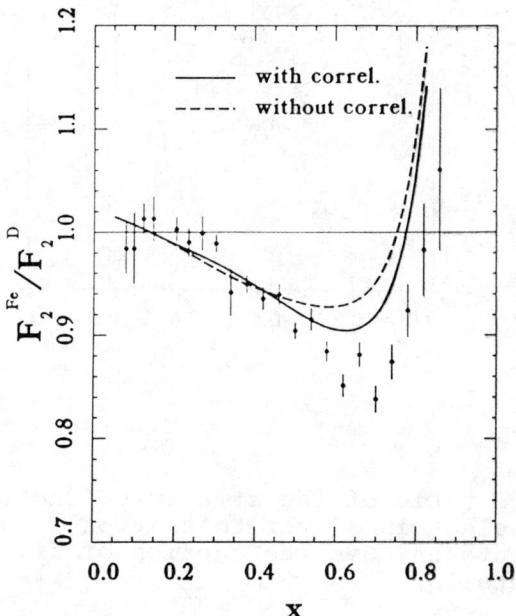

Fig. 15 The ratio of the structure function of a nucleon in iron to that of a nucleon in the deuteron. The solid line and dashed lines denote results with and without NN correlations, respectively.

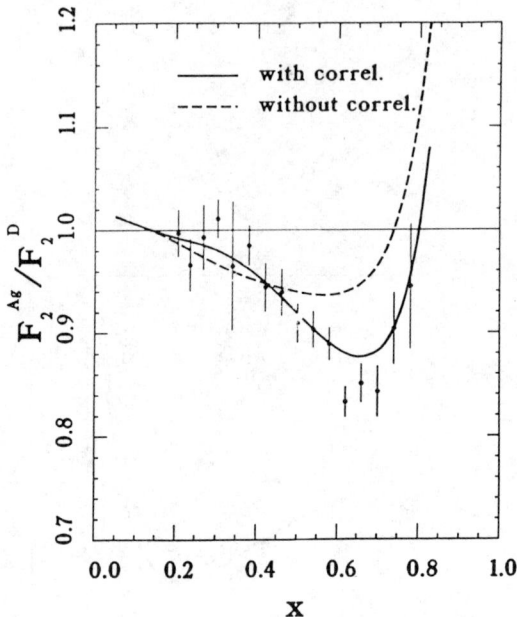

Fig. 16 The ratio of the structure function of a nucleon in silver to that of a nucleon in the deuteron. See the caption of fig. 15 for more details.

The comparison of the results with the data in figures 15 and 16 show that the binding energy of the nucleons $|\varepsilon|$ is important. But it also shows that the Pauli effect and the short range correlations are decisive for the good agreement. They are described by a Quark-Bethe-Goldstone equation on the quark level.

CONCLUSIONS

The NN phase shifts are calculated using the quark model with a QCD inspired quark-quark force. The short range part of the NN force is given by quark and gluon exchange. The long range part is described by π and σ-meson exchange. The data fitted in the model are five values connected with three quarks only: The nucleon mass, the Δ mass, the root mean square radius of the charge distribution of the proton including the pion cloud, the π-N and the σ-N coupling constant at zero momentum transfer. The ^1S and ^3S phase shifts are nicely reproduced. The short range repulsion is decisively influenced by the node in the $[42]_r$ relative wave function. Very important is the colour magnetic quark-quark force which enlarges the $[42]_r$ admixture. We are also able to describe the higher phase shifts.

In the OBEP's the short range repulsion is connected with the exchange of the ω-meson. But to reproduce the short range repulsion one had to blow up the ω-N coupling constant by a factor 2 to 3 compared to flavour SU_3. With quark and gluon exchange the best fit to the ω-N coupling constant lies close to the SU_3 flavour value. This fact strongly supports the notion that we have found the real nature of the short range repulsion of the NN interaction.

In chapter 4 we have applied the same theoretical description to the Λ-nucleon scattering including for the short range interaction quark-gluon exchange and for the long range the exchange of the whole pseudo-scalar octet and singlet mesons (π, K, η, η'). For the elastic Λ-proton cross section we obtain a good agreement with the data.

I would like to thank Dr.'s Bräuer, Fernandez, Shimizu and Dipl.-Phys. Straub with whom I was working on the results reported above.

REFERENCES

1. R. Arndt et al, Phys. Rev. C15, 1002 (1977) and R. Arndt, "Nucleon-Nucleon Interaction", 1977 Vancouver Conference.
2. K. Holinde, Phys. Rep. 68, 191 (1981).
3. D.A. Liberman, Phys. Rev. D16, 1542 (1977).
4. C.E. De Tar, Phys. Rev. D17, 323 (1978).
5. M. Harvey, Nucl. Phys. A352, 301 and 326 (1980).
6. V.G. Neudatchin, I.T. Obukhovsky, V.I. Kukulin, N.F. Golanova, Phys. Rev. C11, 128 (1975).
7. V.G. Neudatchin, Yu.F. Smirnov, R. Tamagaki, Progr. Theor. Phys. 58, 1072 (1977).
8. A. Faessler, F. Fernandez, G. Lübeck, K. Shimizu, Nucl. Phys. A402, 555 (1983).
9. M. Oka, K. Yazaki, Progr. Theor. Phys. 66, 551 and 572 (1981).
10. A. Faessler, F. Fernandez, G. Lübeck, K. Shimizu, Phys. Lett. B112, 201 (1982).
11. A. De Rujula, H. Georgi, S.K. Glashow, Phys. Rev. D12, 147 (1975).
12. A. Faessler, F. Fernandez, Phys. Lett. B124, 145 (1983).
13. K. Holinde, R. Machleidt, Nucl. Phys. A256, 479 (1976).
14. E.G. Lübeck, Diplomarbeit, Universität Tübingen, Februar 1982.
15. R. Arndt et al., Phys. Rev. C15, 1002 (1977); R. Arndt, "NN interaction", 1977 Vancouver Conference, p. 117.
16. G. Strobel, K. Bräuer, A. Faessler, Nucl. Phys. A347, 605 (1985).
    S. Furui, S.B. Khadkikar, A. Faessler, Nucl. Phys. A347, 619 (1985).
17. K. Bräuer, A. Faessler, F. Fernandez, K. Shimizu, Z. Phys. A320, 609 (1985).
18. A. Faessler, U. Straub, Tübingen preprint 1986 (to be published).
19. F. Fernandez, E. Oset, Salamanca preprint 04/85, to be published in Nucl. Phys.
20. A. Faessler, F. Fernandez, K. Bräuer, S. Kuyucak, K. Shimizu, Tübingen preprint, 1985, unpublished
21. G. Alexander et al., Phys. Rev. 173, 1452 (1968).
22. B. Sechi-Zorn et al., Phys. Rev. 175, 1735 (1968)
23. J.A. Kadyk et al., Nucl. Phys B27, 13 (1971).
24. G. Spitz, E.W. Schmid, Few body systems, 1, 37 (1986).
25. Y. Kurihara, A. Faessler, Tübingen preprint September 1986, to be published.
26. J.J. Aubert et al., Phys. Lett. 123B, 275 (1983).
27. R. Arnold et al., Phys Rev. Lett. 52, 727 (1984).
28. B.L. Birbrair et al., Phys. Lett. 166B, 119 (1986).

TIME DEPENDENT APPROACH TO PHOTODETACHMENT IN E, B AND PARALLEL E AND B FIELDS

William P. Reinhardt
Department of Chemistry, University of Pennsylvania, Philadelphia,
Pennsylvania 19104-6323

ABSTRACT

Photodetachment leading to outgoing s-wave and p_z-wave electrons in dc electric and magnetic fields along the z-axis, alone and in combination, is treated using a zero order model which neglects the (weak) interaction between the atom and detached electron. Threshold laws and interferences arising in the mixed field case are illustrated, and qualitatively discussed using a time dependent formulation which gives the total detachment cross section as the fourier transform of the dipole auto-correlation function. First order time dependent perturbation theory is then used to indicate the insensitivity of the results to inclusion of an atomic final state interaction.

I. INTRODUCTION

Photodetachment of negative ions, i.e., processes of the type

$$\hbar\omega + A^- \rightarrow A + e^-$$

has provided a wealth of information on the binding of electrons to neutral species [1] and provided a delicate probe [2, 3] of the Wigner threshold laws [4]. In 1979, Blumberg, Larsen and Itano[5] reported observation of magnetic field induced structures in the photodetachment of S^-, a case of an outgoing s-wave electron. Quite recently Bryant et. al. [6] have seen the first evidence of electric field induced structure in photodetachment of H^-, where the outgoing p-wave electron was oriented parallel to the external E-field. Data from the Bryant experiment is shown in Figure [1] where comparison with a simple model developed by Reinhardt [7, 8] and Overman [9] is made.

It is the purpose of the present note to develop the simple theoretical model of [7, 8, 9] to discuss in somewhat more detail the effect of E, B and combined parallel E and B dc fields on atomic photodetachment. The time dependent formulation of the photodetachment process is introduced in Section II, followed by consideration of photodetachment leading to outgoing s and p waves (the latter polarized parallel to the external fields) in the presence of E or B fields acting separately. A qualitative discussion of field induced changes in the empirically observed threshold laws follows. In Section III the effect of combined parallel E and B is studied

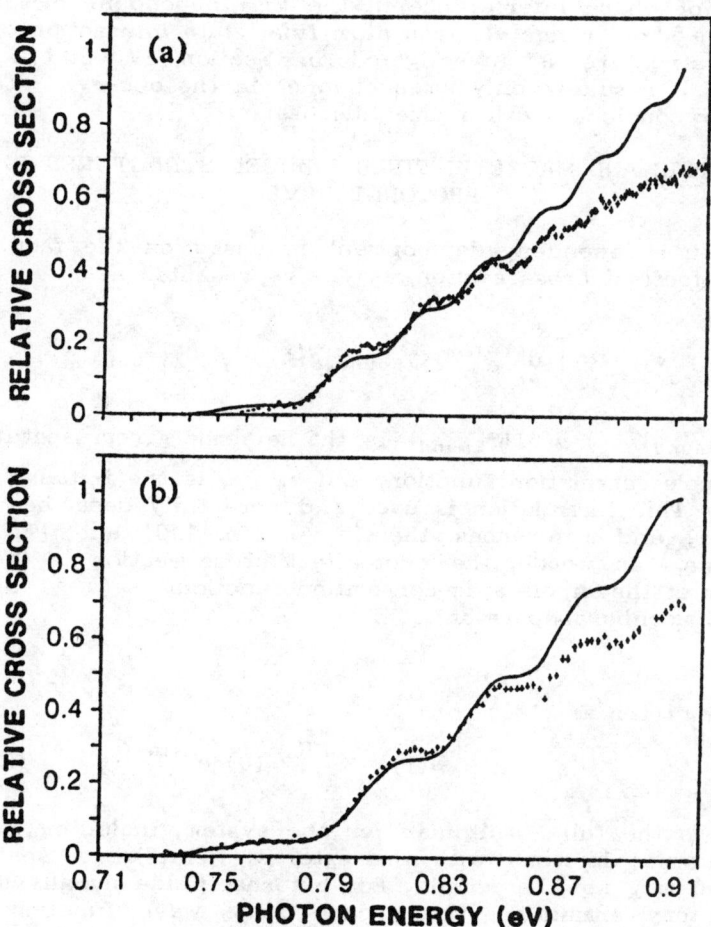

Figure 1. Experimental spectrum for photodetachment of H$^-$ in external electric fields of ca 100,000V/cm (a) and 150,000V/cm (b) with polarization $m_1 = 0$, the quantization axis being defined by the E-field. The solid lines are the result of the zero-order theory discussed here. It is evident that the theory is almost quantitative near the detachment threshold. From Ref.[6], with permission.

numerically, and the observed rich interference structure interpreted in terms of phase interference in the time dependent picture. The effect of a short range electron atom final state interaction on B-field induced structure is investigated in Section IV, and found, as expected, to result in only small changes in the observed structures. The paper concludes with a brief discussion.

II. ELECTRIC AND MAGNETIC FIELD INDUCED STRUCTURES IN ATOMIC PHOTODETACHMENT

The time dependent development is based on the fact that the usual photoeffect cross section may be represented as

$$\sigma(\omega) \propto \int_{-\infty}^{+\infty} dt\, e^{i\omega t} \langle \Psi_{Bound}|\vec{\mu}(t)\cdot\vec{\mu}(0)|\Psi_{Bound}\rangle, \quad (1)$$

where $\langle \Psi_{Bound}|\vec{\mu}(t)\cdot\vec{\mu}(0)|\Psi_{Bound}\rangle$ is the Heisenberg representation dipole-dipole correlation function, and Ψ_{Bound} is the initial state wave function. This formulation is used and more fully described in [7, 8, 9, 10, 11] and references therein. Refs. [10] and [11] contain derivations. In words, the photoeffect cross section is the fourier transform of the dipole auto-correlation function.

The Heisenberg operator

$$\vec{\mu}(t)$$

may be rewritten as ($\hbar=1$)

$$\vec{\mu}(t) = e^{iHt}\,\vec{\mu}(0)\,e^{-iHt} \quad (2)$$

where H is the full Hamiltonian for the system, including interaction of the atomic electrons with the external field(s). A spatial scale separation may now be noted: For external fields usually accessible in laboratory situations, the ground state wave function is little affected by the fields, and thus a reasonable approximation is

$$e^{\pm iHt}|\Psi_{Bound}\rangle = e^{\pm iE_{Bound}t}|\Psi_{Bound}\rangle \quad (3)$$

where E_{Bound} is the atomic ground state eigenvalue in the absence of the field. {Should this assumption break down, the field distorted ground state and energy must be used, necessitating a small additional effort, but changing nothing in principle.} Thus

$$\langle \Psi_{Bound}|e^{iHt}\vec{\mu}(0)e^{-iHt}\vec{\mu}(0)|\Psi_{Bound}\rangle$$

$$= e^{-iE_{Bound}t}\langle \Psi_{Bound}|\vec{\mu}(0)\cdot e^{-iHt}\vec{\mu}(0)|\Psi_{Bound}\rangle \quad (4)$$

and

$$\sigma(\omega) \propto \int_{-\infty}^{+\infty} e^{i(\omega-E_{Bound})t} \langle \Psi_{Bound}|\vec{\mu}(0) \cdot e^{-iHt}\vec{\mu}(0)|\Psi_{Bound}\rangle \quad (5)$$

This is the Schrodinger representation equivalent to Eqn 1, where it is seen that the ground state energy enters as a fourier shift, effectively setting the energy of the detachment threshold as the electron affinity. The Schrodinger correlation function.

$$\langle \Psi_{Bound}|\vec{\mu}(0) \cdot e^{-iHt}\vec{\mu}(0)|\Psi_{Bound}\rangle \equiv \langle \phi(0)|\phi(t)\rangle \quad (6)$$

has a straight forward physical interpretation: $\vec{\mu}(0)|\Psi_{Bound}\rangle$ is the dipole operator acting on the initial state, and $e^{-iHt}\vec{\mu}(0)|\Psi_{Bound}\rangle$ is this non-stationary state time evolved according to the full dynamics. As this "wave packet" spreads to large distances from the atomic core, the effects of even relatively weak external fields become important, and thus affect the time evolution of the overlap (or correlation function) $\langle \Psi(0)|\phi(t)\rangle$ and thus the spectrum. For example in the presence of an electric field, the packet accelerates in the direction of the field, and thus $\langle \phi(0)|\phi(t)\rangle$ is damped, broadening the spectral features; if the electronic packet moves away from the atomic core, but then returns, as it will do if magnetically confined transversely to the B field direction, $\langle \phi(0)|\phi(t)\rangle$ will show a decay followed by recurrence, thus leading to oscillatory structures in the photodetachment spectrum. We now introduce a simple model to illustrate these ideas.

a) Time evolution and Feynman propagators

The simplest approximation allowing investigation of the influence of external fields on photodetachment is to assume: 1) a simple model initial state, of the symmetry appropriate to the outgoing electron (i.e. the symmetry of the dipole operator for the appropriate polarization times the symmetry of the initial state of the detached electron); 2) carry out time propagation analytically using the approximation that the dynamics in the final state are governed by the kinetic energy and potential energies of interaction with the external fields. That is we neglect the electron-atom final state interaction. As negative ions often do not bind states corresponding to the symmetry of the detached electron, this potential is usually weak, and this is a reasonable approximation. For example in detachment of H^- and "s" electron is detached, and the outgoing electron is a p-wave, whose dominant dynamics is determined by the centrifugal potential near the residual H atom, and by the centrifugal potential plus external fields as the electron moves away. The advantage of this approach is that the Feynman propagators in cartesian coordinates may be written in closed form for quadratic Lagrangians, and thus time evolution in the presence of only the external fields is analytic, and especially simple if the initial model state is also separable in cartesian coordinates, namely if the initial state is approximated by a single Gaussian in $r^2 = x^2 + y^2 + z^2$, or by a superposition of such functions. [As in the case of the weak field assumption of Eqn (3), this is a matter of convenience, not necessity. Somewhat more work is involved but the basic ideas are the same, whether or not Gaussians are used to approximate the initial state, their special advantage being that $\exp[-(x^2 + y^2 + z^2)]$ is separable in both the local spherical coordinates appropriate for definition of the initial state, and in the cartesian coordinates, suitable for the longer range interactions with the external field. This no frame transformation [12] need be made in carrying out qualitative studies such as done here].

In cartesian coordinates, free propagation of any initial packet $\Psi(x)$ is carried out via

$$G(xt,x't') = \sqrt{\frac{m}{2\pi i \hbar \tau}} \; \exp\left[\frac{im}{2\hbar\tau}(x-x')^2\right] \tag{7}$$

where $\tau = t - t'$ and

$$\Psi(x,t) = \int_{-\infty}^{+\infty} dx' \; G(x,t,x',0)\Psi(x,0) \tag{8}$$

If a dc electric field is introduced, $G(xt, x't')$ becomes

$$G_E(xt,x't') = \sqrt{\frac{m}{2\pi i \hbar \tau}} \; \exp\left[\frac{i}{\hbar}\left[\frac{m(x-x')^2}{2\tau} - \frac{E}{2}\tau(x+x') - \frac{E^2\tau^3}{24m}\right]\right] \tag{9}$$

or, if there is quadratic confinement due to a magnetic field with cyclotron frequency ω_B

$$G_B(xt,x't') = \sqrt{\frac{m\omega_B}{2\pi i \hbar \sin\omega_B\tau}} \; \exp\left[\frac{im\omega_B}{2\hbar \sin\omega_B\tau}\left[(x^2+(x')^2)\cos\omega_B\tau - 2xx'\right]\right] \tag{10}$$

Three dimensional propagation is carried out as follows: for the normal photoeffect (no external fields)

$$\Psi(x,y,z,t) = \int dx'dy'dz' \; G(xt,x'0') \; G(yt,y'0') \; G(zt,z'0')\Psi(x',y',z',0) \tag{11}$$

For photodetachment in a E field along the z axis

$$\Psi(x,y,z,t) = \int dx'dy'dz' \; G(xtx'0') \; G(yty'0') \; G_E(zt,z'0') \; \Psi(x',y',z',0)$$

For photodetachment in B field along the z axis

$$\Psi(x,y,z,t) = \int dx'dy'dz' \; G_B(xt,x'0') \; G_B(yt,y'0') \; G(zt,z'0')\Psi(x',y',z',0) \tag{12}$$

c) Detachment Leading to a p-wave Final State

We consider the case of photodetachment of an initial electron in an atomic s-orbital, giving an outgoing p-wave, with polarization such that the $m_l = 0$ substate is chosen with respect to the quantization axis defined by the external E and/or B field. That is a p_z-wave is produced, with a node in the x,y plane at $z = 0$. Within the framework of the models considered here such an initial state is represented as

$$z \exp[-(x^2 + y^2 + z^2)/a] \tag{17}$$

where again we choose $a = 1$, as appropriate to detachment of an atomic negative ion. This choice yields a correlation function which is a product of the factor

$$\left[\frac{1}{1+it}\right]^{1/2} \left[\frac{-E^2 t^3}{2(2t-i)} + \frac{1}{1+it} - \frac{E^2 t^2}{4i(t-i)}\right]$$

$$\times \exp\left[\frac{E^2 t^2}{24(2t-i)}(-2i t^2 - 7t + 3i)\right] \tag{18}$$

which corresponds to motion along the z-axis and the factor

$$\left[\frac{1}{\cos(Bt) + i\frac{(4+B^2)}{B}\sin(Bt)}\right] \tag{19}$$

which represents motion along the x and y axies. Figure [3] shows the field free model detachment cross section and the effects of independent E and B fields on the detachment cross section near threshold.

d) Threshold laws

In the absence of external fields the correlation functions are simple and may be analytically fourier transformed. For the outgoing s-wave

$$\left[\frac{1}{1+it}\right]^{3/2} \tag{20}$$

while for the outgoing p-wave

$$\left[\frac{1}{1+it}\right]^{5/2} \tag{21}$$

The general fourier transform of functions of this type is given by [13].

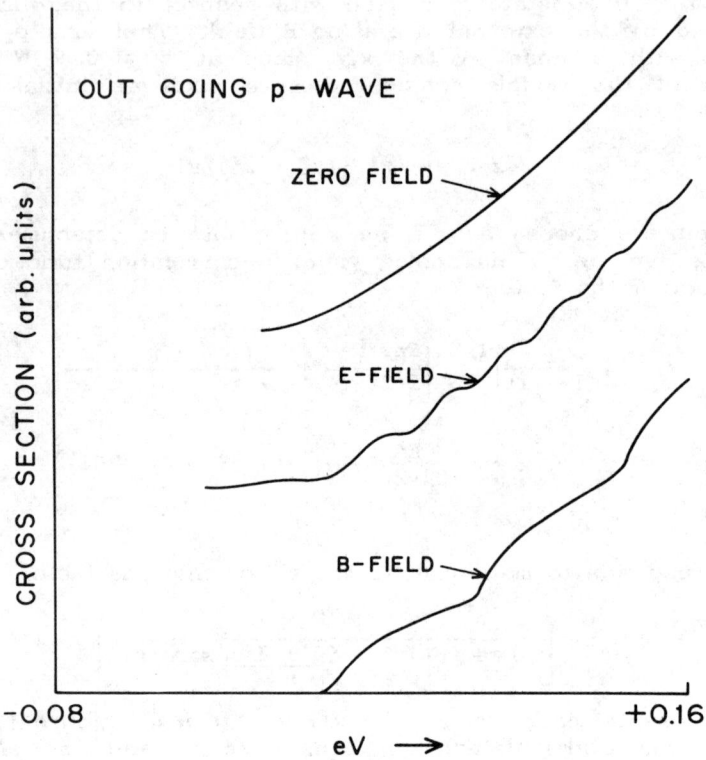

Figure 3. Zero final state interaction model for photodetachment leading to an outgoing p-wave with polarization such that $m_l = 0$, where the quantization axis is that defined by the external E or B field. The fields are those of Figure 2, and the three spectra have been likewise displaced. Comparing the top and botton spectra confirms the fact that the two dimensional confinement of the quadratic Zeeman terms gives rise to a sharp s-wave like $E^{1/2}$ threshold in the field, as opposed to the usual $E^{3/2}$ Wigner law in the absence of the applied field.

$$\sigma(\omega) \sim \int_{-\infty}^{+\infty} dt \, e^{i\omega t} \, \frac{1}{[1 + it]^\nu} \sim \omega^{\nu-1} \exp(-\omega) \tag{22}$$

Thus the models considered here have the expected Wigner law thresholds [4]

$$\sigma(k) \sim k^{2\ell + 1} \tag{23}$$

where ℓ is the angular momentum and k the outgoing momentum of the detached electron.

What is the effect of the external magnetic fields, which confines the motion in the x, and y directions, and thus decay of the correlation function is due only to motion in the z direction? The time dependence of this decay is thus

$$\left[\frac{1}{1 + it}\right]^{1/2} \tag{24}$$

for an out going s-wave, and

$$\left[\frac{1}{1 + it}\right]^{3/2} \tag{25}$$

for the outgoing p-wave. The net effect as correctly foreseen by Blumberg et. al. [5] is to change the Wigner law from

$$\sigma(k) \sim k^{2\ell + 1} \tag{26}$$

to

$$\sigma_B(k) \sim k^{2\ell - 1} \tag{27}$$

in the presence of the confining B field. Thus in the presence of the field we can expect singular k^{-1} threshold behavior for $\ell = 0$, and sudden k^{+1} onsets of the type usually associated with s-wave photodetachment, when a p-wave is produced in the magnetic field. These predictions are clearly born out empirically by the results of Figures [2] and [3].

III. PHOTODETACHMENT IN PARALLEL E AND B FIELDS: INTERFERENCE EFFECTS

The correlation functions of Eqns [14, 15] and Eqns [18, 19] allow modeling of photodetachment in combined paralleled E and B fields. Figure [4] shows detachment leading to an outgoing s-wave in a field of B = 0.001 atomic units (ca 4.7 x 10^6 gauss) and a series of E fields ranging from 0.33 x 10^{-5} through 2 x 10^{-5} atomic units. This

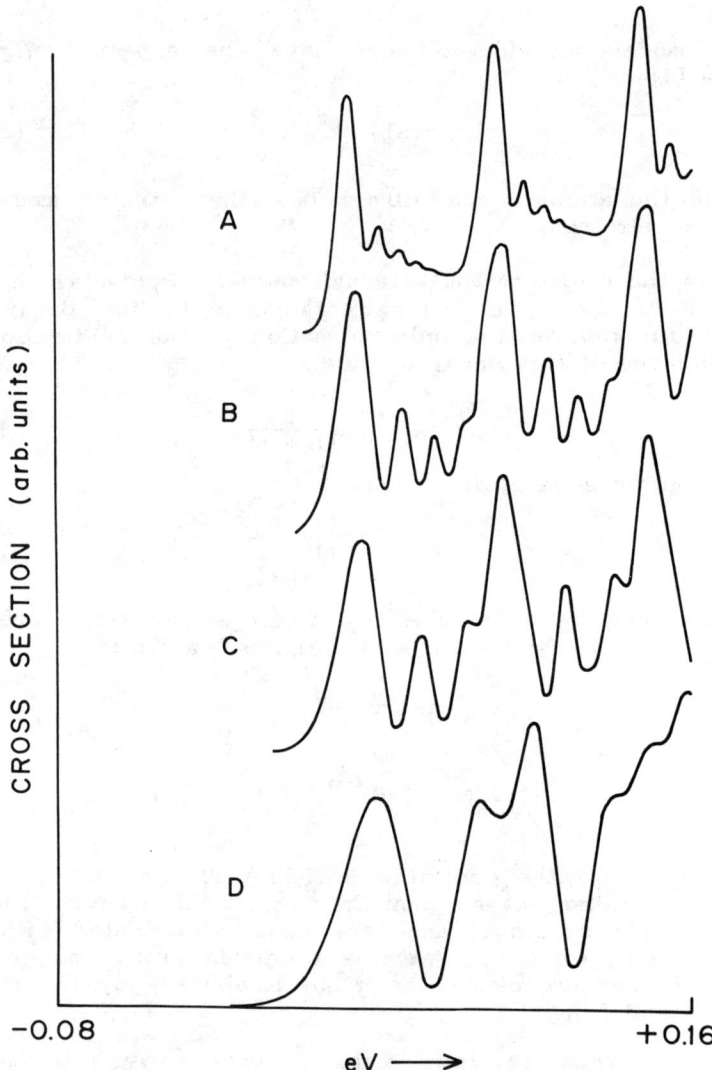

Figure 4. Photoproduction of an s-wave in combined parallel E and B fields. $B = 10^{-3}$ au in all cases, and $E = 0.33 \times 10^{-5}$, 0.67×10^{-5}, 1.0×10^{-5}, and 2.0×10^{-5}, respectively in going from A through D. {See the caption of Figure 2 for conversions to lab units}. A rich interference structure is evident.

series of figures may be compared with the results shown in Fig[2] where the effects of E and B fields alone is illustrated. A similar series for p_z-wave production as a function of a varying E at fixed B is shown in Figure [5]. Here the results should be compared to those of Fig. [3]. It is evident that the effect of the combined fields is to often produce modulations of far greater amplitude, and a higher frequency that if the E or B field were applied separately. What is the origin of this evident interference effect?

In the time dependent picture we must look at the effect of a small electric field on the time evolution of $\langle\phi(t)|\phi(o)\rangle$ in a B-field. Figure [6A] shows the correlation function for s-wave photoproduction in a B-field of 0.04 a.u. (The corresponding detachment cross section appears as Fig [7A].) the effect of a parallel E field is to replace the factor

$$\left[\frac{1}{1 + it}\right]^{1/2} \tag{28}$$

in the overall correlation function by the factor

$$\left[\frac{1}{1 + it}\right]^{1/2} \exp\left[\frac{E^2 t^2}{24(2t - i)} (-2it^2 - 7t + 3i)\right] \tag{29}$$

which introduces a new damping, but, more importantly, a new phase modulation. Comparison of Figs. 6A and 6B shows that the relatively small E field (E = 0.0035 atomic units) gives no major damping at short times, but quite dramatically has changed the phase of the second recurrence of the magnetically confined wave packet. This change at twice the usual Landau recurrence time induces frequency domain structure at half the usual frequency of the magnetic oscillations. A similar phase change at three times the recurrence time would induce interference oscillations at 1/3 of of the usual Landau frequency. Such effects are clearly in evidence in Figs. [4] and [5]. One thinks of the possibility of using field induced creation of such windows to carry out interesting switching processes.

IV. EFFECT OF FINAL STATE INTERACTIONS

How does addition of an approximate attractive interaction between the residual atom and the outgoing detached electron affect the results of the simple model of the previous sections. As a preliminary investigation of the effect of such an improved model, we have carried out a first order correction via time dependent perturbation theory. Namely, with no final state interaction (FSI)

$$\langle\phi(t)|\phi(0)\rangle^o = \left[\frac{1}{1 + it}\right]^{1/2} \frac{1}{\cos(Bt) + i\frac{(4 + B^2)}{B}\sin(Bt)} \tag{30}$$

for photodetachment leading to an s-wave in a B field.

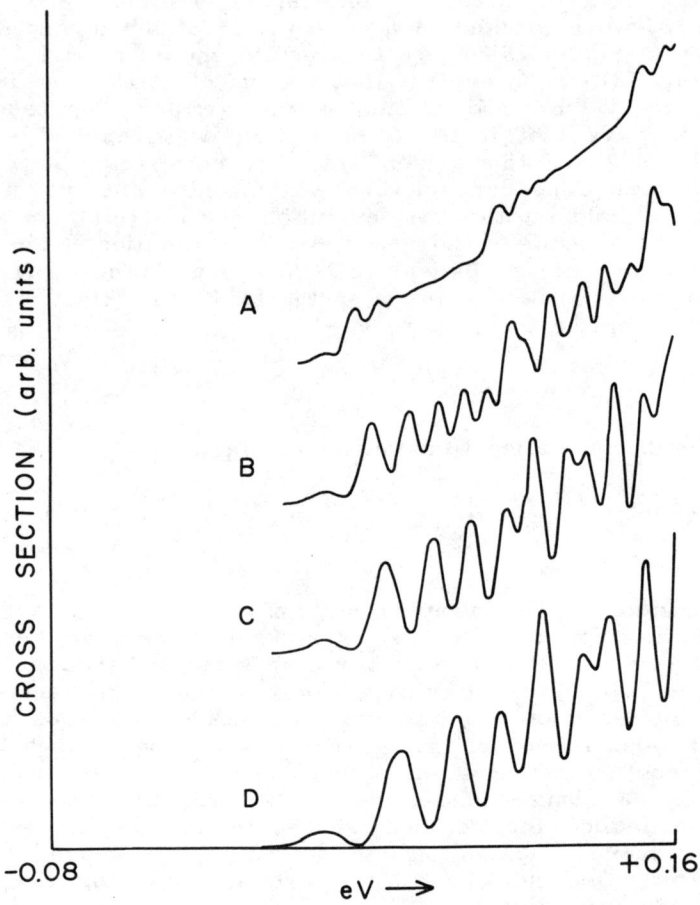

Figure 5. Photoproduction of a p-wave in combined parallel E and B fields. $B = 10^{-3}$ au in all cases and $E = 0.33 \times 10^{-5}$ au through 1.33×10^{-5} au, in steps of 0.33×10^{-5}, in A through D respectively. Polarization is such that $m_l = 0$ in the field defined frame.

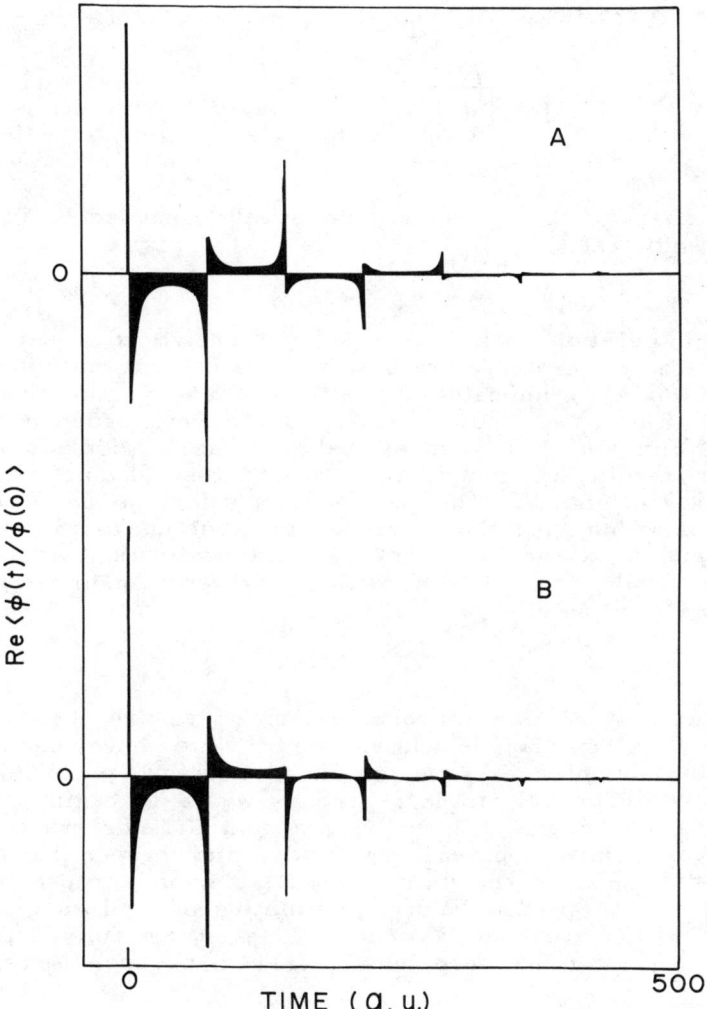

Figure 6A. Real part of the correlation function $\langle\phi(t)|\phi(0)\rangle$ for photodetachment (leading to an outgoing s-wave) in an intense B-field (B = 0.04 au). The field induced recurrences are easily seen, as the area "under" the function above or below the real axis has been shaded.
Figure 6B. A smaller (E = .0035 au) parallel electric field induces modulations in the phase of the correlation function, producing substantial modulations of the detachment spectrum.

A first order correction to this correlation function is obtained by taking (in a mixed notation)

$$\langle\phi(t)|\phi(0)\rangle^1 = \langle\phi(0)|\left[-\frac{i}{\hbar}\int_0^t G_B(\bar{r}t,\bar{r}'t')V(\bar{r}')G_B(\bar{r},'t',\bar{r}''0)d^3r'dt'\right]|\phi(0)\rangle \tag{31}$$

where the effect of the B-field is automatically included to all orders. For the specific case

$$V(\vec{r}) = -\gamma \exp(-r^2/\delta^2) \tag{32}$$

all spatial integrations can be carried out analytically, leaving only the single time integration from 0 to t to be carried out numerically. Figure [7A and 7B] indicates the effect of adding the first order correction with $\delta = 1$ and $\gamma = 0.25$ to the zero order correlation function of Eqn[30]. It is evident that while small modulations of the zero order result are found, the basic threshold and modulation behavior is still intact. This is not surprising, as it is the long range forces which give the dominant contributions to the threshold behavior, and it is the long range B-field confinement which gives rise to the modulations. It is nevertheless reassuring to see that the effects are indeed small.

V. DISCUSSION

Stimulated by the recent observation of electric field induced oscillations in the photodetachment of H^- we have carried out further developments of some simple models of photodetachment leading to production of atomic s- and p_z-waves to begin to catalog the range of phenomena at hand when E and B fields are combined. This is clearly only a preliminary study, and proper inclusion of more realistic modes of the initial atomic state, and proper inclusion of the FSI will be needed before quantitative as well as qualitative comparison with experiment is made. However, we fully expect the basic types of structure seen here to remain as such modifications are made.

ACKNOWLEDGEMENTS

The enthusiastic encouragement of Howard Bryant, and the preliminary work of Lillian Overman are gratefully acknowledged. This work was supported in part by grant CHE84-16459 from the National Science Foundation.

and, finally, for photodetachment in parallel E and B fields along the z axis

$$G_{B,E}(\vec{r}t,\vec{r}'t') = G_B(xt,x't')\, G_B(yt,y't')\, G_E(xt,z't') \quad (13)$$

b) Detachment leading to an s-wave final state

An out going s-wave implies spherical symmetry of $\vec{\mu}|\Psi_{Bound}\rangle$. We take this to suggest a generic approximation

$$|\phi(0)\rangle = \vec{\mu}(0)|\Psi_{Bound}(x,y,z)\rangle = \exp(-(x^2+y^2+z^2)/a^2)$$

for the study of the types of effects to be expected. As the scale size of this initial state is that of the atomic state before photodetachment we choose $a = a_0 = 1$ a.u., giving a scale size of one Bohr radius. The correlation function $\langle\phi(t)|\phi(o)\rangle$ is, analytically given by the simple formula, ($\hbar = e2 = m_e = 1$)

$$e(t) \equiv \left[\frac{1}{1+it}\right]^{3/2} \exp\left[\frac{E^2 t^2}{24(2t-i)}(-2i\,t^2 - 7t + 3i)\right] \quad (14)$$

where E is the field in atomic units (1 a.u. = 5.1×10^9 volts/cm). It is a numerical fourier transform of this function at a field of 0.00002 a.u. (or ca 100,000V/cm) is shown in Figure [2] where it is compared with the field free model detachment. It is seen, that, as discussed earlier in Ref. [7], the usual Wigner $E^{1/2}$ threshold law is modified by field induced oscillations.

Figure [2] also illustrates the effect of a B-field on photodetachment producing a final s-wave electron. The sharp peaks discussed in Refs. [8], and [5] in the experimental studies of S⁻-detachment in an external B-field are easily seen. In this case the correlation function is

$$\left[\frac{1}{\cos(Bt) + i\left[\frac{(4+B^2)\sin(Bt)}{B}\right]}\right]^{3/2} \quad (15)$$

which reduces (as does the correlation function of Eqn 14) to

$$\left[\frac{1}{1+it}\right]^{3/2} \quad (16)$$

in the absence of the external field. We will return to discussion of this reduction in the discussion of threshold laws in part d) of this section.

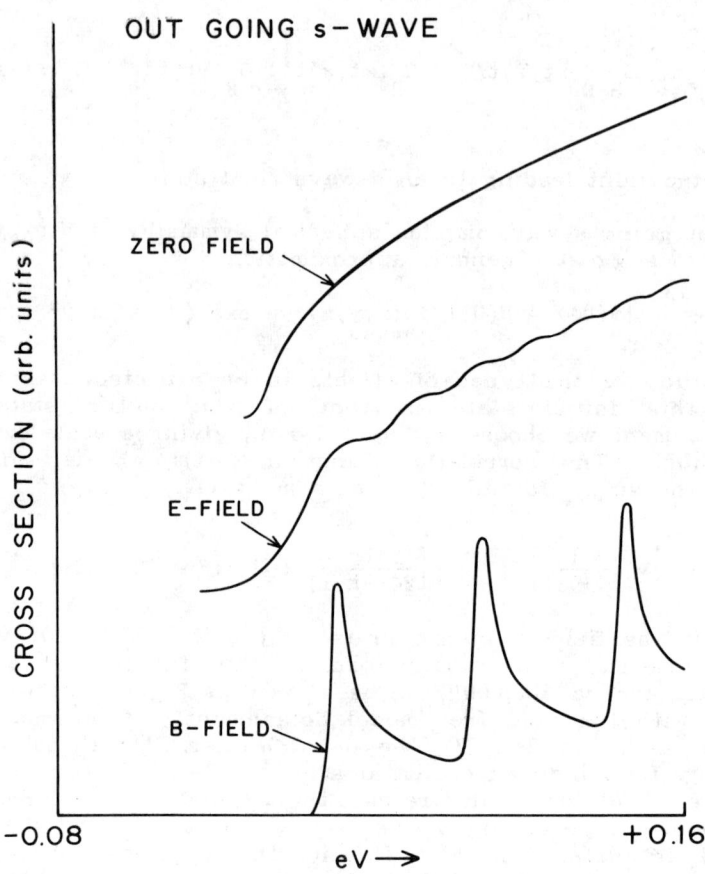

Figure 2. Model for photodetachment leading to an outgoing s-wave as implemented with zero external field, in an E-field of 2×10^{-5} au, and in a B-field of 10^{-3} au. A small broadening has been introduced in this and all other spectra shown here for realistic comparison with experiment. Figure 4 shows the result of mixed (parallel) E and B fields of the same magnitude as shown here. The three spectra have been displaced on the vertical axis to simplify comparison. The conversions appropriate to the results presented here are (approximately) 1 au E-field is 5.1×10^9 V/cm and 1 au B-field is 4.7×10^9 gauss.

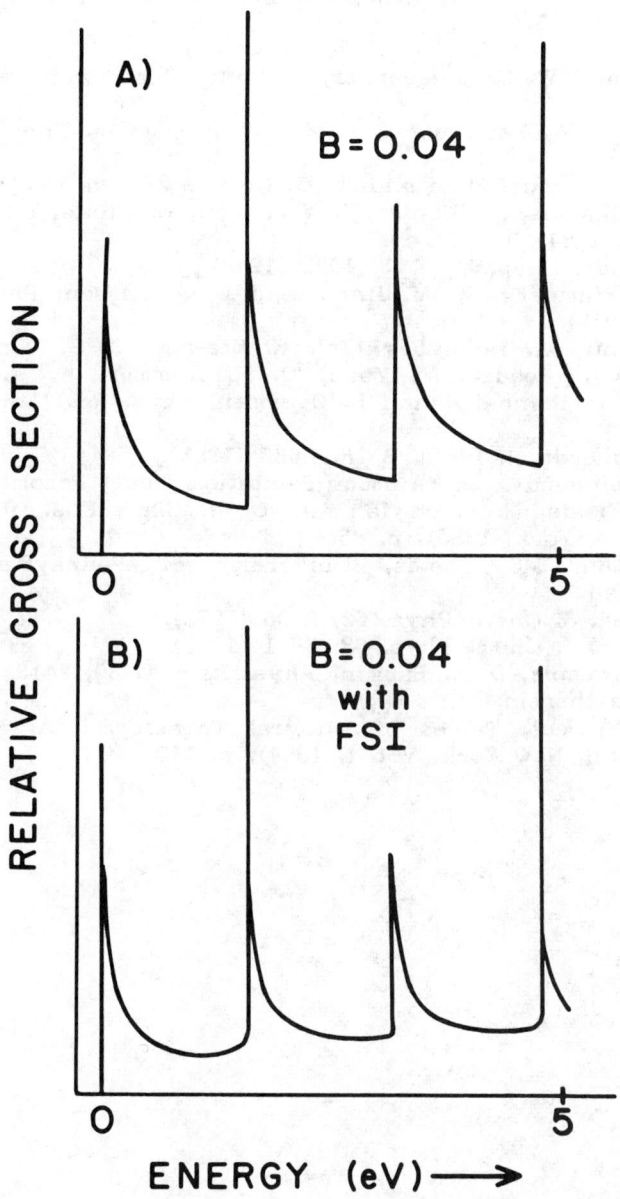

Figure 7. Effect of a short range attractive final state interaction (FSI) on B-field induced modulations near an s-wave photodetachment threshold. As discussed in the text the FSI was included via first order time-dependent perturbation theory.

REFERENCES

1. H. Hotop and W. C. Lineberger, J. Phys. Chem. Ref. Data, 14, 731 (1985).
2. H. Hotop, T. A. Patterson and W. C. Lineberger, Phys. Rev. A 8, 762, (1973).
3. R. D. Mead, K. R. Lykke and W. C. Lineberger, in Electronic and Atomic Collisions, J. Eichler, I. V. Hertel, N. Stolterfoht, (Elsever, New York, 1984), p. 721.
4. E. P. Wigner, Phys. Rev. 73, 1002 (1948).
5. W. A. M. Blumberg, W. M. Itano, and D. R. Larson, Phys. Rev. A. 19, 139 (1979).
6. H. C. Bryant, A. Mohagheghi, J. E. Stewart, J. B. Donahu, C. R. Quick, R. A. Reeder, V. Yuan, C. R. Hummer, W. W. Smith, S. Cohen, W. P. Reinhardt and L. Overman, Phys. Rev. Letts 58, 2412 (1987).
7. W. P. Reinhardt, J. Phys. B 16, L635 (1983).
8. W. P. Reinhardt, in Atomic Excitation and Recombination in External Fields, M. H. Nayfeh and C. W. Clark Eds., (Gordon and Breach, New York, 1985), p. 85.
9. L. Overman, MS Thesis, University of Pennsylvania, 1985, unpublished.
10. E. J. Heller, J. Chem. Phys. 68, 2006 (1978).
11. E. J. Heller, J. Chem. Phys. 68, 3891 (1978).
12. See, for example, D. A. Harmin, Phys. Rev. A. 30, 2413 (1984), and references therein.
13. Bateman Project, Tables of Integral Transforms, A. Erdelyi Ed. (McGraw-Hill, New York, Vol. I, 1954), p. 118.

DEUTERON–NUCLEUS COLLISIONS: REDUCTION OF A MANY–FERMION COLLISION SYSTEM TO AN EQUIVALENT THREE–BODY MODEL

R. Kozack
Los Alamos National Laboratory, Los Alamos, NM 87545

F. S. Levin
Brown University, Providence, RI 02912

ABSTRACT

Elastic scattering and breakup of deuterons by nuclei are strongly coupled processes usually treated together theoretically by means of a three–body model. This model consists of the neutron (n) and proton (p) forming the deuteron, and a structureless core (A) representing the unexcited target nucleus. Their collision dynamics is governed by an empirical Hamiltonian H_M which consists of the kinetic energy operators, the neutron–proton interaction V_{np} which binds the deuteron, plus neutron–nucleus and proton–nucleus absorptive interactions. These latter are the relevant optical potentials evaluated at half the bombarding energy. Although H_M has been successful in fitting data up to deuteron energies of 400 MeV, only very recently have some of the underlying theoretical problems been solved. The most fundamental problem is the reduction of the deuteron–nucleus, many–fermion Hamiltonian H_{A+2} to an equivalent three–body Hamiltonian H_3 which describes those collisions in which the target nucleus remains unexcited. Approximations to H_3 should then yield H_M. In making the $H_{A+2} \to H_3$ reduction, Pauli-principle exchange effects must be properly accounted for. It has been argued that inclusion of such effects will lead to a Pauli–blocked V_{np} as the neutron–proton interaction in H_3 and thus in H_M, where Pauli-blocking means that n or p cannot be in already occupied orbitals. Kozack and Levin, however, have shown that no such Pauli-blocking occurs in an exact reduction of H_{A+2}. As shown herein, the resolution of this inconsistency is the existence of two different definitions of the equivalent three–body model. The definition used by Kozack and Levin leads to an H_3 containing a non–Pauli-blocked V_{np} and the nucleon–nucleus optical potentials in the form originally predicted by Austern and Richards; this H_3 leads by standard approximation to H_M. The other definition yields a Hamiltonian H'_3, which contains a Pauli-blocked V_{np}, although not in the previously expected form, as well as Pauli-blocked optical potentials. These latter seem not to have been previously anticipated. It is concluded that modifying H_M to include a Pauli-blocked V_{np} is correct only if the optical potentials are also modified away from the standard forms currently used in data analyses.

© American Institute of Physics 1987

INTRODUCTION

As a general rule, the energy level spectra of quantal systems contain a series of excited, particle–stable states lying above the ground state. The deuteron — i.e., the nucleus formed from one neutron and one proton — is an intriguing exception to this rule, since its spectrum contains no excited bound states. Furthermore, on the scale of typical nuclear two–fragment dissociation energies, the deuteron ground state is very weakly bound, requiring an injection of only 2.225 MeV to break it up into a free neutron and proton. An important consequence of this low dissociation threshold is that in deuteron–induced collisions in which the bombarding energy is not too low, the probability of projectile breakup is significant. As a result, deuteron elastic scattering can be strongly influenced by deuteron breakup, i.e., these two elastic processes are strongly coupled. This strong coupling suggests that the theoretical description of deuteron elastic scattering should be different than that typically used to describe the elastic scattering of either single particles or strongly–bound, composite–particle projectiles.

The typical description of elastic scattering is based on a property that makes this collision process unique, viz, no matter how many particles comprise the target and the projectile, elastic scattering can always be analyzed as if it is a one–body scattering event. Because of this, one can construct an equivalent one–body Hamiltonian for elastic scattering in which the effects of all the non–elastic processes, of the numbers of particles, of particle identity, etc., are represented by an effective potential. This effective potential — known as an optical potential or as a complex potential well — is energy dependent, absorptive, and non–local. One of the best known and most elegant discussions of this potential is that of Feshbach, who examined the scattering of a single particle by a structured target, basing his analysis on the use of projection operators.[1,2] This analysis is easily extended to the case of composite–particle projectiles.[1] The optical model has been widely used for data analysis, e.g., in electron–atom and nucleon–nucleus scattering, where by "nucleon" is meant a neutron or proton.

Two aspects of the optical potential formulation of elastic scattering are worth commenting on here. First, the optical potential description represents a reduction of the many–particle Hamiltonian for the full collision system to an equivalent one–body Hamiltonian describing elastic scattering only. The second point concerns the complexity of the optical potential: because it is an effective interaction depending on the full many–particle dynamics, it is too difficult to evaluate exactly. As a consequence, an empirical optical model potential is used in fitting data.[3]

Although deuteron–nucleus elastic scattering can also be formulated as an equivalent one–body problem, the ease with which breakup can occur suggests that in this case elastic scattering and elastic breakup of the projectile should be treated together. Instead of a one–body description involving the relative coordinate or relative momentum between the deuteron and the target nucleus CM's, the inclusion of the channel with a free neutron and a free proton (breakup) means that a three-body description is required. These three bodies are the neutron (n) and proton (p) forming the deuteron (d) plus the unexcited and unexcitable target nucleus (A), represented by a structureless mass.

Since this three-body description is to be equivalent to the full deuteron-nucleus system when it is restricted to the elastic scattering — (d,d) — and breakup — (d,np) — events, then there must be a reduction of the deuteron-nucleus many-particle Hamiltonian H_{A+2} to an equivalent three-body Hamiltonian H_3 from which the elastic scattering and breakup amplitudes can in principle be calculated. The effective interactions appearing in H_3 are too complicated to evaluate exactly and, just as in the pure elastic scattering case, analyses of (d,d) and (d,np) data have been based on empirical (three-body) model Hamiltonians.[4] In the model that has been most widely used in analyzing data, the Hamiltonian H_M has the form[4]

$$H_M = H_0 + V_{np} + \mathcal{V}_n + \mathcal{V}_p, \qquad (1)$$

where H_0 is the sum of the kinetic energy operators, V_{np} is the neutron-proton interaction that binds the deuteron and $\mathcal{V}_n(\mathcal{V}_p)$ is the neutron-nucleus (proton-nucleus) elastic scattering optical potential evaluated at half of the incident deuteron's kinetic energy. The relevant Schrödinger equation is

$$(E - \varepsilon_0 - H_M)\psi_M(n,p) = 0, \qquad (2)$$

where ε_0 is the energy of the target nucleus.

Because of the facts that the empirical Hamiltonian H_M should be an approximation to H_3 and the nucleon-nucleus optical potentials are present in (1), one expects that the nucleon-nucleus optical potentials should somehow be incorporated in H_3. Unfortunately, the derivation of H_3 via a standard projection operator approach does not yield expressions that show how the optical potentials are incorporated, if indeed they are. The need to understand the structure of H_3, so that the structure of the empirical Hamiltonian(s) can then be justified if not validated, is evident.

Lack of insight via a standard projection operator approach has not blocked progress in this field. Using as their starting point a simple ansatz in which the identity of the nucleons n and p in the deuteron with those in the target was ignored, Austern and Richards[5] seem to have been the first to infer a form for H_3, denoted H_{A-R}. Their Hamiltonian is a sum of five terms, viz, the relevant kinetic energy operators; the (bare) neutron-proton interaction which binds the deuteron; the two optical potentials describing elastic scattering of n and of p each with the target nucleus, but modified by subtracting the kinetic energy operator of the other or "spectator" nucleon from the energy appearing in the optical potential (see Eq. (20) for a mathematical statement of this); and an absorptive three-body interaction involving in an unspecified manner the three labels n, p and A. It is not difficult to see on physical grounds that H_{A-R} is an example of the most general H_3 one can postulate, viz, a Hamiltonian consisting of the kinetic energy operators plus four unspecified potentials (three pair interactions and one three-body interaction). However, not only did Austern and Richards propose specific forms for the absorptive n-A and p-A interactions, they also suggested that the spectator kinetic energy operators appearing in the optical potentials could be replaced by half of the incident deuteron's kinetic energy. This prescription plus neglect of the absorptive three-body interaction changes H_{A-R} to H_M, Eq. (1) above.

The empirical three-body Hamiltonian, fleshed out with specific forms for the interactions, has been reasonably successful in its applications.[4] However, neither it nor the Austern-Richards Hamiltonian from which it is obtained are wholly well-founded theoretically. Three of the questions which cannot be answered within the framework of the Austern-Richards approach are: Are there effects due to the Pauli principle that are not contained in H_{A-R} or in the empirical Hamiltonian? Would an exact reduction of H_{A+2} (not employing an ansatz for the deuteron-nucleus wave function) yield n-A and p-A absorptive interactions in the same optical potential form as originally deduced for H_{A-R}? What is the form of the three-body absorptive interaction (which is neglected in the empirical Hamiltonian and which would provide one source of corrections to it)?

Since Austern and Richards used an ansatz which did not take into account the identity of the incident neutron and proton with the nucleons in the target, the answer to the first question is yes. Furthermore, the structure of H_{A-R} suggests how to modify the n-A and p-A optical potentials appearing in it so as to include the effects of the Pauli principle: Just replace the "distinguishable-particle" optical potentials by those taking full particle identity into account. In the empirical Hamiltonian this has meant using the optical potentials that are used to fit neutron-nucleus and proton-nucleus elastic scattering data.

This means of imposing the Pauli principle is trivial to carry out and has been an accepted procedure even though it was not rigorously derived by a formally exact reduction of H_{A+2}. A much less trivial affirmative response to the first question has also been given. Starting with the early work of Johnson and Soper, various authors have advocated altering the bare neutron-proton interaction V_{np} appearing in H_{A-R} so as to take account of Pauli blocking effects.[6] Pauli blocking here refers to the impossibility of either of the nucleons in the deuteron being able to occupy an orbital initially filled by one of the target nucleons. In analogy to the treatment of a pair of particles in Bethe-Goldstone theory,[7] it was proposed that V_{np} be changed to $(1-Q)V_{np}$, where $(1-Q)$ is the Pauli blocking operator. A variety of theoretical and/or numerical studies have been carried out on the effects of replacing V_{np} by $(1-Q)V_{np}$, but applications to experimental data seem to have been made only in cases where there was prior disagreement between experiment and calculations based on use of the unmodified V_{np}. Modifying V_{np} has not achieved the desired agreement, although some improvements in the calculated cross sections have been found.[8]

The Pauli principle modifications just described seem to be reasonable responses to the first question posed above, even though the usefulness of the $V_{np} \to (1-Q)V_{np}$ replacement is not established. However, answers to the second and third questions can only be obtained via a rigorous reduction of H_{A+2}, and this has not been achieved within the framework of the standard projection operator approach. Since this is due to the complexity associated with this approach, an alternate procedure, if one can be found, may be able to provide the desired reduction. Such a procedure has been recently developed and applied to the deuteron–nucleus collision problem.[9] It realizes the reduction by first constructing the fully antisymmetrized, deuteron–nucleus scattering eigenstate via a multi–particle collision theory. This eigenstate is then projected onto the target nucleus ground state. The result of this projection is an antisymmetric three–body wave function $\psi(n,p)$ which yields the properly symmetrized elastic scattering and breakup amplitudes. Since $\psi(n,p)$ obeys a Schrödinger equation with effective interactions, the alternate procedure yields H_3.

The most intriguing aspect of the H_3 obtained this way is that it is in the form of the original Austern–Richard Hamiltonian H_{A-R}: ab initio inclusion of the Pauli principle does not lead to the modified neutron–proton interaction $(1-Q)V_{np}$. That is, while Pauli blocking effects are present in H_3, they do not alter the isolated V_{np} interaction. Thus, in this approach, the deuteron–nucleus collision problem is seen not to be analogous to the behavior of a pair in the Bethe–Goldstone formulation of the structure (bound–state) problem. This has been a surprising result; not surprisingly, it has also been a controversial one, since it seems to violate the intuition developed from many years of research on identical fermion systems. Such a situation almost automatically provokes questions. An obvious one concerns the validity of the analysis: could the form of H_3 derived in Ref. 9 be ambiguous? And, if this form is unique, should or could it have been anticipated? Further, does this result have implications that intuition based on previous work has not suggested?

One of the main purposes of the present review is to provide answers to these questions. In particular we shall discuss qualitatively the origin and meaning of the results of Ref. 9 as well as some of their implications. In order to do so, it is helpful to recall some of the standard approaches that yield an effective three–body Hamiltonian in which the Pauli blocking operator $(1-Q)$ does appear. This is done in the next two sections. Following that is a discussion of the alternate procedure.

The key point which will be developed and explored is that even though the antisymmetrized (A+2)–particle wave function is unique, the identity of the particles means that there is not a unique way of defining the three–body elastic/breakup wave function.

Although this ambiguity is an inevitable consequence of working with fully antisymmetric collision wave functions, it is not the familiar one associated with, e.g., a Slater determinant representation of the ground state.[10] And it does allow for at least a partial resolution of the V_{np} vs $(1-Q)V_{np}$ problem in that it provides for two different equivalent Hamiltonians, H_3, containing V_{np}, and H_3', containing a modified V_{np}. That this may only be a partial resolution is due to several unresolved problems concerning the Hamiltonian H_3'. First, despite the analyses in Refs. 4 and 6, it is not clear that in H_3', V_{np} should be modified by subtracting from it the term QV_{np}: there are good theoretical reasons for expecting the modification to be of the form $V_{np}(1-Q)$, not $(1-Q)V_{np}$. Second, this modification seems to apply to the three-body wave function itself, so that the energy and the kinetic energy operators in H_3' are also changed from the usual form. Finally, since H_3 contains both an unmodified V_{np} and the nucleon-nucleus optical potentials in the expected form that leads via standard approximation to the $\mathcal{V}_n$ and $\mathcal{V}_p$ of Eq. (1), we do not expect approximations to H_3' also to yield $\mathcal{V}_n$ and $\mathcal{V}_p$. That is, H_3' should not lead to an empirical Hamiltonian which differs from H_M only by a Pauli-blocked V_{np}.

In order to understand the origin and meaning of these remarks, we must turn to a more quantitative description of deuteron-nucleus collisions. The next four sections are devoted to this topic.

ANTISYMMETRIZATION: WAVE FUNCTION AMBIGUITIES

Because the deuteron-nucleus collision system involves identical fermions, the Schrödinger wave function must be antisymmetric. By ignoring the small neutron-proton mass difference the neutron and proton can be considered as two different states of a single particle, the nucleon.[11] Nucleon states are characterized by the values of a dichotomic quantum number denoted the isospin.[11] For the purposes of this discussion, the only relevant aspect of this characterization is the use of the nucleon as the fundamental ingredient in nuclei. Since the nucleon is a spin-one-half fermion and all nucleons are identical, then the total wave function must be fully antisymmetric in the nucleon labels. Hence, if a nucleus contains nucleons 1,2,...,N, then a state $\Phi(1,2,\ldots,N)$ of this N-nucleon nucleus must obey $(1+P_{ij})\Phi(1,2,\ldots,N) = 0$, all i and j, where P_{ij} is the two-particle transposition operator. Because of the convenience of working with only one kind of particle, the nucleon concept is employed herein.

The nucleon labels that we will use for the deuteron-nucleus system are n,p,1,2,...,A; A of these A+2 identical nucleons form the target nucleus. For simplicity, only two-particle interactions are included in the (non-relativistic) Hamiltonian H_{A+2} governing the behavior of the system:

$$H_{A+2} = \sum_i K_i + \sum_{i<j} V_{ij}, \qquad (3)$$

where K_i denotes a kinetic energy operator and V_{ij} a pair interaction.

The antisymmetrized (A+2)-nucleon wave function describing the deuteron–nucleus collision system is denoted $\Psi^A(n,p,1,\ldots,A)$ such that

$$(1 + P_{ij})\Psi^A(n,p,1,\ldots,A) = 0, \quad \text{all } i,j \in \{n,p,1,\ldots,A\}. \tag{4}$$

This Ψ^A is the solution of

$$(E - H_{A+2})\Psi^A(n,p,1,\ldots,A) = 0, \tag{5}$$

obeying the asymptotic boundary conditions that it: (a) is generated by a deuteron incident on the target nucleus in its ground state Φ_0 and (b) contains asymptotically only outgoing scattered waves in all energetically open channels.

The wave function Ψ^A describes the behavior of the system and thus contains a portion corresponding to deuteron elastic scattering and breakup. In order to clarify the ambiguity associated with the definition of a three–body wave function describing the elastic and breakup processes, we first consider pure elastic scattering, since the same ambiguity arises there as well. In elastic scattering, both the deuteron and the target nucleus are observed (asymptotically) in their ground states ϕ_d and Φ_0, respectively. One means of defining a one–body, elastic scattering wave function which makes explicit the antisymmetry property is via a projection operator P' such that:[1,2]

$$P'\Psi^A = \mathcal{A}\{F(n+p)\phi_d(n-p)\Phi_0(1,2,\ldots,A)\}. \tag{6}$$

Here, $\mathcal{A}$ is an antisymmetrizer, $\phi_d(n-p)$ depends on the relative coordinate or momentum between the nucleons n and p, and the one–body, elastic scattering wave function $F(n+p)$ depends on the relative coordinate or momentum between the CM of the n - p pair and that of the remaining nucleons $(1,2,\ldots,A)$ which form the target. Using standard procedures,[1,2] a formal equation for F can be derived which contains an antisymmetrized, deuteron optical potential. The asymptotic, coordinate–space boundary condition used to determine F from this equation is that it contain a plane wave and an outgoing spherical wave.

The foregoing approach accomplishes two goals. First, it yields a definition of an elastic scattering state which explicitly incorporates the Pauli principle. Second, it ultimately leads from the (A + 2)-particle Schrödinger equation to a one–body equation from which F can in principle be calculated (although its detailed structure is difficult to elucidate). This method is not, however, the only way to define an elastic scattering wave function. An equally valid procedure is to project $\Psi^A(n,p,1,\ldots,A)$ onto an unsymmetrized product of the deuteron and target nucleus ground states. To simplify notation let [i,j] mean all labels $n,p,1,\ldots,A$ but i and j. Then an alternate elastic scattering wave function is [2]

$$\chi(n+p) = (\phi_d(n-p)\Phi_0([n,p])|\Psi^A(n,p,1,\ldots,A)), \tag{7}$$

where the symbology n + p in $\chi(n+p)$ has the same meaning as in $F(n+p)$ above. Because of the antisymmetry of Ψ^A, any pair of labels i and j can be used in place of n and p: the only effect on χ will be to multiply it by the same phase factor (equal to plus or minus unity) as appears when going from $F(n+p)$ to $F(i+j)$. No matter which nucleons are in ϕ_d and which are in Φ_0, the alternate procedure (7) will yield, to within a phase factor, the same elastic scattering amplitude as obtained using (6). A proof of this follows by projecting $P'\Psi^A$ onto $\phi_d(n-p)\Phi_0([n,p])$ and then allowing the coordinate R_{np} between the deuteron and nucleus CM's to become asymptotic.[2]

Although $F(n+p)$ and $\chi(n+p)$ are equal asymptotically, they differ for finite values of R_{np}, as the above projection shows:

$$\chi(n+p) = F(n+p) + \text{short range contributions.} \qquad (8)$$

This is not the only difference associated with these two elastic scattering functions. For example, the projection of Ψ^A onto $\phi_d(n-p)\Phi_0([n,p])$ singles out a specific labelling, so that $\chi(n+p)$ is not defined by an explicitly antisymmetric procedure. Furthermore, suppose we try to derive an equation for $\chi(n+p)$ directly from the Schrödinger equation by using the projection operator $P' = |\phi_d(n-p)\Phi_0([n,p])><\phi_d(n-p)\Phi_0([n,p])|$. Then it is trivial to show that the resulting equation for $\chi(n+p) = (\phi_d(n-p)\Phi_0([n,p])|P'\Psi^A)$ is the same one as would occur if n and p were distinguishable from nucleons $1, 2, \ldots, A$.[1] Feshbach dealt with this kind of problem by using, in effect, the relation (8) and the P' projection operator to derive[2] an equation for χ whose form is so similar to that of the equation for F that there seems to be no essential structural differences these two equations. This apparent lack of an essential difference is a crucial point that will be examined below.

The inability to derive directly from (5) an equation for χ in which the effects of full antisymmetry are manifest is a major reason why F and not χ has been regarded as the more fundamental quantity in describing elastic scattering of identical particles.[12] On the other hand, if the antisymmetry requirement were somehow imposed on Ψ^A via a secondary set of equations, then this secondary set could be an alternate basis for describing elastic scattering. Such a procedure is not only feasible, it has been developed and used by Adhikari, Kozack and Levin to derive an equation for any elastic scattering wave function defined as is $\chi(n+p)$.[13] This equation contains the complete effect of antisymmetry by means of a fully antisymmetrized optical potential, i.e., a one–body operator which, when acting on the elastic scattering wave function, leads to exchange effects. For properties of this optical potential and the associated elastic scattering wave function, plus a comparison with more (and less) conventional approaches, see Ref. 14. The relevance of the analysis of Ref. 13 to the present discussion is that it underlies the $H_{A+2} \to H_3$ reduction of Ref. 9: what works for elastic scattering also carries over to the deuteron elastic scattering/breakup problem.

Let us now return to the coupled elastic scattering/breakup problem, and examine the analogous ambiguity in the definition of the three–body wave function. When elastic scattering and breakup are to be considered together, the physical requirement is that the target nucleus be detected in its ground state. The standard projection operator approach to this situation involves a new projector $\mathcal{P}$ such that

$$\mathcal{P}\Psi^A = A\{u(n,p)\Phi_0(1,2,\ldots,A)\}, \qquad (9)$$

where the antisymmetric three-body wave function u(n,p) depends on two vector variables and in coordinate space asymptotically yields both the elastic scattering and the breakup amplitudes. In the standard way,[2] a formal equation for $\mathcal{P}\Psi^A$ or u(n,p) can be derived. This equation, and approximations to it, will be considered below. We shall refer to this approach as the method of antisymmetrized projection. The alternate approach, denoted the method of unsymmetrized projection and formulated in analogy with the pure elastic scattering case, leads to the three-body wave function $\psi(n,p)$ defined by

$$\psi(n,p) = (\Phi_0([n,p])|\Psi^A(n,p,1,2,\ldots,A)). \qquad (10)$$

This wave function also depends on two vector variables (in the barycentric system) and yields the same elastic scattering and breakup amplitudes as does u(n,p). It obeys the following Schrödinger equation:[9]

$$(E - \varepsilon_0 - H_3)\psi(n,p) = 0, \qquad (11)$$

where ε_0 is the target nucleus ground state energy and H_3 is the reduced Hamiltonian describing all collision events in which the target nucleus remains unexcited. Properties of $\psi(n,p)$ and H_3 will be discussed after our consideration of u(n,p), to which we turn next.

REDUCTION VIA ANTISYMMETRIZED PROJECTION

The projection operator $\mathcal{P}$ has so far been defined by its action on Ψ^A, Eq. (9). Following Feshbach,[2] it can be expressed in terms of a non-local integral kernel function K, denoted $K(r_1r_2;r_nr_p)$ in a coordinate representation, an expression for which is given in the Appendix. As in the case of the kernel term for pure elastic scattering,[2] it is straightforward to show that: K and 1-K are each Hermitian and semi-positive definite, TrK is finite, and the eigenvalues $\{\mu_\alpha\}$ of K obey $0 \leq \mu_\alpha \leq 1$. If K' is the part of K with the unity eigenvalue portions removed and $\mathcal{Q} = 1 - \mathcal{P}$, then the same analysis as in the pure elastic scattering case[2] leads to the following equations for u(n,p) and $\psi(n,p)$:

$$E(1-K')u(n,p) = (\Phi_0([n,p])|H_{eff}|\mathcal{A}\{u\Phi_0\}) \qquad (12)$$

and

$$E\psi(n,p) = (\Phi_0([n,p])|H_{eff}|\mathcal{A}\{\Phi_0\frac{1}{1-K'}\psi\}), \qquad (13)$$

with[2]

$$H_{eff} = H_{A+2} + H_{A+2} Q \frac{1}{E + i0 - Q H_{A+2} Q} Q H_{A+2}, \qquad (14)$$

and $u = (1 - K')^{-1}\psi$.

Equations (12) and (13) are remarkably similar in structure, and also have the same forms as their counterparts for the pure elastic scattering case, although the details are obviously different (e.g., in pure elastic scattering, $P \to P'$ and $K(r_1 r_2; r'_1 r'_2)$ is replaced by a simpler kernel, $K(\mathbf{r}, \mathbf{r}')$). The identical fermion nature of the collision system is clearly expressed by (12) and (13). Let us now concentrate on Eq. (12). It is clear that (12) can in principle be rewritten in the general form $H'_3 u = (E - \varepsilon_0)u$, although the full structure of this H'_3 is difficult to deduce directly from (12). Because of this, we are aware of no direct transformations from (12) to the H'_3 form which establish that the antisymmetrized nucleon–nucleus optical potentials actually appear in H'_3: it appears necessary to use indirect means, as in Eqs. (23) – (25) below. On the other hand, the presence or absence in H'_3 of a Pauli-blocked neutron–proton binding interaction is easy to determine. This may be done by approximating H_{eff} by its first term. It is helpful to use the decomposition

$$H_{eff} \cong H_{A+2} = H_0 + V_{np} + V_{nA} + V_{pA} + H_A, \qquad (15)$$

where H_0 is the sum of the kinetic energy operators for the (n–p) internal motion and for the motion between the CM's of the (n + p) and $(1,\ldots,A)$ systems, V_{np} is the n – p binding interaction, $V_{nA}(V_{pA})$ is the interaction between n(p) and the nucleus formed by nucleons $1,\ldots,A$, while $(\varepsilon_0 - H_A)\Phi_0([n,p]) = 0$. Then substitution of (15) into (12) leads to

$$(E - \varepsilon_0 - H_0 - V_{np})(1 - K')u(n,p) = (\Phi_0([n,p])|V_{nA} + V_{pA}|\mathcal{A}\{u\Phi_0\}). \qquad (16)$$

For later comparison, we note here that use of (15) in (13) yields

$$E\psi(n,p) = (\varepsilon_0 + H_0 + V_{np})(\Phi_0([n,p])|\mathcal{A}\{\Phi_0 \frac{1}{1-K'}\psi\})$$

$$+ (\Phi_0([n,p])|V_{nA} + V_{pA}|\mathcal{A}\{\Phi_0 \frac{1}{1-K'}\psi\}). \qquad (17)$$

Equation (16) is of particular interest in the present context because $1 - K'$ is precisely the Pauli blocking factor $1-Q$ advocated in Ref. 6, although here the modified V_{np} term is $V_{np}Q$ rather than QV_{np}. A simple way to see that $1-K'$, or equivalently $1-K$, is the Pauli blocking operator is to let Φ_0 be given by a Slater determinant of occupied orbitals $\{\phi_1 \cdots \phi_A\}$. Then, using the Appendix, a short calculation yields

$$Ku(n,p) = -\sum_{i=1}^{A}[\phi_i(p)(\phi_i(p')|u(n,p')) + \phi_i(n)(\phi_i(n')|u(p,n'))]$$
$$+ \sum_{i<j}^{A} \phi_i(n)\phi_j(p)(\phi_i(n')\phi_j(p')|u(n',p')),$$
$$\equiv Qu(n,p), \qquad (18)$$

which realizes the standard form for Q.[4,6]

The structure of (16) may be contrasted with the analogous equations that have been derived from specific models by Austern[6] and by Tostevin, Lopes and Johnson.[6] In these latter, QV_{np} rather than $V_{np}Q$ appears, and the operator $E - \varepsilon - H_0$ acts directly on the equivalent to u(n,p) rather than on (1-Q)u(n,p) as in (16). A crucial ingredient in the derivations of Austern and of Tostevin et al. is the requirement that each of their u(n,p)'s be orthogonal to the occupied orbitals (i.e., in these works, Φ_0 is essentially a Slater determinant). Since they approximate Ψ^A by an antisymmetrized ansatz of the form (9), that is, since they use $\Psi^A \simeq \mathcal{P}\Psi^A$, there is the freedom in their u(n,p) to impose such a requirement. Thus, in each of these articles, use of this additional requirement leads to an approximate result containing a Pauli–blocked V_{np} in the QV_{np} form. On the other hand, in (12) and in (16), it is the $V_{np}Q$ form that occurs, without any a priori requirement that u(n,p) be orthogonal to the occupied orbitals. Notice also that if a resonating group method (RGM) ansatz[15] were used, whose simplest form would be $\Psi^A \simeq \mathcal{P}\Psi^A$ and $(\Phi_0([n,p])|E - H_{A+2}|\mathcal{P}\Psi^A) = 0$, then the result is again Eq. (16), which contains $V_{np}Q$.

We therefore see that in an exact treatment such as the Feshbach formulation and also in an RGM, $\Psi^A \to \mathcal{P}\Psi^A$ approximation to it, Pauli blocking is associated with V_{np} via $V_{np}Q$, while in other approximate treatments employing an orthogonality condition QV_{np} occurs; which approximate form is to be preferred is not known. Additionally, one may interpret the RHS of (16) as containing the lowest order terms in the antisymmetrized n – A and p – A optical potentials. It is thus tempting to extrapolate from these terms and hypothesize that the full, antisymmetrized, nucleon–nucleus optical potentials (evaluated at "energies" $E - \varepsilon_0 - K_p$ and $E - \varepsilon_0 - K_n$, as in Eq. (20) below), will be present in the H'_3 introduced above. While this is an attractive hypothesis, it has not only been proved, we shall demonstrate later in this article that H'_3 contains modified nucleon–nucleus optical potentials plus $V_{np}Q$.

The foregoing analysis suggests that the reduced Hamiltonian H'_3 will not yield H_M when standard approximations are used in it, a point we expand on in more detail below. We may therefore ask if (13) would do so. Working first with Eq. (17), Feshbach's analysis gives[2]

$$(E - \varepsilon_0 - H_0)\psi(n,p) = (\Phi_0([n,p])|V_{np} + V_{nA} + V_{pA}|\mathcal{A}\{\Phi_0 \frac{1}{1-K'}\psi\}). \quad (19)$$

Although this relation is more like (2) than is (16) and (12), it and (13) do not allow for a straightforward delineation of the structure of the corresponding H_3, essentially because of the presence of the operator $(1-K')^{-1}$. That is, the portion of (19) with V_{np} now does not contain a simple QV_{np} or $V_{np}Q$ form, and the other terms in (19) plus the higher-order ones in (13) are not readily seen to contain the nucleon-nucleus optical potentials in what may be regarded as the "expected" form.[5] However, this complexity in (19) and (13) is only apparent. As discussed in the next section, the modified V_{np} term in (19) is restored to an unmodified V_{np} form by higher order contributions from (13). In addition, the nucleon-nucleus optical potentials do occur in H_3 in the expected form.[9] The fact that the structure of the reduced Hamiltonian governing $\psi(n,p)$ is that of H_{A-R} implies that the Hamiltonian H_3' governing $u(n,p)$ will not be of this latter form, a point discussed in more detail below.

REDUCTION VIA UNSYMMETRIZED PROJECTION

In the preceding section, the antisymmetrized projector $\mathcal{P}$ was employed as an intermediary in obtaining an impractical equation for the three-body wave function $\psi(n,p)$. Use of $\mathcal{P}$ can be avoided only if one works with an explicit set of equations whose solution is the fully antisymmetrized scattering solution Ψ^A. It is through such a set that the alternate projector P can be used to define $\psi(n,p)$ in a practical way: $\psi(n,p) = (\Phi_0([n,p])|P\Psi^A) = (\Phi_0([n,p])|\Psi^A)$.

A considerable effort has been devoted to deriving sets of equations from which the scattering solution to the N-particle Schrödinger equation can be uniquely determined. While many such sets have been obtained,[16] very few have been formulated in a way that facilitates a description of nuclear reactions, in particular the $H_{A+2} \rightarrow H_3$ reduction of the deuteron-nucleus collision problem. One such formulation is that of Ref. 9, based on the general approach given in Ref. 13. A crucial ingredient in this approach is ab initio inclusion of the Pauli principle in the defining sets of equations: they do yield fully antisymmetrized solutions.

Let us recall next why a set of integral equations is needed.[17] Although scattering solutions to the one-particle Schrödinger equation are specified by the Lippmann-Schwinger (LS) integral equation,[18] this latter equation is not appropriate for the deuteron-nucleus system, since it does not yield unique solutions when the collision system contains three or more particles.[19] While unique solutions can be specified if the inhomogeneous LS equation is supplemented by a set of homogeneous LS equations[20] (the entire group of which can be easily antisymmetrized[21]), this particular set of integral equations is inappropriate to use as a calculational tool because its kernel is not mathematically well–behaved. Among the sets of equations that have been developed as mathematically well–behaved replacements for the set of LS equations, that used in Ref. 9 not only is easily antisymmetrized but it has been shown to lead via a lengthy analysis to the desired H_3. Due to the complexity of the analysis, neither the structure of the set of equations nor the reduction procedure used in Ref. 9 is appropriate in a qualitative review like this. For our purposes, it is sufficient to state the form of H_3 which has been derived, viz,

$$H_3 = H_0 + V_{np} + \mathcal{V}_n^{opt}(E - \varepsilon_0 - K_p) + \mathcal{V}_p^{opt}(E - \varepsilon_o - K_n) + W_{np}$$
$$\equiv H_{A-R}. \tag{20}$$

As indicated, the form of H_3 is precisely that proposed by Austern and Richards.[5] The meaning of the terms in (20) is as follows: $\mathcal{V}_i^{opt}(E - \varepsilon_0 - K_j)$ is the antisymmetrized nucleon–nucleus optical potential between nucleon i = n or p and the target nucleus, but evaluated at the shifted "energy" $E - \varepsilon_0 - K_j$, where K_j is the kinetic energy operator for the other or spectator nucleon (j = p or n). The operator version of this potential is the one that occurs in the Lippmann–Schwinger equation for the antisymmetrized, nucleon–nucleus, elastic scattering transition operator, as shown in Ref. 9. These two optical potential terms and V_{np} are each two–body interactions, i.e., in them, only two of the three bodies n, p and A are mutually interacting. The final term in (20), W_{np}, is an absorptive three–body interaction. Although this term is too complex to be evaluated in closed form, relatively simple, low–order approximations can be introduced to help understand its general structure.[9]

Let us next recall that a Pauli–blocked V_{np} term appears in Eq. (16) and has been advocated in Refs. 4 and 6, while none appears in H_3 or H_{A-R}. Since QV_{np} and $V_{np}Q$ are each three–body interactions and W_{np} cannot be evaluated in closed form, it is conceivable that W_{np} could contain a portion of the form (Q–1) V_{np} or V_{np}(1–Q), which, when combined with the isolated V_{np} term in (20), would then yield a Pauli–blocked V_{np}. The similarity in structure between Eqs. (16) and (17) suggests that W_{np} might contain such a portion. This point has been investigated in Ref. 9, where it has been shown that W_{np} contains no such terms. Instead, the lowest order terms in W_{np} in which V_{np} occurs are matrix elements of the following products of three factors: non–V_{np}–interactions times a Green's function times V_{np}. (See also the discussion following Eq. (21).) This latter structure cannot be reduced to either a (Q–1)V_{np} or a V_{np}(Q–1) form. Thus, for the H_3 of (20), Pauli blocking occurs in $\mathcal{V}_i^{opt}$ and in W_{np}, but does not occur at the level of the isolated V_{np} term, in contrast to any expectations fostered by Eqs. (12) and (13) or by model calculations employing various approximate wave functions.[4,6] Furthermore, the absence of Pauli blocking at the isolated V_{np} level means that for the elastic–type processes, the deuteron–nucleus collision system is not analogous to the description of a pair of identical particles in the

Bethe–Goldstone formulation of the structure problem.[7,4] This dissimilarity with Bethe–Goldstone theory may now be seen to be unsurprising, since in the latter description the Bethe–Goldstone pair wave function is introduced in the same way as is u of Eq. (9).

Because of the relationship (7) for the pure elastic scattering case on the one hand and the structural similarities between Eqs. (12) and (13) on the other, the form taken by H_3, especially its lack of a QV_{np} or $V_{np}Q$ term, was unexpected, as noted in the Introduction. In view of the preceding analyses and remarks, it is thus appropriate to consider two of the questions raised near the end of the Introduction, viz, Could the $\psi(n,p)/H_3$ formulation, obtained via unsymmetrized projection, be ambiguous?, and if not, Could the form of H_3 have been anticipated?

The answer to the first question is no. To see this, note that $\psi(n,p)$ and $u(n,p)$ are related in a manner similar to Eq. (8) and that by construction, $\psi(n,p)$ is unique. Uniqueness follows from the fact that once the complete set of asymptotic boundary conditions are imposed, a solution Ψ^A to the Schrödinger equation that satisfies these boundary conditions is unique. By construction, the set of integral equations used in Ref. 9 to define Ψ^A does specify the complete set of boundary conditions. Furthermore, this set of integral equations can be converted to differential equations, which ultimately reduce to $(E-H)\Psi^A = 0$. Finally, the kernel of the set of integral equations used in Ref. 9 is connected,[22] implying that the solution to the set is unique as long as the two-body interactions are assumed to be well-behaved; this is an assumption intrinsic to our analysis. The combination of a connected kernel and well-behaved interactions connotes compactness, and it is this latter condition which is required for uniqueness of the solution.[23] Mathematically rigorous proofs of compactness have been given for the three-particle Faddeev equations[24] and for the Chandler–Gibson equations.[25] Since these sets of equations have connected kernels, as does the set used in Ref. 9 (and also in Ref. 13), it is to be expected that this latter set will yield unique solutions. As this is a standard assumption, it is one we make herein.

The answer to the second question is that the non-QV_{np} expression (20) for H_3 could indeed have been anticipated. The argument is as follows. Suppose that the requsite Ψ^A has already been obtained. Since it will be a solution of Eq. (5), then the projection operators P and $Q_p = 1-P$ can be applied to it in the standard way.[1,2] If the decomposition (15) is used for H_{A+2}, then projection with P leads to

$$[E - \varepsilon_0 - H_0 - V_{np} - (\Phi_0([n,p])|V_{nA} + V_{pA}|\Phi_0([n,p]))]\psi(n,p)$$

$$= (\Phi_0([n,p])|V_{nA} + V_{pA}|Q_p\Psi^A). \qquad (21)$$

An analogous equation for $Q_p\Psi^A$ in terms of $P\Psi^A$ also exists. In contrast to the distinguishable particle case, this latter equation cannot be straightforwardly solved to obtain a formal solution for $Q\Psi^A$ by the usual means involving a Green's function.[1] Were this possible, then the resulting equation for $\psi(n,p)$ would be the same as if n and p were distinguishable from nucleons 1,2,...,A. Instead, the equation for $Q_p\Psi^A$ must be formally solved (i.e., inverted) as a sum in which Pauli principle exchange effects are explicitly included.

While (21) does not serve as a means of defining $\psi(n,p)$, this is of no concern here, because $\psi(n,p)$ is assumed known. Equation (21) is therefore an equation which $\psi(n,p)$ must satisfy rather than one which generates it. Nevertheless, (21) is a key to understanding why a Pauli–blocked V_{np} term does not occur in H_3, i.e., why W_{np} of (20) does not contain a portion of the specific form $(Q-1)V_{np}$ or $V_{np}(Q-1)$. Comparison of Eqs. (13), (15) and (21) readily shows that

$$W_{np}\psi(n,p) = (\Phi_0([n,p])|V_{nA} + V_{pA}|Q_p\Psi^A)$$
$$+(\Phi_0([n,p])|V_{nA} + V_{pA}|\Phi_0([n,p])\psi(n,p)$$
$$- \left[\mathcal{V}_n^{opt}(E - \varepsilon_0 - K_p) + \mathcal{V}_p^{opt}(E - \varepsilon_0 - K_n)\right]\psi(n,p). \quad (22)$$

Only the first term on the RHS of (22) can give rise to the three–body interaction W_{np}. Since $Q_p\Psi^A$ will be given by a sum, each term of which is necessarily a product of at least one Green's function and one two–body interaction,[26] then the contributions to W_{np} that contain the fewest number of interactions and Green's functions will have the generic form $(V_{nA} + V_{pA})$ times Green's function times interaction. This is precisely the form stated above for the lowest order contributions to W_{np}, thus verifying our claim that the absence of an isolated, Pauli–blocked V_{np} interaction in H_3 could have been anticipated.

The preceding discussion has been intended to clarify aspects of the $H_{A+2} \to H_3$ reduction. We conclude this section reviewing the two approximations that convert H_3 into H_M of Eq. (1), and thus $\psi(n,p)$ into $\psi_M(n,p)$ of Eq. (2). Since H_M contains only two–body interactions, one of these approximations is that W_{np} is negligible. This is an approximation of convenience and is unsupported, to our knowledge, by any estimates of the actual importance of W_{np}. It seems possible to obtain estimates based on the approximate forms of W_{np} derived in Ref. 9, but doing so is far from a trivial exercise. It is plausible that W_{np} is small, a conclusion we reach by making a comparison with the nuclear shell model, where a reduction from the A–particle Hamiltonian to the shell–model Hamiltonian involves many–body forces in principle but only one– and two–body operators in practice. Verification that W_{np} may be ignored remains for the future.

The other approximation is that $\mathcal{V}_i^{opt}(E_d/2) = \mathcal{V}_i$ of Eq. (1), where $E_d = E - \varepsilon_0$. The physics underlying this replacement is that on average, nucleon j, the spectator in $\mathcal{V}_i^{opt}(E - \varepsilon_0 - K_j)$, carries half of the deuteron kinetic energy E_d. Among other things, this approximation ignores the 2.225 Mev nuclear binding energy in the initial state and so requires $E_d \gg 2.225$ MeV, say $E_d \gtrsim$

20 MeV. The approximation also implies that in the breakup process, neither n nor p is detected at very large scattering angles relative to the direction of the incident deuteron, so that each will have about the same energy. In the breakup analyses of Refs. 4, this condition has meant restriction to scattering angles of less than or roughly equal to about 20°. The application of H_M to this latter situation by the Kyushu group has been a deliberate (and successful) one.[27]

CONCLUSIONS/IMPLICATIONS

The major result of Ref. 9, described in the preceding section, is that the method of unsymmetrized projection leads to an $H_{A+2} \to H_3$ reduction in which H_3 contains both a non–Pauli–blocked V_{np} interaction and the antisymmetrized nucleon–nucleus optical potentials evaluated at the shifted energies. With inclusion of the three–body term W_{np}, this H_3 has exactly the form of the Austern–Richards Hamiltonian H_{A-R}. Use of what are by now standard approximations in H_3, viz, $E - \varepsilon_0 - K_j \to E_d/2$ and $W_{np} \cong 0$, then yields the standard model Hamiltonian H_M, employed in many data analyses.[4] From this we draw two conclusions. First, the standard model H_M is an approximation to the H_3 derived from the method of non–symmetrized projection. Second, because H_M approximates this H_3, one should not modify H_M by adding to it a Pauli–blocked term in the form $(Q-1)V_{np}$, as advocated in Ref. 6 and some of Refs. 4. In other words, the reduction which yields the $\mathcal{V}_i^{opt}(E - \varepsilon_0 - K_j)$ of H_3 also yields a non–Pauli–blocked V_{np}; to add Pauli–blocking to the V_{np} term without also modifying the optical potential term is not only inconsistent, it is also theoretically unfounded.

In order to clarify and assess the full implications of this latter remark, we compare Eqs. (11) and (20) with Eqs. (13) and (14). Since (11) and (13) define the same quantity $\psi(n,p)$, then Eqs. (11) and (13) must be transformable into each other. This makes it clear that the factor $(1-K')^{-1}$ in (13) does not lead to a Pauli–blocked V_{np} in H_3, even though Eq. (12) for $u(n,p)$ contains such a term and $\psi(n,p)$ is related to $u(n,p)$ via $(1-K)u(n,p) = \psi(n,p)$, an equation derived in Ref. 2. Thus, appearances are deceiving. Although we have not succeeded in directly transforming Eq. (13) into Eq. (11), it is evident that the absence of the Pauli–blocked V_{np} in H_3 of (11) must be a result of cancellations between the low order terms in (17) and the higher–order ones in (13). As a consequence, we re-iterate our previous point, viz, not only do these cancellations lead to a non–Pauli–blocked V_{np}, they also produce precisely $\mathcal{V}_n^{opt}$ and $\mathcal{V}_p^{opt}$ as the other two–body interactions.

The implications of this are simply stated. If Eq. (12) is transformed into

$$(E - \varepsilon_0)u(n,p) = H_3' u(n,p), \qquad (23)$$

then H_3' contains two–body absorptive interactions in addition to the optical potentials $\mathcal{V}_n^{opt}(E - \varepsilon_0 - K_p)$ and $\mathcal{V}_n^{opt}(E - \varepsilon_0 - K_n)$ of H_3 and H_{A-R}. To see this, first substitute the relation[2] $(1-K)u(n,p) = \psi(n,p)$ into (11):

$$(E - \varepsilon_0 - H_3)(1 - K)u(n,p) = 0. \tag{24}$$

Using (20) in (24) and comparing with (23) yields

$$H_3' = H_3 - V_{np}K + (E - \varepsilon_0 - H_0 - \mathcal{V}_n^{opt} - \mathcal{V}_p^{opt} - W_{np})K, \tag{25}$$

where the energy dependence in $\mathcal{V}_i^{opt}$ is again suppressed and the modification to V_{np} is made explicit. The extra two–body terms in H_3' are obtained from the relevant two–body portions of K appropriately combined with the individual terms in $(E - \varepsilon_0 - H_0 - \mathcal{V}_n^{opt} - \mathcal{V}_p^{opt})$; the former are the first two terms on the RHS of Eq. (A3). Equations (23) and (25) clearly demonstrate that if H_3' is used as a basis for deriving an equivalent model Hamiltonian H_M' describing (d,d) and (d,np) processes, then because H_M' contains a Pauli–blocked V_{np} it will necessarily contain modified nucleon–nucleus optical potentials. To use an H_M' with the replacement $V_{np} \to V_{np}(1-Q)$ without modifying $\mathcal{V}_i^{opt}$ is inconsistent. The form of these latter modifications is so far not investigated; it seems to us to be a necessary investigation on the part of anyone who would prefer to use a model Hamiltonian containing a Pauli–blocked neutron–proton interaction.

ACKNOWLEDGEMENT

The authors thank N. Austern and R. C. Johnson for their helpful comments and/or correspondence. This work has been supported in part by the U.S. Department of Energy and in part by the U.S. National Science Foundation.

APPENDIX

The kernel $K(\mathbf{r}_1\mathbf{r}_2; \mathbf{r}_n\mathbf{r}_p)$ is defined through the relation

$$(\Phi_0([n,p]) | \mathcal{A}\{u(n,p)\Phi_0([n,p])\})$$
$$= \int d^3r_1 d^3r_2 [\delta(\mathbf{r}_1 - \mathbf{r}_n)\delta(\mathbf{r}_2 - \mathbf{r}_p) + K(\mathbf{r}_1\mathbf{r}_2; \mathbf{r}_n\mathbf{r}_p)]u(\mathbf{r}_1\mathbf{r}_2), \tag{A1}$$

where spin indices are suppressed. K can be determined once the antisymmetrizer $\mathcal{A}$ is specified. It is straightforward to show that, apart from normalization,

$$\begin{aligned}
\mathcal{A}\{u(n,p)\Phi_0([n,p])\} &= u(n,p) \\
&- \sum_{i=1}^{A} u(i,p)\Phi_0(1,2,\ldots,i-1,n,i+1,\ldots A) \\
&- \sum_{j=1}^{A} u(n,j)\Phi_0(1,2,\ldots,j-1,p,j+1,\ldots A) \\
&+ \sum_{\substack{j>i \\ i=1}}^{A} u(i,j)\Phi_0([i,j]).
\end{aligned} \tag{A2}$$

Substitution of (A2) into (A1) then yields

$$\begin{aligned}K(r_1 r_2; r_n r_p) &= A\delta(r_2 - r_n) \int d^3 r_2 \ldots d^3 r_A \Phi_0^\dagger(r_1, \ldots r_A) \Phi_0(r_p, r_2, \ldots r_A) \\ &+ A\delta(r_2 - r_p) \int d^3 r_2 \ldots d^3 r_A \Phi_0^\dagger(r_1, \ldots r_A) \Phi_0(r_n, r_2, \ldots r_A) \\ &+ A' \int d^3 r_3 \ldots d^3 r_A \Phi_0^\dagger(r_1, r_2, r_3, \ldots r_A) \Phi_0(r_n, r_p, r_3, \ldots r_A), \quad (A3)\end{aligned}$$

where $A' = (A-1)(A-2)/2$. Note that the ordering of labels in (A3) is chosen for consistency with the signs in Eqs. (A1) and (A2). Equation (A3) may be easily re-expressed in terms of one- and two-particle density matrices.

REFERENCES

1. H. Feshbach, Ann. Phys. (NY) $\underline{5}$, 357 (1958); ibid $\underline{19}$, 287 (1962).

2. H. Feshbach in Reaction Dynamics, by F. S. Levin and H. Feshbach, (Gordon and Breach, New York, 1973) pp. 171 ff, and references cited therein.

3. See, e.g., G. R. Satchler, Direct Nuclear Reactions (Oxford, New York, 1983).

4. Some representative studies of (d,d) and (d,np) processes using empirical Hamiltonians can be found, e.g., in G. H. Rawitscher, Phys. Rev. C$\underline{9}$, 2210 (1974); J. P. Farrell, C. M. Vincent and N. Austern, Ann. Phys. (NY) $\underline{96}$, 333 (1976), and $\underline{114}$, 93 (1978); B. A. Anders and A. Linder, Nucl. Phys. A$\underline{296}$, 77 (1978); M. Kawai, M. Kamimura and K. Takesako, Proc. 1978 INS International Symposium, Fukuoka (unpublished), p. 710. A comparative study of results based on different approximations is given in M. Yahiro, M. Nakano, Y. Iseri and M. Kamimura, Prog. Theor. Phys. $\underline{67}$, 1467 (1982). A comprehensive review by N. Austern, Y. Iseri, M. Kamimura, M. Kawai, G. Rawitscher and M. Yahiro of both theoretical and numerical work on this topic, especially that of the so-called Continuum-Discretized Coupled-Channel (CDCC) approach is to appear soon in Physics Reports.

5. N. Austern and K. C. Richards, Ann. Phys. (NY) $\underline{49}$, 309 (1968).

6. See, e.g., R. C. Johnson and P. J. R. Soper, Phys. Rev. C$\underline{1}$, 976 (1970) and P. J. R. Soper, Ph.D. thesis, University of Surrey (unpublished); N. Austern, Phys. Lett. $\underline{46}$B, 49 (1973); W. S. Pong and N. Austern, Ann. Phys. (NY) $\underline{93}$, 369 (1975); R. C. Johnson, N. Austern and M. H. Lopes, Phys. Rev. C$\underline{26}$, 348 (1982); Q. Ma and N. Austern, Nucl. Phys. A $\underline{463}$, 620 (1987); J. H. Tostevin, M. H. Lopes and R. C. Johnson, Nucl. Phys. A (in press).

7. H. A. Bethe and J. Goldstone, Proc. R. Soc. London, Ser. A 238, 551 (1957); J. Goldstone, ibid, 239, 267 (1957).

8. See Johnson, Austern and Lopes, and Tostevin, Lopes and Johnson, Ref. 6.

9. R. Kozack and F. S. Levin, Phys. Rev. C34, 1511 (1986).

10. The familiar ambiguity arises on adding one (or more) of the occupied orbitals to the elastic scattering wave function, since the process of fully antisymmetrizing this ansatz (i.e., the Pauli principle) eliminates this orbital as a contributor.

11. The nucleon/isospin representation is discussed in many texts. See, e.g., A. deShalit and H. Feshbach, Theoretical Nuclear Physics, vol I (Wiley, New York, 1974); H. Frauenfelder and E. M. Henley, Nuclear and Particle Physics (Benjamin, Reading, 1975); M. A. Preston and R. K. Bhaduri, Structure of the Nucleus (Addison–Wesley, Reading, 1975).

12. See, for example, p. 202 in Ref. 2, where such a comment is explicitly made (there, our F and χ are denoted u and U).

13. S. K. Adhikari, R. Kozack and F. S. Levin, Phys. Rev. C29, 1628 (1984). The various sets of equations considered in this reference are for components of Ψ^A, a term defined therein. For the case of deuteron–nucleus collisions examined in Ref. 9, each component of Ψ^A turns out, for technical reasons, to be equal to Ψ^A itself. This point is not essential for the discussion of the present article.

14. For comments on the terminology and the properties of "optical potentials", see Ref. 13 and literature cited therein, plus K. L. Kowalski, in Few Body Problems in Physics, vol I, ed. by B. Zeitnitz (North Holland, Amsterdam, 1984), p. 465, and Nuovo Cimento A92 289 (1986).

15. For references on the resonating group method and the cluster model, see e.g., Prog. Theor. Phys. Supp. 62 (1977); K. Wildermuth and Y. C. Tang, A Unified Theory of the Nucleus (Academic, New York, 1978); and Y. C. Tang, M. LeMere and D. R. Thompson, Phys. Rep. 47, 167 (1978).

16. The literature on N–particle collision theory is extensive. References to much of it can be found in the review articles of F. S. Levin, Nucl. Phys. A353, 143c (1981) and K. L. Kowalski, Ref. 14.

17. Integral equations are used since they incorporate the asymptotic boundary conditions.

18. The Lippmann–Schwinger equation is treated in many graduate texts on quantum mechanics. For a more detailed discussion, see, e.g., R. G. Newton, Scattering Theory of Waves and Particles (Springer, New York, 1982).

19. See, e.g., L. L. Foldy and W. Tobocman, Phys. Rev. 105, 1099 (1957).

20. W. Glöckle, Nucl. Phys. A141, 620 (1970); W. Sandhas, in Few Body Nuclear Physics (IAEA, Vienna, 1978), p. 3. The validity of the triad of LS equations in the three–particle case has been the subject of recent work, e.g., F. S. Levin and W. Sandhas, Phys. Rev. C29, 1617 (1984); P. Benoist–Gueutal, Phys. Rev. C33, 412 (1986); F. S. Levin and W. Sandhas, Phys. Rev. C34, 1140 (1986); and E. Gerjuoy, ibid, p. 1143.

21. Details of the antisymmetrization procedure are given in Ref. 13.

22. For a discussion of connectivity and references to some of its applications, see, e.g., W. N. Polyzou, J. Math Phys. 21, 506 (1980).

23. Discussions of compact operators are given, e.g., in T. Kato, Perturbation Theory for Linear Operators (Springer, New York, 1966) and M. Reed and B. Simon, Methods of Modern Mathematical Physics, III: Scattering Theory (Academic, New York, 1979). Application to the three–particle scattering problem is given in Ref. 24.

24. L. D. Faddeev, Mathematical Aspects of the Three–Body Problem in the Quantum Scattering Theory (Davey, New York, 1965).

25. See, e.g., C. Chandler and A. G. Gibson, J. Math Phys. 25, 1841 (1984) and references cited therein.

26. That $Q_p \Psi^A$ is necessarily a sum of products as stated in the main text follows from the general structure of the projected equations, as in Ref. 1. See also Ref. 13.

27. Private communications from M. Kawai and M. Yahiro.

COINCIDENCE REACTIONS AND THE
THREE-BODY STRUCTURE OF ^{6}Li

D.R. Lehman[*]

The George Washington University, Washington, D.C. 20052

ABSTRACT

This paper concerns two of the three virtual disintegration amplitudes of ^{6}Li that are possible below the ^{3}He-^{3}H threshold: ^{6}Li $\to$ α + d and ^{6}Li $\to$ p + (nα). These amplitudes embody the three-body nature of ^{6}Li at low-excitation energies and thereby serve as a testing ground for three-body models of ^{6}Li. The current status of a three-body approach to the dynamics of these amplitudes is given. Emphasis is placed on the physics underlying the three-body approach and the results obtained. The theoretical predictions for the ^{6}Li $\to$ α + d and ^{6}Li $\to$ p + (nα) momentum distributions are compared to new data from the kinematically optimized coincidence reaction ^{6}Li(e,e'd)^{4}He and existing data from ^{6}Li(p,2p)nα, respectively.

INTRODUCTION

Below the ^{3}He-^{3}H threshold, there are three possible virtual-disintegration amplitudes or vertices for ^{6}Li:

(a) ^{6}Li $\to$ α + d   (1.472 MeV),

(b) ^{6}Li $\to$ α + (np)   (3.697 MeV),

and (c) ^{6}Li $\to$ p + (nα) [or $\to$ n + (pα)]   (3.697 MeV),

where a pair of particles enclosed in parentheses, e.g., the neutron and alpha-particle in (nα), means that they are unbound but interacting, i.e., in a scattering state. These amplitudes embody the three-body nature of ^{6}Li at low energies; therefore, they can serve as a test for three-body models of ^{6}Li (alpha-particle plus two nucleons). Such tests are possible because three-body models of ^{6}Li give unequivocal predictions as to the structure of these vertices. Moreover, such predictions can be compared with experimental data from kinematically-optimized coincidence reactions like ^{6}Li(e,e'd)^{4}He, ^{6}Li(e,e'α)pn, ^{6}Li(e,e'p)nα, ^{6}Li(p,2p)nα, etc., thus testing the validity of the three-body hypothesis. In this work, we focus on the current status of theory and experiment for amplitudes (a) and (c).

THREE-BODY MODELS OF ^{6}Li

The three-body model of the A=6 system has as its basic interactions the nucleon-nucleon (NN) and alpha-nucleon (αN) interactions. Since ^{6}Li is an isospin singlet, as is the alpha particle, the NN interaction is limited to the isospin-zero component. Therefore,

---

[*]Work supported in part by The U.S. Department of Energy, Grant No. DE-FG05-86-ER40270

we take the NN interaction to be present in the $^3S_1$-$^3D_1$ partial waves. For the αN interaction, we include the partial-wave components that are dominant at low energies, i.e., $S_{\frac{1}{2}}$, $P_{\frac{1}{2}}$, and $P_{3/2}$. All the interactions are represented by separable potentials: 1.) The NN potential is the standard Yamaguchi-Yamaguchi form[1]; and 2.) The αN potentials are phenomenologically constructed from the low-energy phase shifts.[2] The Schrödinger equation for the three-body system is solved in momentum space with neglect of the Coulomb interaction between the proton and alpha-particle.[2,3]

Ideally, it would be nice to be less phenomenological with respect to the αN interaction, e.g., set up the effective αN interaction from an underlying microscopic theory that accounts for the compositeness of the alpha-particle. However, for the initial examination of the physics that concerns us in this work, it is quite adequate and justified from the resonating-group approach to use the phenomenological αN potentials.[4] Nevertheless, some subtlety is involved in dealing with the $S_{\frac{1}{2}}$ partial-wave αN interaction, that partial-wave which makes manifest the Pauli exclusion principle between the nucleons in the alpha-particle and the remaining nucleon. The $S_{\frac{1}{2}}$ partial-wave phase shift can be reproduced by means of either a purely repulsive potential or a purely attractive potential, but the latter potential leads to an αN bound state of 12 to 13 MeV binding energy. Of course, such a bound state does not exist in nature and it can be viewed as a 'Pauli-forbidden' state that must be removed without altering the on-shell properties of the $S_{\frac{1}{2}}$ αN t-matrix. This has been accomplished.[5] So, in this work, we consider two representations of the $S_{\frac{1}{2}}$ αN interaction: 1.) Repulsive $S_{\frac{1}{2}}$; and 2.) Attractive-Projected $S_{\frac{1}{2}}$.

Within the above framework, several ^{6}Li wave functions are generated: 1.) With and without the NN tensor force; and 2.) With either the Repulsive $S_{\frac{1}{2}}$ or the Attractive-Projected $S_{\frac{1}{2}}$ αN interaction.[3,5] The calculated three-body ^{6}Li binding energy is sensitive to the NN tensor force, but is insensitive to the representation of the $S_{\frac{1}{2}}$ αN interaction. When the NN tensor force has a strength such that a 4% D-wave component is generated in the deuteron, the ^{6}Li binding energy turns out to be ~0.3 MeV greater than the experimental value. This implies that if the Coulomb interaction is included, underbinding can be expected. Nevertheless, these wave functions lead to good results for the static properties of ^{6}Li, e.g., magnetic moment[6], charge radius[7], and asymptotic normalization constants.[8-10] These results are tabulated in Table I (next page) for the Attractive-Projected $S_{\frac{1}{2}}$ model with 4% NN tensor force. In conjunction with equivalent wave functions for ^{6}He, good results are obtained for ^{6}He β-decay (last line of Table I).[9,11] Consequently, strong evidence already exists for validity of a three-body model of the A=6 system.[12]

^{6}Li → α + d VIRTUAL DISINTEGRATION

The amplitude for ^{6}Li → α + d virtual disintegration, as depicted in Fig. 1, is given by

$$A(\vec{q}) = < \vec{q} \; \phi^{[1]}_{m_d} | V_{\alpha n} + V_{\alpha p} | \psi^{[1]}_{m_6} > \;. \tag{1}$$

Table I. Static Properties of ^{6}Li
Three-Body Model: Attractive-Projected $S_{\frac{1}{2}}$ with 4% NN tensor force

Property	Theory	Experiment
Three-body binding energy (MeV)	3.903	3.697
L-S wave function components (%)	$^3S_1$ - 91.47 $^3P_1$ - 0.48 $^1P_1$ - 4.65 $^3D_1$ - 3.40	
Charge radius (fm)	2.434	2.56 ± 0.05
Magnetic moment (n.m.)	0.8417	0.82202
Quadrupole moment (fm^2)	?	-0.0644 ± 0.0007
Asymptotic normalization constants (^{6}Li → α + d)	C_0 = 2.034 C_2 = 0.0112 D_2 = 0.0617 fm^2	1.86 ± 0.05 0.01 ± 0.03 ?
^{6}He → ^{6}Li + e- + $\bar{\nu}$ ft-value (sec)	835 ± 13	813 ± 2

Ψ represents the ^{6}Li ground state, Φ represents the deuteron, $\vec{q}$ is the relative momentum of the alpha-particle and the deuteron, and the potentials are those that are excluded in the final state. By means of Schrödinger's equation, $A(\vec{q})$ can be written in terms of the overlap of the αd state with the ^{6}Li ground state:

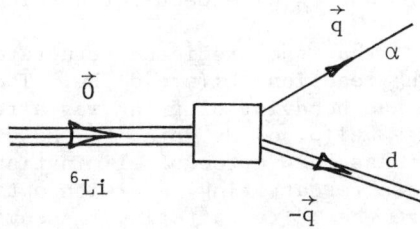

Fig. 1. ^{6}Li → α + d vertex.

$$A(\vec{q}) = -\left(\frac{q^2}{2\mu_{\alpha d}} + B_{\alpha d}\right) < \vec{q} \; \Phi^{[1]}_{m_d} | \Psi^{[1]}_{m_6} > , \qquad (2)$$

where $\mu_{\alpha d}$ is the αd reduced mass and $B_{\alpha d}$ is the binding energy of the αd within ^{6}Li. The αd overlap, or equivalently, the αd momentum-distribution amplitude, has only two possible partial-wave components since the ^{6}Li ground state has $J^\pi = 1^+$. Therefore, it can be expressed as

$$< \vec{q} \; \Phi^{[1]}_{m_d} | \Psi^{[1]}_{m_6} > = \sum_{\ell=0,2 \atop m_\ell} f_\ell(q) < \ell \; m_\ell \; 1 \; m_6 | 1 \; m_d > \sqrt{4\pi} \; Y^{[\ell]}_{m_\ell}(\hat{q}), \quad (3)$$

where $f_\ell(q)$ is the ℓ-th partial-wave momentum-distribution amplitude. The physics of the ^{6}Li → α + d vertex is contained in the $f_\ell(q)$, where $f_0(q)$ has contributions from the $^3S_1$ and $^3D_1$ components of the ^{6}Li ground-state wave function and $f_2(q)$ from the $^3S_1$, $^3P_1$, and $^3D_1$ components. The $^1P_1$ component plays no role in the structure of the ^{6}Li → α + d vertex due to the deuteron spin projection.

The f_ℓ have been calculated with the three-body wave functions described above and they are displayed in Figs. 2 and 3.[8,9] $f_0(q)$ is seen to change sign around $q \sim 0.7$ fm^{-1}. The value of q for which $f_0(q)$ changes sign is connected to the underlying $S_{1/2}$ αN interaction, i.e., the Pauli exclusion principle. This latter fact is demonstrated with a model that does not have the $S_{1/2}$ component of the αN interaction.[8] Such a model yields an f_0 that does not fall off as rapidly with q as those shown in Fig. 2, so the sign change occurs at a value of $q \gtrsim 1$ fm^{-1}. Since $f_0(q)$ is the momentum-space effective αd wave function for ^{6}Li, its configuration-space form is found to have a node (near 1.5 fm) due to the change of sign in momentum space.[8,9] The configuration-space effective αd wave function for ^{6}Li is of 2s form -- consistent with Pauli exclusion. Moreover, as can be seen from Fig. 2, $f_0(q)$ is sensitive to the representation of the $S_{1/2}$ αN interaction, but insensitive to the presence or absence of the NN tensor force. The Attractive-Projected $S_{1/2}$ αN model predicts the zero of $f_0(q)$ to occur for a slightly smaller value of q and the magnitude of $f_0(q)$ to the right of the zero to be significantly larger compared to the Repulsive $S_{1/2}$ αN model. On the other hand, as can be seen in Fig 3, $f_2(q)$ is sensitive to both the NN tensor force and the $S_{1/2}$ αN interaction, but in magnitude is more than a factor of 10 smaller than $f_0(q)$ except for q values near the zero of $f_0(q)$.

It was suggested[13] that the predicted structure of $f_0(q)$ might be observed through the reaction ^{6}Li(e,e'd)^{4}He. The small q ($0 < q < 0.3$ fm^{-1}) magnitude and behavior of $f_0(q)$ was already known to be consistent with 670 MeV ^{6}Li(p,pd)^{4}He data.[8,9] However, an electron-coincidence experiment has the virtue of eliminating projectile initial- and final-state rescattering, and with optimized kinematics can be made to minimize the αd rescattering. Assuming no αd rescattering and deuteron-pole dominance (the virtual photon interacts

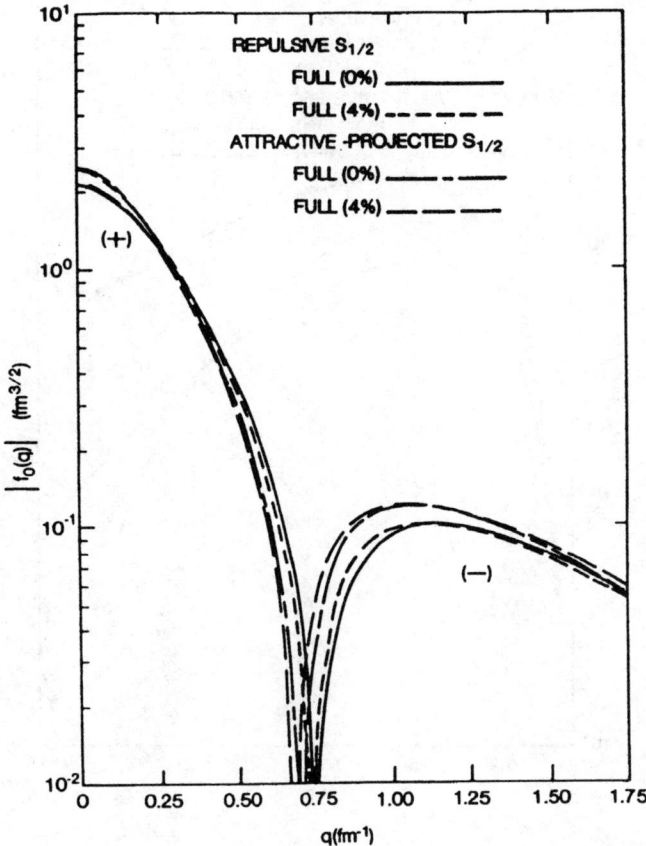

Fig. 2. ^{6}Li $\to$ α + d S-wave momentum-distribution amplitude.

only with the ejected deuteron), the ^{6}Li(e,e'd)^{4}He cross section is

$$\frac{d^3\sigma}{d\Omega_e d\Omega_d dE_e} = (k.f.) \left.\frac{d\sigma}{d\Omega}\right|_{ed} \rho(q) \quad , \tag{4}$$

where $f_2(q)$ has been neglected, $\rho(q) = [f_0(q)]^2$, (k.f.) is a known kinematic factor, and the cross section on the right-hand side is for elastic ed scattering (taken on-shell in the CM). Rescattering effects can be incorporated approximately by computing the 'distorted' momentum distribution which depends on the ejected deuteron's momentum p, call it $\rho^D(q,p)$. The αd rescattering is obtained from an optical potential.

The ^{6}Li(e,e'd)^{4}He experiment has been carried out at NIKHEF-K (The Netherlands) and the results are shown in Fig. 4.[14] The different values of energy for the data correspond to the relative αd energies in the final state and p_m=q is the recoil momentum of the

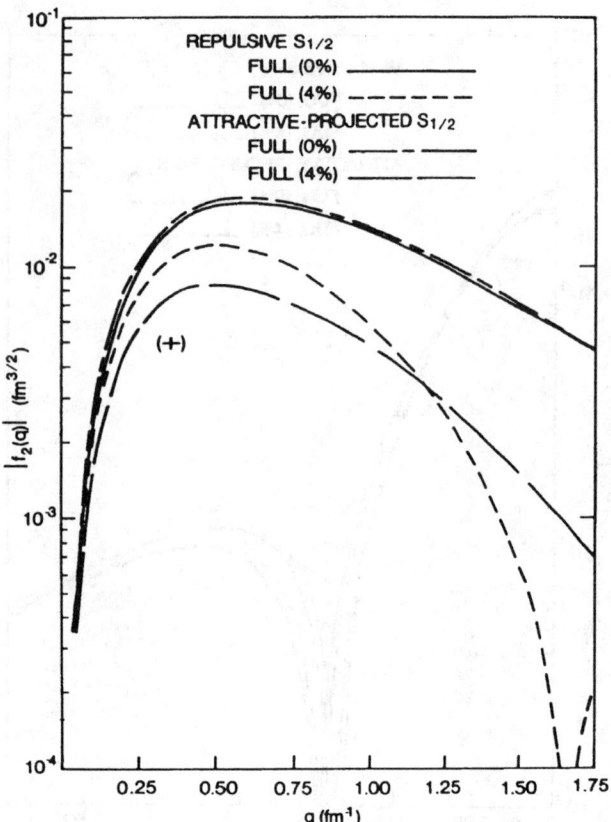

Fig. 3. ^{6}Li → α + d D-wave momentum-distribution amplitude.

α-particle. The curves are the results of calculations in which the Repulsive $S_{\frac{1}{2}}$ αN model of ^{6}Li with 4% tensor force was used. The Attractive-Projected model gives an almost equivalent though slightly poorer description of the data. The deuteron-pole dominance (PWIA) curve shows the features of $f_0(q)$ seen in Fig. 2. The DWIA curves are the results for the relevant deuteron energies. Clearly, rescattering effects are relatively small at low p_m. This can be understood from the fact that for small p_m only the asymptotic region of the αd wave function is probed where the optical-model potential is weak. At larger p_m, the minimum of the PWIA is almost completely filled in because contributions from the interior of the αd wave function, which are responsible for the minimum, are strongly suppressed by the absorptive part of the optical-model potential. The overall agreement in shape between the data and the DWIA calculations is rather good.

Another point should be made: the 2s form of the effective αd wave function is essential for reproduction of the slope change seen

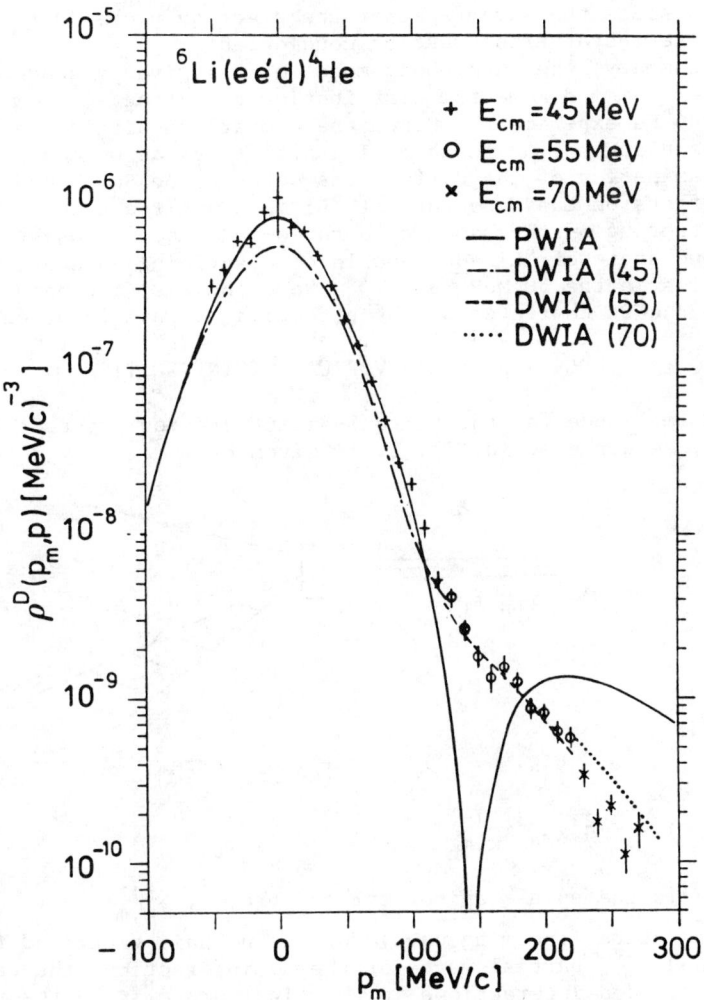

Fig. 4. Measured ^{6}Li → α + d momentum distribution from the reaction ^{6}Li(e,e'd)^{4}He. The theoretical curves are from the Repulsive $S_{1/2}$ αN model (with 4% NN tensor force) of ^{6}Li.

in the data of Fig. 4. Assumption of a 1s form for the effective αd wave function yields a momentum distribution that decreases smoothly in the region of p_m~150 MeV/c, in contrast with the data. This is in contradistinction to the reaction ^{6}Li(p,pd)^{4}He (out of the plane detection of the deuteron and incident proton energies in the 100 to 200 MeV range) where the various rescattering effects completely mask the 1s or 2s character of the effective αd wave function.[15]

This emphasizes the advantages of the electron probe in that only αd final-state rescattering must be considered.

In summary, the three-body model of ^{6}Li gives a good description of the ^{6}Li → α + d momentum distribution as extracted from a ^{6}Li(e,e'd)^{4}He experiment. Since the comparison with the data involves handling the αd rescattering effects by an optical potential, finer examination of the differences between the $S_{\frac{1}{2}}$ αN models and their effects on the structure of ^{6}Li must await a calculation of the αd rescattering on the basis of a three-body model consistent with the ground state of ^{6}Li and capable of describing αd scattering for CM energies in the 50 MeV range.[16] Nevertheless, the data clearly indicate the 2s character of the effective αd wave function.

$$^6\text{Li} \rightarrow p + (n\alpha) \text{ VIRTUAL DISINTEGRATION}$$

The amplitude for the virtual-disintegration process ^{6}Li → p + (nα), as depicted in Fig. 5, is given by

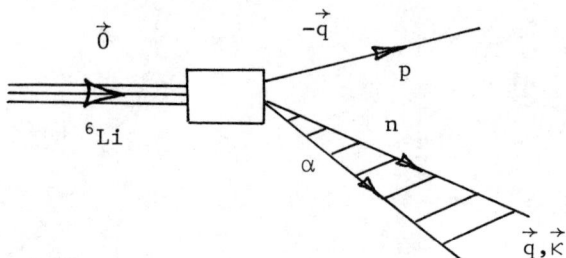

Fig. 5. ^{6}Li → p + (nα) vertex.

$$M(\vec{q},\vec{\kappa}) = <-\vec{q}\ \chi_{m_p}^{[\frac{1}{2}]} \phi_{\vec{\kappa},m_n}^{(-)} | V_{np} + V_{p\alpha} | \psi_{m_6}^{[1]}> \quad , \quad (5)$$

where $\chi_{m_p}^{[\frac{1}{2}]}$ is the spin-½ spinor for the proton, $\phi_{\vec{\kappa},m_n}^{(-)}$ is the outgoing scattering state (incoming wave) for the nα pair generated from the $S_{\frac{1}{2}}$, $P_{\frac{1}{2}}$, and $P_{3/2}$ partial waves of the nα interaction, the potentials are the excluded interactions in the final state, $\vec{\kappa}$ is the relative momentum of the nα pair, and $\vec{q}$ is the relative momentum of the proton with respect to the nα CM. Again, through Schrödinger's equation, the key element is the overlap amplitude:

$$M(\vec{q},\vec{\kappa}) = -\left(\frac{q^2}{2\mu_{p-(n\alpha)}} + \frac{\kappa^2}{2\mu_{n\alpha}} + B_3\right) <-\vec{q}\ \chi_{m_p}^{[\frac{1}{2}]} \phi_{\vec{\kappa},m_n}^{(-)} | \psi_{m_6}^{[1]}> , \quad (6)$$

where B_3 is the three-body (αnp) binding energy of ^{6}Li. The absolute value squared of the overlap amplitude divided by 4π, summed and averaged over spins, and then integrated over the solid angles of $\vec{\kappa}$ and $\vec{q}$, can be extracted from kinematically-optimized ^{6}Li(e,e'p)nα measurements. Specifically, for proton-pole dominance (the virtual photon interacts only with the ejected proton), the ^{6}Li(e,e'p)nα cross section is

$$\frac{d^4\sigma}{dE'_e d\Omega'_e d\Omega_p dE_\kappa} = (\text{k.f.}) \frac{d\sigma}{d\Omega_{ep}}\bigg|_{lab} S(q,E_\kappa) , \qquad (7)$$

where $E_\kappa = \kappa^2/2\mu_{n\alpha}$, (k.f.) is a known kinematic factor, the cross section on the right-hand side is for elastic ep scattering, and $S(q,E_\kappa)$ is the ^{6}Li → p + (nα) <u>spectral function</u> given by

$$S(q,E_\kappa) = \mu_{n\alpha} \kappa V(q,\kappa) \qquad (8)$$

with

$$V(q,\kappa) = \frac{1}{3} \sum_{m_p, m_n, m_6} \frac{1}{4\pi} \int d\Omega_q \int d\Omega_\kappa |M^0(\vec{q},\vec{\kappa};m_p,m_n,m_6)|^2 . \qquad (9)$$

M^0 in Eq. (9) represents the overlap amplitude on the right-hand side of Eq. (6). $V(q,\kappa)$ is the ^{6}Li → p + (nα) joint momentum distribution: the probability of finding within ^{6}Li a proton moving with momentum q relative to the CM of an interacting nα-pair that has relative momentum κ per unit momentum-space volume per unit momentum cubed. Due to the completeness of the p-(nα) states and the normalization of the ^{6}Li ground state, $V(q,\kappa)$ satisfies a sum rule consistent with its probability interpretation:

$$4\pi \int q^2 dq \int \kappa^2 d\kappa \, V(q,\kappa) = 1. \qquad (10)$$

The spectral function gives the probability of finding within ^{6}Li a proton moving with momentum q relative to the CM of an interacting nα pair that has excitation energy E_κ per unit momentum-space volume per unit energy. As a consequence of Eq. (10), the spectral function satisfies the sum rule

$$4\pi \int q^2 dq \int dE_\kappa \, S(q,E_\kappa) = 1. \qquad (11)$$

Equations (10) and (11) are used as checks on the numerical calculations.

The spectral function for the repulsive $S_{\frac{1}{2}}$ αN model with 4% tensor force is displayed in Fig. 6.[17] The predominant features are the large peak centered at $E_\kappa \sim 0.8$ MeV and the nonzero value of the spectral function along q=0. The large peak comes from the nα pair rescattering in the resonant $P_{3/2}$ partial wave. For q~0.3 fm^{-1} and $E_\kappa \sim 0.8$ MeV, this contribution dominates. However, for q ≳ 0, the nonzero value of the spectral function arises from interference between the $S_{\frac{1}{2}}$ component of the plane wave and the $S_{\frac{1}{2}}$ scattering term of the nα-pair rescattering. In fact, for $E_\kappa \lesssim 1.5$ MeV, $V(q,\kappa)$ has a secondary maximum in this region which is damped by the κ factor in the spectral function. The Attractive-Projected model gives a qualitatively similar, but quantitatively different, spectral func-

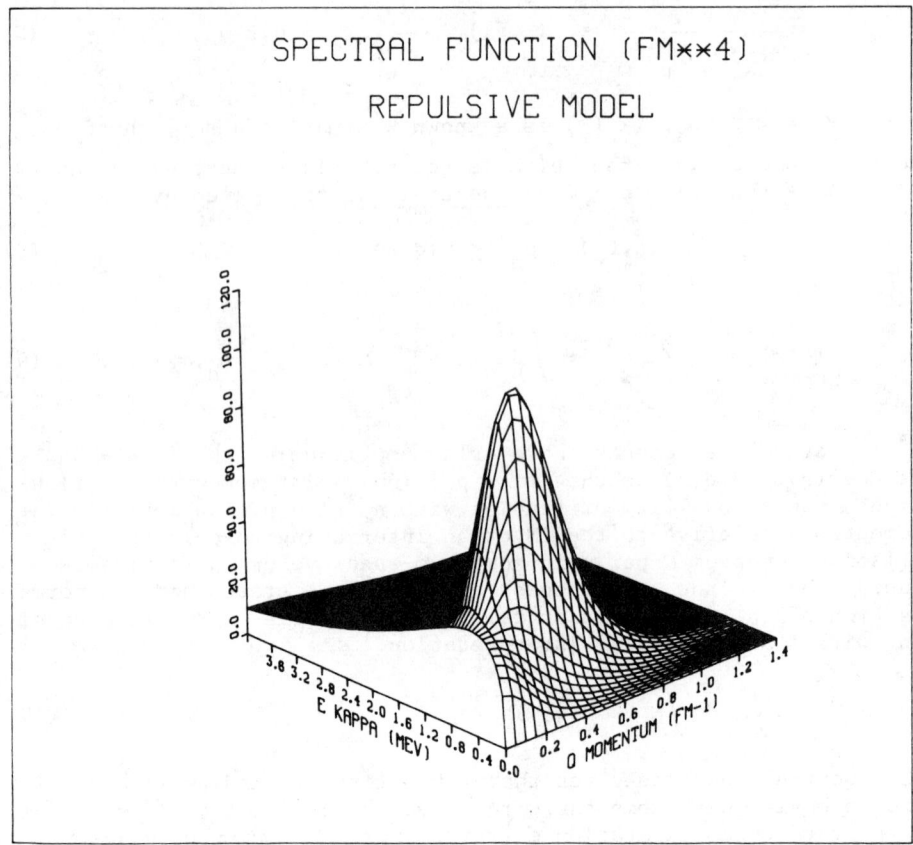

Fig. 6. ^{6}Li → p + (nα) spectral function for the Repulsive $S_{1/2}$ αN model (with 4% NN tensor force) of ^{6}Li.

tion.

Another quantity of interest is the ^{6}Li → p + (nα) momentum distribution defined as

$$\rho(q) = \int_{E_{K_{min}}}^{E_{K_{max}}} dE_K \, S(q, E_K) \quad , \qquad (12)$$

which can be extracted from experiment provided the (k.f.) and the ep cross section (See Eq. (7)) are approximately constant within the range of E_K integration. Usually, E_K is taken over the $P_{3/2}$ resonance peak. So far, $\rho(q)$ has been looked at in ^{6}Li(p,2p)nα experiments and interpreted with ^{6}Li αd cluster models.[18] The fact that $\rho(q)$ is observed to be nonzero with a minimum at q~0, whereas zero

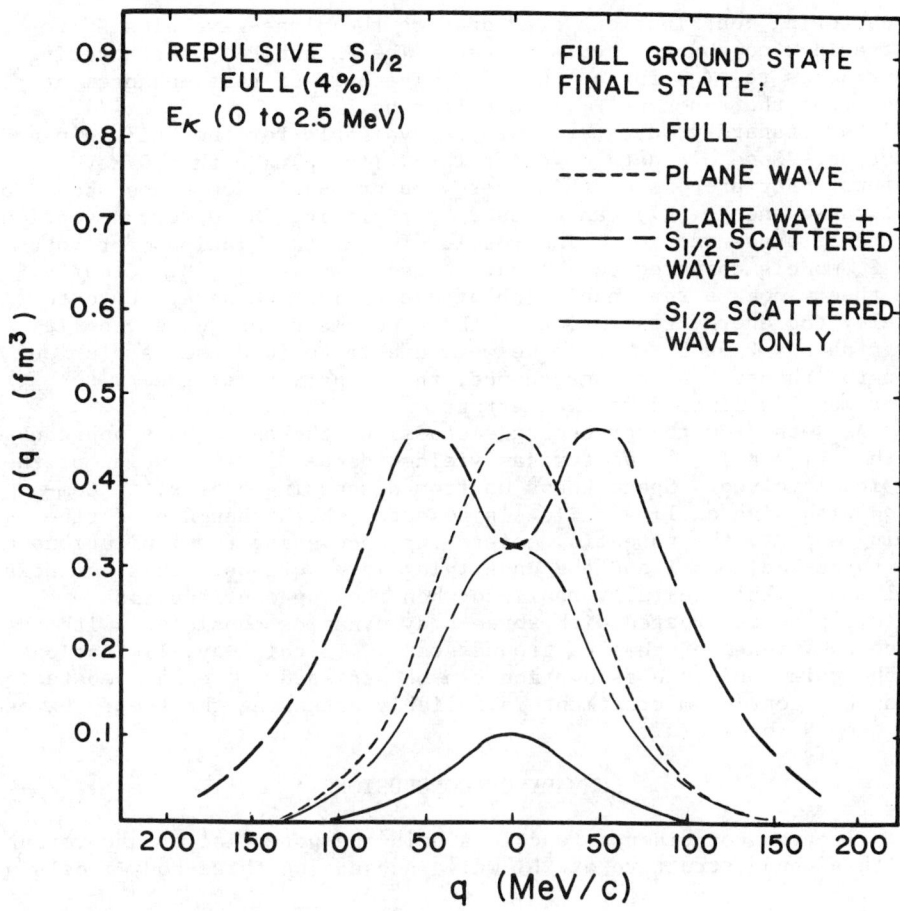

Fig. 7. ^{6}Li → p + (nα) momentum distribution for the Repulsive $S_{1/2}$ αN model (with 4% NN tensor force) of ^{6}Li.

is expected based on a pure p-shell (valence nucleons) shell model of ^{6}Li, is argued to be due to the presence of the $S_{1/2}$ αn interaction.[19] From the three-body viewpoint, the interpretation is different.[17] Consider Fig. 7. When no interactions are present between the n and α in the final state, the plane-wave component of $\Phi_{\vec{K},m_n}^{(-)}$ describes their motion and leads to a ρ(q) with a <u>maximum</u> at q=0. When only the $S_{1/2}$ partial-wave component of the nα-rescattering is included, there is <u>destructive</u> interference between the plane-wave and $S_{1/2}$ rescattering contribution. If the plane-wave contribution is dropped and only the $S_{1/2}$ rescattering piece retained, ρ(q) is small but has a maximum at q=0. It is as though this latter contribution was taken away from the plane-wave (of course, it is more complicated since the rescattering contribution has real and imaginary parts, whereas the plane-wave piece is real). Finally, when the $P_{1/2}$ and $P_{3/2}$

rescattering contributions are added to the plane-wave plus $S_{1/2}$ rescattering, no change occurs at q=0. The $P_{1/2}$ and $P_{3/2}$ rescattering contributes to $\rho(q)$ for $q > 0$ only. The bulk of this enhancement comes from the resonant $P_{3/2}$ rescattering.

Unfortunately, data are not yet available for the ^{6}Li(e,e'p)nα reaction. Some old data do exist for ^{6}Li(p,2p)nα with 460 MeV protons. For protons of this energy, proton-pole dominance should be applicable and Eq. (7) can be used by replacing the ep cross section by the pp cross section. The result of such calculations for both the $S_{1/2}$ models compared to the data[20] is shown in Fig. 8. Clearly, the theory does a reasonable job of describing the data. Unfortunately, the energy resolution of the experiment was not given (it was probably 2 or 3 MeV), so we were unable to fold the resolution into the theory. As a consequence, the comparison of theory to experiment is limited to that extent.

As seen from the preceding discussion, the three-body approach to the ^{6}Li → p + (nα) vertex has yielded deeper understanding of the physics involved. Space keeps us from discussing others.[21] Combined with high quality ^{6}Li(e,e'p)nα data, which should be forthcoming soon[22], the potential exists for uncovering the limitations of the three-body model and the underlying interactions. This potential will most likely be fully realized when the αnp continuum in ^{6}Li(e,e'p)nα is treated with three-body dynamics consistent with the three-body model of the ^{6}Li ground state. In this way, limitations of the pole-dominance assumption can be assessed. The main obstacle to such a continuum treatment is reliably extending the theory beyond the ^{3}He-^{3}H threshold.[16]

SUMMARY OF CONCLUSIONS

Electron coincidence reactions hold the potential of uncovering the three-body structure of ^{6}Li while subjecting three-body models of

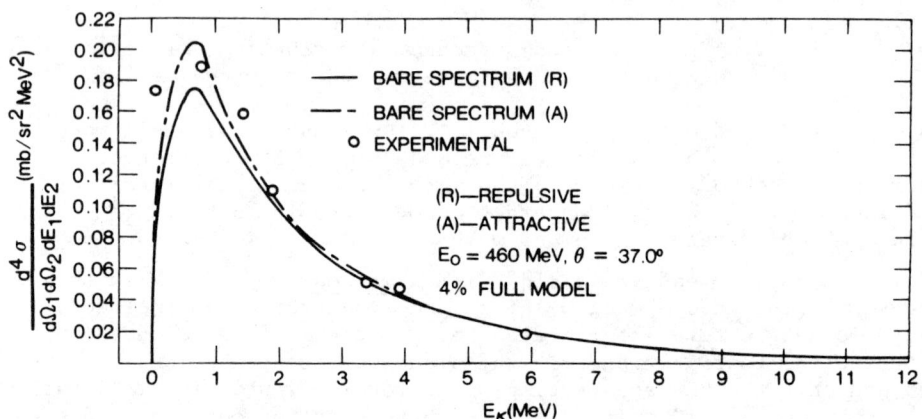

Fig. 8. ^{6}Li(p,2p)nα coincidence cross section.

^{6}Li to severe tests. Good data for ^{6}Li(e,e'd)^{4}He already exists and it points to the 2s nature of the ^{6}Li effective αd wave function as predicted by three-body models. The theory generally reproduces the data, but there is a need for a three-body continuum theory to eliminate reliance on the approximate distorted-wave (optical potential) treatment of the final-state rescattering. Such a theory must be consistent with the ground state and applicable to αd scattering at ~50 MeV in the CM. Data for ^{6}Li(e,e'p)nα is not yet available, but is forthcoming. Nevertheless, three-body treatment of the ^{6}Li $\rightarrow$ p + (nα) vertex has shed new light on the dynamics underlying the spectral function, especially the role of the $S_{\frac{1}{2}}$ partial-wave component of the nα rescattering. As in the ^{6}Li(e,e'd)^{4}He reaction, a proper three-body treatment of the αnp continuum in ^{6}Li(e,e'p)nα is essential for understanding any limitations inherent to three-body models of the ^{6}Li $\rightarrow$ p + (nα) vertex.

ACKNOWLEDGMENTS

The author gratefully acknowledges his collaborators on different aspects of this work: Carol T. Christou and William C. Parke. The author would like to thank William J. Briscoe and Barry L. Berman, both of GWU, and Henk P. Blok and Eddy Jans, associated with NIKHEF, for their keen interest in the ^{6}Li(e,e'd)^{4}He experiment. He would also like to thank Peter K.A. deWitt Huberts and Gerard van der Steenhoven, both of NIKHEF, for correspondence and discussions concerning the ^{6}Li(e,e'p)nα experiment.

REFERENCES

1. Y. Yamaguchi and Y. Yamaguchi, Phys. Rev. 95,1635(1954).
2. A. Ghovanlou and D. R. Lehman, Phys. Rev. C9,1730(1973).
3. D. R. Lehman, M. Rai, and A. Ghovanlou, Phys. Rev. C17,744(1978).
4. K. Hahn, E. W. Schmid, and P. Doleschall, Phys. Rev. C31,325 (1985).
5. D. R. Lehman, Phys. Rev. C25,3146(1982).
6. D. R. Lehman and W. C. Parke, Few-Body Systems 1,193(1986).
7. A. Eskandarian, D. R. Lehman, and W. C. Parke, unpublished.
8. D. R. Lehman and M. Rajan, Phys. Rev. C25,2743(1982).
9. W. C. Parke and D. R. Lehman, Phys. Rev. C29,2319(1984); ibid. C34,E1496(1986).
10. D. R. Lehman and W. C. Parke, Phys. Rev. C31,1920(1985).
11. W. C. Parke, A. Ghovanlou, C. T. Noguchi, M. Rajan, and D. R. Lehman, Phys. Lett. 74B,158(1978).
12. See also, Y. Koike, Prog. Theor. Phys. 59,87(1978); Nucl. Phys. A301,411(1978); ibid. A337,23(1980); V. I. Kukulin, V. M. Krasnopol'sky, V. T. Voronchev, and P. B. Sazonov, Nucl. Phys. A417,128(1984); ibid. A453,365(1986).
13. C. T. Christou, C. J. Seftor, W. J. Briscoe, W. C. Parke, and D. R. Lehman, Phys. Rev. C31,250(1985).
14. R. Ent, H. P. Blok, J. F. A. van Hienen, G. van der Steenhoven, J. F. J. van den Brand, J. W. A. den Herder, E. Jans, P. H. M. Keizer, L. Lapikás, E. N. M. Quint, P. K. A. deWitt Huberts,

B. L. Berman, W. J. Briscoe, C. T. Christou, D. R. Lehman, B. E. Norum, and A. Saha, Phys. Rev. Lett. $\underline{57}$,2367(1986).

15. R. E. Warner, R. S. Wakeland, J-Q. Yang, D. L. Friesel, P. Schwandt, G. Caskey, A. Galonsky, B. Remington, and A. Nadasen, Nucl. Phys. $\underline{A422}$,205(1984); R. E. Warner, J-Q. Yang, D. L. Friesel, P. Schwandt, G. Caskey, A. Galonsky, B. Remington, A. Nadasen, N. S. Chant, F. Khazaie, and C. Wang, Nucl. Phys. $\underline{A443}$, 64(1985).

16. K. Miyagawa, Y. Koike, T. Ueda, T. Sawada, and S. Takagi, Prog. Theor. Phys. $\underline{74}$,1264(1985); K. Miyagawa, T. Ueda, T. Sawada, and S. Takagi, in Book of Contributions to the Eleventh International IUPAP Conference on Few Body Systems in Paritcle and Nuclear Physics, 1986, edited by T. Sasakawa, K. Nisimura, S. Oryu, and S. Ishikawa (Supplement to Research Report of Laboratory of Nuclear Science, Tohoku University, 1986), Vol. 19, p.272.

17. C. T. Christou, Ph.D. dissertation, The George Washington University, July 1986, unpublished.

18. R. K. Bhowmik, C. C. Chang, P. G. Roos, and H. D. Holmgren, Nucl. Phys. $\underline{A226}$,365(1974).

19. S. Saito, J. Huira, and H. Tanaka, Prog. Theor. Phys. $\underline{39}$,635 (1968).

20. H. Tyren, S. Kullander, O. Sundberg, R. Ramachandran, P. Isacsson, and T. Berggren, Nucl. Phys. $\underline{79}$,321(1966).

21. C. T. Christou, D. R. Lehman, and W. C. Parke, to be published.

22. P. K. A. deWitt Huberts, private communication.

THREE-BODY THEORY OF ELECTRON CAPTURE

Joseph Macek
Freiburg Universität, 7800 Freiburg i. Br.[*]

ABSTRACT

Capture of a free electron by a positive ion is forbidden by energy-momentum conservation. Capture of a bound electron does occur since the nucleus of the target atom can recoil and thereby absorb the energy and momentum needed to maintain energy-momentum conservation. Electron capture reactions therefore involve at least three particles all playing essential dynamical roles, that is, electron capture is an inherently three-body problem. Multiple scattering theories do incorporate much of the relevant three-body dynamics but require special care to avoid singularities peculiar the the Coulomb potential. Some specific formulations will be reviewed with emphasis on observed features including the Thomas double collision peak, the continuum electron capture cusp and impact parameter dependent capture probabilities.

INTRODUCTION

The motion of three charged particles, two of which are of like charge, is central to our understanding of a wide variety of physical phenomena. Many such systems are being studied intensively, both experimentally and theoretically. Because theoretical concepts can be closely confronted with high-quality experimental data, the whole field is progressing rapidly. On the theoretical side, progress is less noticeable, primarily because we deal with the long range electrostatic potential. The long range of the potential appears to require concepts which are not expected on the basis of our fairly complete understanding of how two particles interact. I will concentrate

---

[*] Permanent address: Dept. of Physics and Astronomy, University of Nebraska, Lincoln NE

© American Institute of Physics 1987

on the specific example of electron capture at high velocity as an example of a basic three-body problem involving three charged particles. Here much progress has been made in confronting theory and experiment. Atomic units, $m_e = h = e = 1$, are used, where m_e is the electron mass.

IMPORTANCE OF ELECTRON CAPTURE

Electron capture is important technologically. When energy is concentrated in a small region of space filled with atoms or molecules, highly charged ions are invariably produced. One thing such highly charged ions like to do is capture electrons. For that reason capture is interesting for applications. The emphasis here, however, is on the another aspect, namely, the intrinsic three-body nature of capture reactions.

Recall that the process

$$P^{Z_P+} + e^- \to P^{(Z_P-1)+} \quad (1)$$

where P^{Z_P+} represents a projectile ion of charge $+Z_P$, cannot occur owing to energy-momentum conservation. A third particle is required to take up some recoil momentum. Call the third particle the target ion T^{Z_T+} of charge $+Z_T$. Then we have the generic, non-radiative, capture reaction

$$P^{Z_P+} + T^{(Z_T-1)+} \to P^{(Z_P-1)+} + T^{Z_T+} \quad (2)$$

Because three particles all play essential roles in capture reactions, it follows that capture is an intrinsically three-body process. Even at high energies it cannot be approximated by the interaction of two particles. In contrast, excitation and ionization of targets is simply understood, at high energies, in terms of the transfer of energy in a two-body collision of P^{Z_P+} and e^-. For this reason a high energy theory for excitation and ionization in the form of the first Born approximation[1] was developed several decades ago. Only recently has our understanding of capture at high energies progressed to the same level. In this talk I will recount the experimental and theoretical basis for our current understanding of capture at high energies, stressing experimental tests of basic concepts.

THE CAPTURE SPECTRUM

It is well recognized that the a fast charged particle excites both discrete and continuum states. The recognition that capture also extends to the continuum (in a sense to be made precise below) is less well known and was achieved only in the early 1970's. To see this, consider that almost all theories predict cross sections for capture to high n'ℓ'm' Rydberg states, where n'ℓ'm' represent principal, angular momentum and magnetic quantum numbers respectively, vary with n' according to

$$\sigma_{n'\ell'm'} = \bar{\sigma}_{\ell'm'}(Z_P^2/n'^3). \qquad (3)$$

Eq. (3) applies only to capture to bound states, but what happens as n' → ∞? By continuity, we know that bound states of high n' join smoothly with continuum states of low energy E'. This must also hold for matrix elements. The experimental significance of this statement is recognized by noting that the binding energy of a Rydberg state is given by

$$E_{n'} = -Z_P^2/2n'^2, \qquad (4)$$

so that

$$dE_{n'} = (Z_P^2/n'^3)\, dn'. \qquad (5)$$

If we define $\sigma_{n'\ell'm'}\, dn' = d\sigma_{\ell'm'}$, then Eq. (3) takes the form

$$d\sigma_{\ell'm'} = \bar{\sigma}_{\ell'm'}\, dE' \qquad (6)$$

In this form Eq. (6) applies for both positive and negative E', thus we see that "capture" to projectile continuum states is expected.

Eq. (6) applies for each partial wave e' of the continuum electron in a <u>frame moving with the projectile P</u>, i.e. the primed frame. It implies that the differential, in angle, ionization cross section in the moving frame is given by

$$d\sigma' = \frac{\bar{\sigma}}{4\pi} dE' d\Omega' \qquad (7)$$

Upon transforming to the lab frame one finds

$$d\sigma = \frac{\bar{\sigma}}{4\pi} \frac{v}{|\vec{v}-\vec{v}_f|} dE\, d\Omega \qquad (8)$$

where $\bar{v}_f$ is the final velocity of P^{Z_P+}, $\bar{v}$ is the electron velocity in the lab frame and unprimed quantities refer to the lab frame. We immediately see that Eq. (8) predicts a strong enhancement in the energy and angular distributions of electrons ejected from targets by fast projectiles.

The cusp shaped feature characteristic of Eq. (8) was observed by Crooks and Rudd[2] and has been subsequently studied in great detail. A typical 0° electron distri-

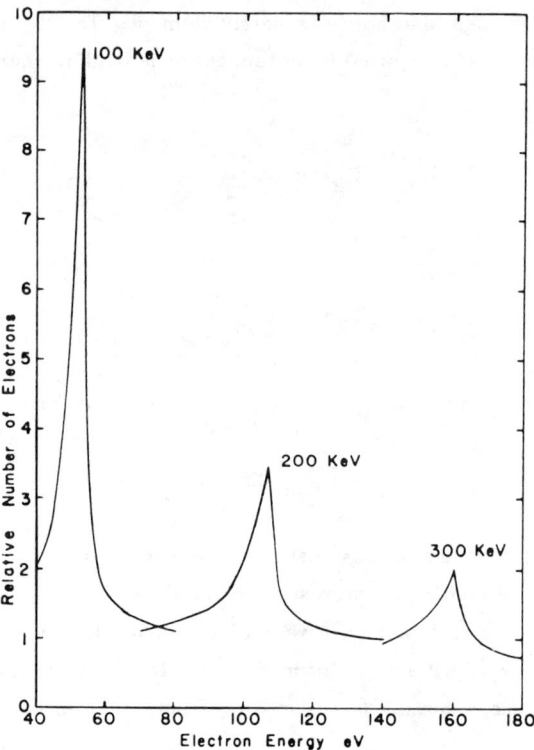

Fig. 1. Typical continuum electron capture cusp. From Ref. 2.

bution, taken from Ref. 2 is shown in Fig. 1. The $\bar{v} = \bar{v}_f$ peak is quite prominent[3] and confirms the extension into the continuum of capture to bound states[4].

It might seem that we obtain something for nothing by the transformation from Eqs. (7) to (8), since no physical interactions are involved in the purely kinematic change of reference frames. Actually we have employed the electrostatic attraction of e^- and P^{Z_P+} in the form of the $1/n^3$ law of Eq. (3). Without this, the constant $\bar{\sigma}$ would be identically zero at $\bar{v} = \bar{v}_f$ and Eq. (8) would not predict a cusp.

Extension of capture into the continuum is conceptually important since it shows that secondary electron production cannot be completely understood without also understanding the, seemingly unrelated, process of electron capture. For this reason we now know that finding the capture mechanism at high energies is of wide significance.

THE CAPTURE MECHANISM.

The first Born approximation for capture, referred to as the Brinkman-Kramers approximation[5], gives a cross section σ^{BK} for capture which decreases with incident particle velocity v_i as v_i^{-12}. In contrast, a classical calculation by Thomas[6], who computed the cross section assuming that the projectile P^{Z_P+} struck the electron e^- which then scattered from the target ion T^{Z_T+} in a direction favorable for capture as illustrated in Fig. 2, gave a v_i^{-11} behaviour. Classical and quantum theories were reconciled

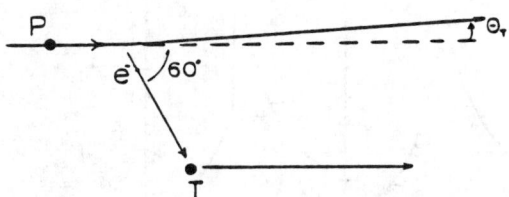

Fig. 2. The Thomas double scattering mechanism.

when it was recognized that the Thomas double collision mechanism correspond to a second-Born theory[6]. Second Born computations gave the high velocity result for 1s - 1s transfer

$$\sigma = \sigma^{BK}\left[0.295 + 5\pi 2^{-11} v_i (Z_T+Z_P)^{-1}\right] \quad (9)$$

where

$$\sigma^{BK} = (\text{const}) \, v_i^{-12} \quad (10)$$

The second Born theory was satisfactory in many regards, in particular the double collision mechanism provided a means of transfering momentum to the recoil ion, and the quantal theory gave a v_i^{-11} term in the cross section. But this term is so small that it has never been observed directly. Confrontation of theory and experiment required more indirect means.

Shakeshaft and Spruch[6] pointed out that the v^{-11} term is more important for high ℓ' states so that capture to 3d states at 5 MeV in proton-hydrogen collisions has a large second-Born component. Even this is not observable, but Shakeshaft and Spruch[6] suggested an ingeneous observation exploiting capture to continuum states.

The shape of the cusp predicted by Eq. (8) depends upon a coherent sum of partial waves for electrons with E' ≈ 0. Normally, one expects only ℓ = 0 waves so that $\bar{\sigma}$ is continuous at the point $\vec{v} = \vec{v}_f$. But if odd partial waves contribute at E' ≈ 0, then they may add destructively with even waves for Θ' = 0, corresponding to

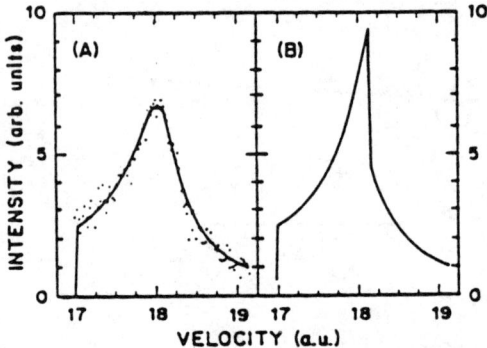

Fig. 3. Step function discontinuity in the continuum electron capture cusp. Fig. 3a is the data folded with the cusp shape in 3b. From Ref. 7.

$v > v_f$, and constructively for $\Theta' = \pi$, corresponding to $v > v_f$. If this happens σ has a step function discontinuity at $v = v_f$. The discontinuity was predicted to be observable for ions of high Z_p. Measurements by Brening et al.[7], shown in Fig. 3 for $Z_p = 14$, found the step-discontinuity and provided the first clear evidence for second Born contributions in electron capture in high ℓ' states.

A second way to confirm the Thomas mechanism was suggested by Briggs and co-workers[8]. It had been known that the double collision mechanism gave rize to a peak in the angular distribution of the scattered ion $P^{(Z_p-1)+}$ at an angle $\Theta_T = \sqrt{3/2}\, m_e/M_p$ where M_p represents the mass of P in atomic units. Plane-wave-Born calculations predicted, however, that this peak would not show up under experimentally accessible conditions[6]. More sophisticated calculations employing Coulomb waves for the electron in the intermediate state predicted a peak at the relatively low energy of 2 Mev/amu for protons. The challenge of measuring the peak was successfully met by Pederson[9] and co-workers. Their angular distribution data is shown in Fig. 4. Dots represent experimental data, theory denoted by the salid curve employs

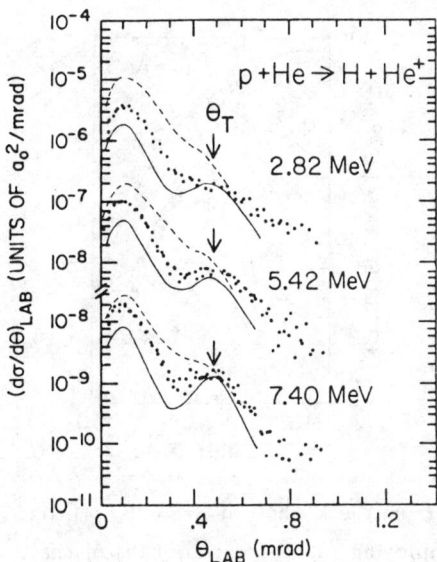

Fig. 4. Differential cross section for electron capture in proton-helium collisions showing the Thomas peak at Θ_T. From Ref. 9.

Coulomb waves as intermediate states while the theory denoted by the dashed curve employs plane waves[10]. Not only is the peak observable, confirming the double collision mechanism, but it suggests that Coulomb rather than plane waves are needed for intermediate states.

The need for more realistic intermediate states is apparent if the ratio Z_p/Z_T is small. Then Z_p/Z_T is a small expansion parameter and if the amplitude is expanded in powers of Z_p/Z_T one obtains the expression[11]

$$T_{fi} \approx \langle \psi_f | V_{Te} + V_{Te} G_T V_{pe} | \psi_i \rangle \qquad (11)$$

for the capture amplitude. Here V_{Te} and V_{pe} denote the $e^- - T^{Z_T+}$ and $e^- - P^{Z_P+}$ interactions respectively and

$$G_T = (E - KE - V_{Te})^{-1}. \qquad (12)$$

Eq. (11) employs a standard transformation, valid when $(M_p + M_T)/M_p M_T \ll 1$, to eliminate the P-T potential. Actually, as pointed out by Dewangen and Eichler[12],

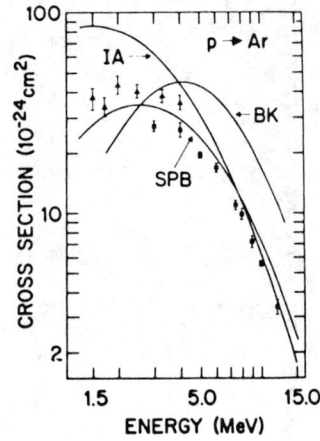

Fig. 5. Electron capture from the K-shell of Argon by protons. The SPB corresponds to calculation employing $\overline{T}_{fi}(E)$ from Eq. (11) of the text. Data from Ref. 14.

the amplitude in Eq. (11) contains a divergent term which is eliminated by taking the proper limits[13].

The amplitude in Eq. (11) employs target intermediate states rather than plane waves. Experimental measurements[14] show quite good agreement with the predictions of Eq. (11) (Fig. 5) suggesting that an expansion in the small parameter Z_p/Z_T is a good guide to finding the appropriate intermediate states.

TOWARDS A MORE COMPLETE THEORY

Fig. 6 summarizes the picture that emerges for the charge transfer mechanism at high energy: The projectile ion P^{Z_p+} excites virtual ionization states of the target with electron momentum $\vec{k}$ centered around the projectile velocity $\vec{v}$ (recall that $m_e = 1$ in atomic units). These states overlap favorably with bound or continuum states of the projectile leading to electron capture.

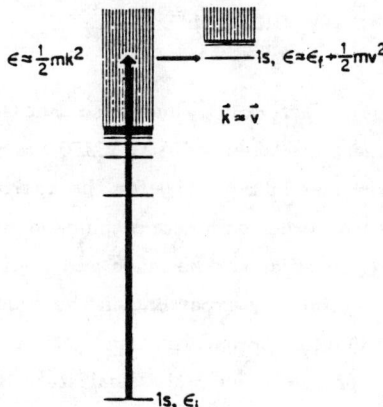

Fig. 6. Electronic energy-level diagram showing the two-step mechanism: virtual ionization and attachment to the moving ion.

This picture is adequate provided intermediate states are indeed represented by virtual states of the target. More generally, one expects that both virtual target and projectile states might play a role. This leads to some varient of the Faddeev-Watson[15] multiple scattering expansion, however there are still unsolved problems employing this expansion with Coulomb interactions. These problems relate to the Dewangen-Eichler singularities which can be delt with when there are only two charged systems in the final state[16]. Then we can define $\overline{T}_{fi}(E)$ by the limit procedure:

$$\overline{T}_{fi}(E) = \lim_{\substack{E' \to E \\ E'' \to E}} g_f(E',E) \langle E'f | V_f + V_f G(E) V_i | E''i \rangle g_i(E'',E) \qquad (13)$$

where V_f, V_i and $G(E)$ have the usual meanings, and g_i and g_f are known functions which include logarithmically divergent factors

$$|E'-E|^{i\nu_f} \text{ and } |E''-E|^{i\nu_i}, \qquad (14)$$

where ν_f, and ν_i are Sommerfeld parameters c_i/v, c_f/v depending upon the long range Coulomb interaction between the bound systems in initial and final states. The singularities of

$$T_{fi}(E',E,E'') = \langle E'f | V_f + V_f G(E) V_i | E''i \rangle \qquad (15)$$

are exactly cancelled by the singular factors in g_f and g_i, so that the limits in Eq. (13) are well defined. In any approximate treatment of $T_{fi}(E',E,E'')$ the nature of the singularities may change, and then the limit in Eq. (13) for the approximate amplitudes usually does not exist. If appropriate order parameters can be identified the expansion of all factors in Eq. (13) to a given order can be made and then the limits exist, to that order. Indeed, Eq. (11) is obtained by expanding in the order parameter ν_i for the model with V_{PT} removed by a unitary transformation. This gives a consistant expansion, but is only rigorously applicable when the final state is bound. Extension to continuum states is not apparent since then there are three charged particles in final states so that Eq. (13) cannot be an appropriate expression for $\overline{T}_{fi}(E)$.

More immediately, when there are only two bodies in the final states, it is possible to apply multiple scattering expansions to $T_{fi}(E',E,E'')$, but because $T_{fi}(E',E,E'')$ is not exact we may not know how to take the limits in Eq. (13) consistantly. We can summarize the situation by saying that various multiple scattering theories for $T_{fi}(E',E,E'')$ incorporate known mechanisms for capture at high energies, but we do not have a general prescription for taking the limits $E' \to E$, $E'' \to E$. For capture to bound states the limits exist for the exact amplitude, but there remains the practical problem of performing the limits consistantly for approximate amplitudes. This problem has been only partially solved for certain expansions, including the expansion represented by Eq. (11), but has not been solved for the Faddeev-Watson expansion.

For capture to continuum states the situation is much worse. We can give formal expressions for $T_{fi}(E',E,E'')$ but do not know how to take the limits, even for the exact amplitude, let alone for approximate amplitudes that are used for interpretation of experiment. Much progress has been made by deferring these formal questions and by proceeding in analogy with capture to bound states. There are now many new effects being discovered experimentally, that appear to require a comprehensive theory[17]. In these circumstances it is difficult to further defer these formal questions.

ACKNOWLEDGEMENTS

Support for this research by NSF grant PHYS 82-03400 and the Deutsche Forschungsgemeinschaft is gratefully acknowledged.

REFERENCES

1. H.A. Bethe, Ann. Physik $\underline{5}$, 32 (1930).
2. J. Crooks and M.E. Rudd, Phys. Rev. Lett. $\underline{25}$, 1599 (1970).
3. A.J. Salin, J. Phys. B $\underline{2}$, 631 (1969).
4. J. Macek, Phys. Rev. A $\underline{1}$, 235 (1970).
5. N.F. Mott and H. Massey, Theory of Atomic Collisions (Cambridge University Press, Cambridge, 1968) p. 428.
6. R. Shakeshaft and L. Spruch, Rev. Mod. Phys. $\underline{151}$, 369 (1979), and Phys. Rev. Lett. $\underline{41}$, 1037 (1978).
7. M. Brenig, S. Elston, I. Sellin, L. Liljeby, R. Thoe, C.R. Vane, H. Gould, R. Marrus and R. Laubert, Phys. Rev. Lett. $\underline{45}$, 1689 (1980).
8. J.S. Briggs, P.T. Greenland and L. Kocbach, J. Phys. B $\underline{15}$, 3085 (1982).
9. E. Horsdal-Pederson, L. Cocke and M. Stockli, Phys. Rev. Lett. $\underline{50}$, 1910 (1982).
10. R. Shakeshaft A $\underline{17}$, 1011 (1978).
11. J.S. Briggs, J. Macek and K. Taulbjerg, Comments on Atomic and Molecular Physics $\underline{12}$, 1 (1982).
12. P. Dewangen and J. Eichler, J. Phys. B $\underline{18}$, L65 (1985).
13. J. Macek, J. Phys. B $\underline{18}$, L71 (1985); J.H. McGuire, J. Phys. B $\underline{18}$, L75 (1985).
14. J.R. MacDonald, C.L. Cocke and M. Stockli, Phys. Rev. Lett. $\underline{32}$, 648 (1974).
15. J. Burgdorfer and K. Taulbjerg, Phys. Rev. A $\underline{33}$, 2959 (1986).
16. J. Macek, Phys. Rev. To be published.
17. W. Meckbach, P.J. Focke, A.R. Goni, S. Suàrez, J. Macek and M.G. Menendez, Phys. Rev. Lett. $\underline{57}$, 1587 (1986).

THE TIME-DEPENDENT HARTREE-FOCK METHOD AND CHAOS IN NUCLEAR DYNAMICS*

M. R. Strayer
Oak Ridge National Laboratory, Oak Ridge, TN 37831

ABSTRACT

The long-time classical behavior of many-body systems is examined using time-dependent Hartree-Fock (TDHF) theory. Detailed computer simulations for the formation of a nuclear molecule in low-energy $^{12}C+^{12}C$ collisions are discussed at length. The representation of effective classical coordinates, and their correspondence to physical observables for systems which undergo large-amplitude quasiperiodic and chaotic motion is briefly summarized. The relationship between these calculations and new methods, which allow the calculation of quasiperiodic solutions to the TDHF equations under more general circumstances, is summarized.

INTRODUCTION

In this talk I shall examine some of the problems that arise in microscopic theories of large-amplitude collective motion. I principally want to describe the results of a numerical experiment using TDHF calculations to simulate the formation of a nuclear molecule in low-energy $^{12}C+^{12}C$ collisions.[1] These simulations extensively used supercomputers, and required about 3,000 mflop hours of time. The molecular structure is believed to be related to the observation of a class of narrow resonances in different $^{12}C+^{12}C$ reaction channels,[2] and may correspond to the formation of shape isomers of the ^{24}Mg system. Betts[3] and Rae[4] have discussed the question of clustering and shape isomeric structures in these reactions; however, we shall see that TDHF brings a whole new dimension to this problem.

In addressing this problem in terms of TDHF, we are led to the long-time classical behavior of a strongly coupled many-body system, a nonlinear Hamiltonian system, and we shall have to concern ourselves with the classification of the TDHF motion in terms of the concepts of classical chaos.[5] I shall briefly discuss a simple three-level model as an example of a system with a well-defined classical limit which undergoes stochastic and quasiperiodic motion.

The more difficult question of how to impose time periodic, or quasiperiodic, boundary conditions on the THDF equations[6] will be discussed briefly. The style of presentation of this material will be informal, descriptive, and nonrigorous.

---

*Research sponsored by the Division of Nuclear Physics, U.S. Department of Energy under contract DE-AC05-84OR21400 with Martin Marietta Energy Systems, Inc.

The derivation of the TDHF equations which clearly displays their classical properties is given in a paper by Kerman and Koonin.[7] The derivation follows from the minimization of the many-body action,

$$\mathscr{S} = \int dt \, \langle \Phi(t) | (i\partial_t - H) | \Phi(t) \rangle \qquad (1)$$

where the many-body Hamiltonian is H, and the integrand on the right-hand side of Eq. (1) is the Lagrangian in the Schrödinger picture. Here the Lagrangian is not in canonical form. The additional assumption that the time-dependent wave function is a Slater determinant of single-particle states ϕ_λ is needed. The action is minimized with respect to ϕ_λ, resulting in the TDHF equations. Conjugate single-particle states, ψ_λ and π_λ are defined in terms of ϕ_λ and $\phi_\lambda^\dagger$ and have the interpretation of classical fields,

$$\phi_\lambda = (\psi_\lambda + i\pi_\lambda)/2. \qquad (2)$$

In terms of these fields, the TDHF equations can be written as a set of Hamilton's equations for the fields

$$\begin{aligned} \dot{\psi}_\lambda &= \delta E/\delta \pi_\lambda \\ \dot{\pi}_\lambda &= -\delta E/\delta \psi_\lambda. \end{aligned} \qquad (3)$$

In Eq. (3) E is the total energy in the state $\Phi(t)$. These equations reflect our present understanding of TDHF as a semiclassical collision theory of macroscopic systems.[8] Let me summarize in a general way our current interpretation of TDHF. This is shown schematically in Fig. 1. TDHF is a fully microscopic dynamical theory which describes both short- and long-time phenomena in low-energy, heavy-ion collisions. The short-time phenomenon is the deep-inelastic part of the collision, and the long-time behavior subdivides into fusion and fission phenomena and periodic and quasiperiodic motion. For fission and fusion, the heavy-ion trajectories do not recur; the entrance channel configurations are very much different than the final configurations, whereas the periodic motion is recurring.

Blocki and Flocard[9] have calculated quasiperiodic motion associated with giant resonances using the TDHF approximation. In Fig. 2 are some of their results obtained for monopole and quadrupole isoscalar resonances. Their method of calculation is as follows. For monopole vibrations, they prepare an initial state which is near the Hartree-Fock minimum, but which has a larger radius, as shown in Fig. 2. They use this state as the initial time-boundary condition and solve the TDHF equations. The proton and neutron radii undergo nearly periodic motion, as shown at the bottom of Fig. 2. The amplitude of the motion is approximately the amount of the initial displacement away from the Hartree-Fock minimum, r_0. The Fourier time transform of this motion gives the frequencies as shown in the figure. In the same way, initializing a nucleus with a nonzero value of the quadrupole moment yields quadrupole vibrational frequencies. This method gives a rapid and simple way to calculate giant-resonance frequencies for a variety of nuclei.

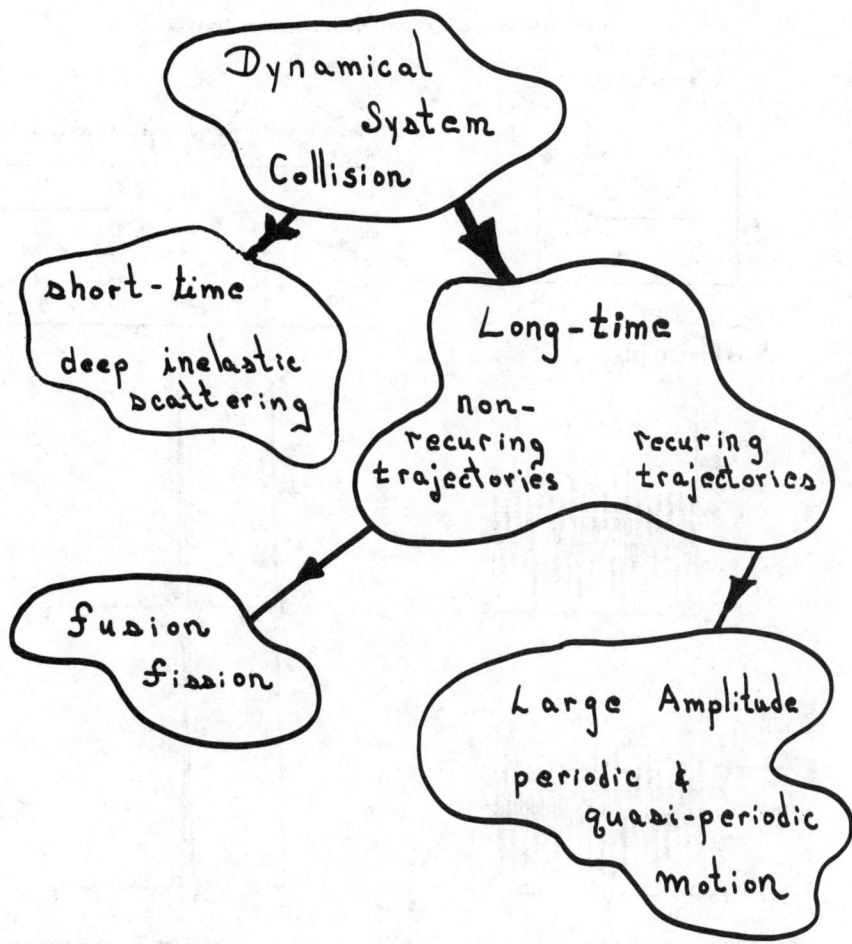

Fig. 1. A schematic representation of the possible classes of physical problems addressed in TDHF.

The suggestion that the focusing of heavy-ion reactions into regions of phase where there exist isomeric states with new shell structure was suggested by Kerman more than 20 years ago.[5,10] In terms of the low-energy $^{12}C+^{12}C$ reactions, this would entail potential minima in the energy surface of ^{24}Mg giving rise to transient vibrational motion. The energy surface for ^{24}Mg computed with the Strutinsky method[11] is shown in Fig. 3. These are potential contours in polar coordinates, where the variables ε (and ε_2) measure the degree of deformation and where γ and also ε_3 and ε_5 display the shape; $\gamma = 0$ for prolate, $\gamma = 60°$ for oblate, and triaxial shapes otherwise. The contours of the potential as a function of ε and ε_3 are given at the bottom of the figure. The ground state of ^{24}Mg is the triaxial minimum with coordinates (0.42, 20°). The triaxial isomer at large

CALCULATION OF GIANT RESONANCES WITH TIME-DEPENDENT HARTREE-FOCK

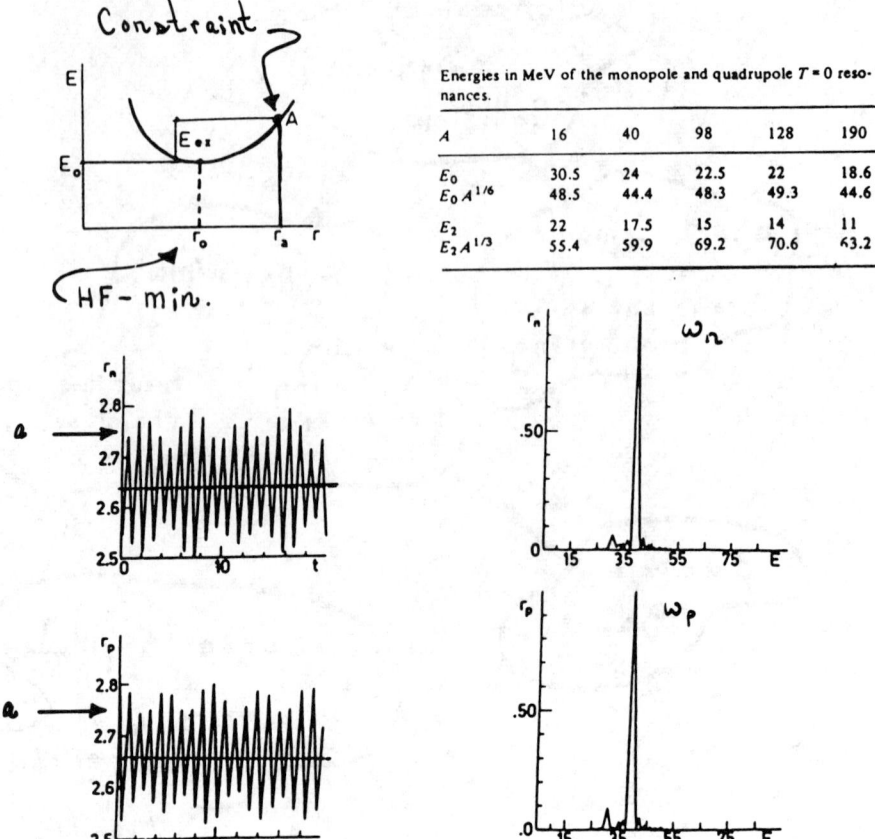

Energies in MeV of the monopole and quadrupole $T = 0$ resonances.

A	16	40	98	128	190
E_0	30.5	24	22.5	22	18.6
$E_0 A^{1/6}$	48.5	44.4	48.3	49.3	44.6
E_2	22	17.5	15	14	11
$E_2 A^{1/3}$	55.4	59.9	69.2	70.6	63.2

Fig. 2. Isoscalar monopole and quadrupole giant resonances calculated using TDHF.

deformation has a six-α cluster structure of two "touching" ^{12}C nuclei.[12] Along the prolate axis there is another isomer with non-zero octupole deformation which corresponds to an α-^{16}O-α cluster structure. The latter cluster configuration has about the right deformation properties for building up the fine structure resonances observed in the ^{12}C+^{12}C reaction; however, the puzzle remains of how the system evolves from the entrance channel configuration shown on the figure to the region near this structure.

This question has been partially answered by the TDHF evolution of the ^{12}C+^{12}C system.[1] In Fig. 4 is shown the energy of a ^{12}C nucleus as a function of its mean quadrupole deformation, β. The ground state, as shown in the figure, is slightly oblate and the spherically symmetric configuration, A, is a reasonable approximation to this state. However, the prolate configuration, B, is an isomer which is a fully self-consistent solution of the static Hartree-Fock equations. The corresponding contours of the matter density are

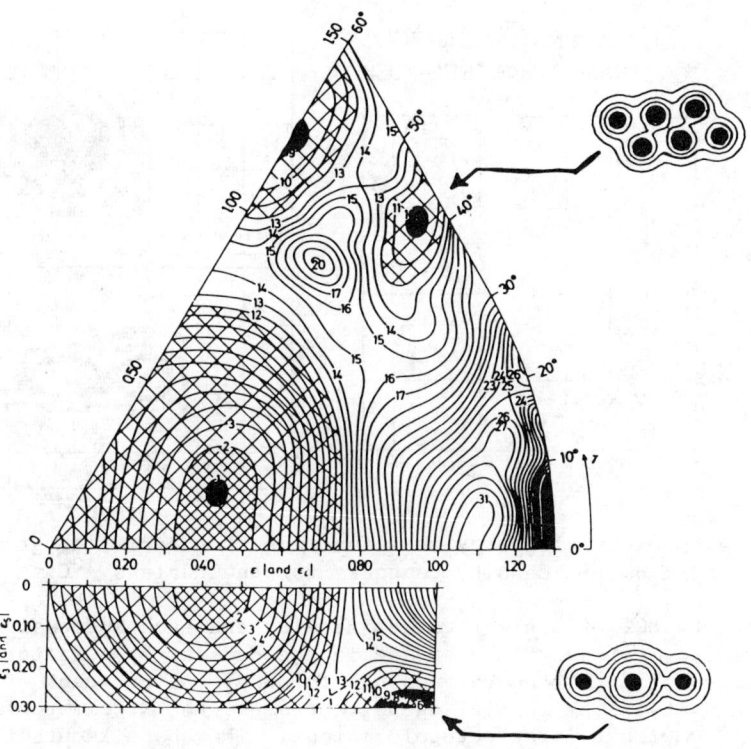

Fig. 3. The potential energy of ^{24}Mg computed using the Strutinsky method as a function of the variables describing the nuclear shape. Two different cluster configurations are indicated by the insets.

shown to the right of the curve in Fig. 4. Note that the isomer density appears to have the structure of a cluster of three α particles arranged in a linear chain. Low-energy collisions of ground-state configurations, A, lead to nonrecurring trajectories, which are usually interpreted as fusion. However, it is known from magnetic transition data[13] that the true ground state of ^{12}C must be a combination of these two configurations, approximately,

$$|^{12}C\rangle = .98 |A\rangle - .21 |B\rangle \qquad (4)$$

Thus, a simulation of the $^{12}C+^{12}C$ collision should also include collisons between configurations A and B as shown in the figure. Note that this would represent an initial configuration which does not appear on the energy surface in Fig. 3.

The results of a collision of nucleus A and nucleus B at a bombarding energy of 15 MeV are given in Figs. 5 and 6. In Fig. 5, the mean isoscalar density contours in the collision plane are given for eight different times during the collision. The interior dark areas

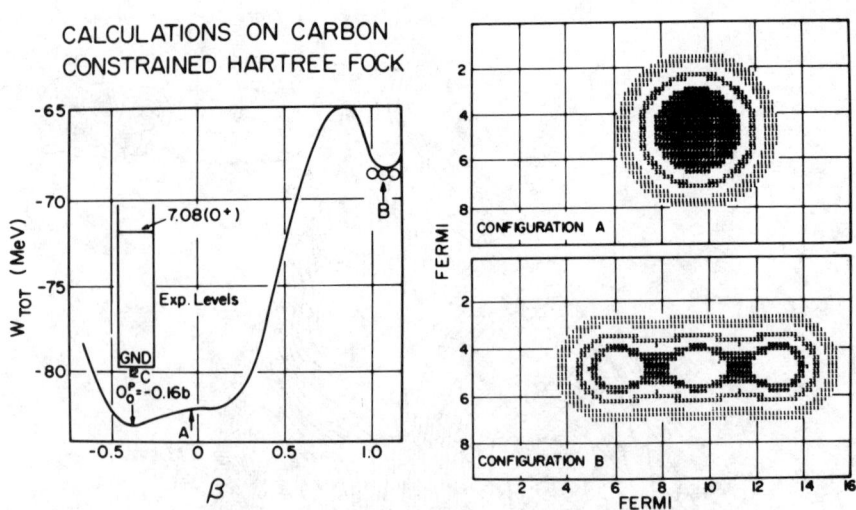

Fig. 4. Potential energy as a function of deformation ground and isomeric density contours for the nucleus ^{12}C.

correspond to 80% of nuclear matter density and each successive shaded region represents a decrease in the density by a factor of 2.3. The outer dark band corresponds to about one per cent of the central density. The initial shape represents an extreme configuration of ^{24}Mg with a large octupole moment. The shape recurs throughout the time evolution with an approximate period of about 600 and 750 fm/c, or equivalently, a classical frequency of between 1.5 and 2 MeV. The isoscalar density is the sum of the proton and neutron densities; equivalently the isovector density, the difference between the proton and neutron densities, is shown in Fig. 6. These densities are displayed at the same collision times as those in the previous figure, however, the scale is very different. The darkest contour corresponds to about +0.5% of the central density, the grey contours circled by the darkest contour to about +1.5%, the isolated grey contours to -3%, the white regions to -2.5%, and the grey contours encircled by the white region to -6% of the central density. The motion here appears very rapid, and has large multipole configurations with no smooth oscillatory motion as noted in the previous figure.

Even though we are solving for the dynamical evolution of 24 single-particle wave functions, we can appeal to the Kerman and Koonin work and view this motion in the following classical sense. As the system evolves in time, there is a corresponding potential energy surface associated with its motion. This is a type of dynamical energy surface, different than Fig. 3, which corresponds to minimizing the Hartree-Fock energy while keeping the density fixed. Thus the minimum energy is a function of the density during the evolution. This energy surface is displayed in Fig. 7 as a function of the q_2 q_3, respectively, the quadrupole and the octupole moments of the isoscalar density. The time evolution begins at the coordinates (q_2, q_3) $\sim (1.7 q_{20}, -2.5 q_{30})$ and "moves" the system towards the local minimum

163

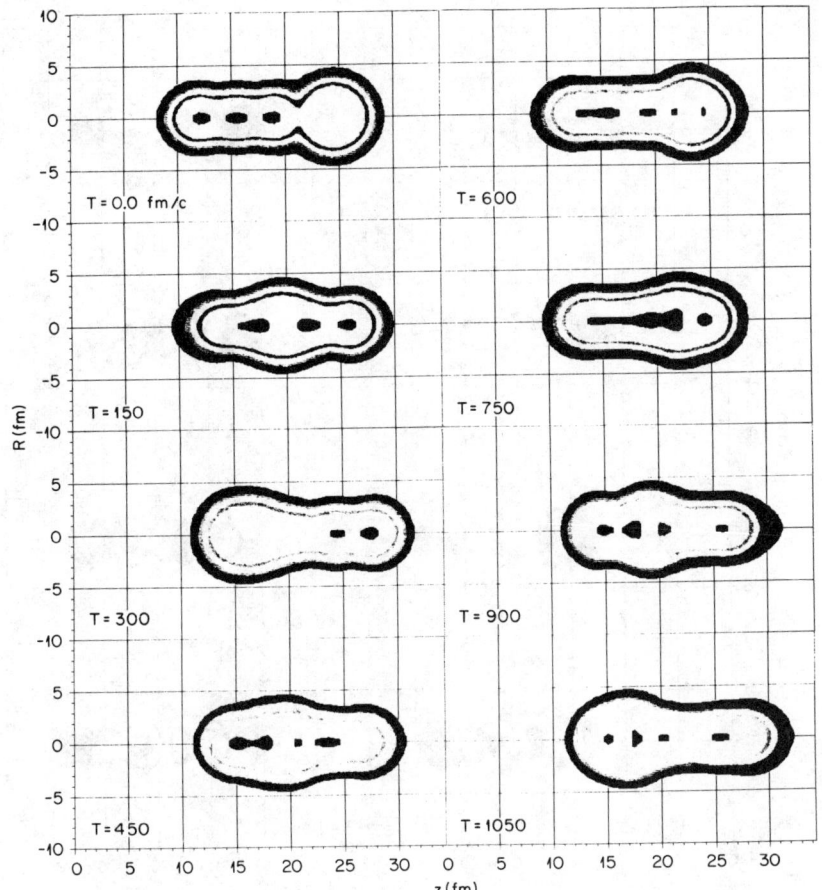

Fig. 5. Isoscalar density contours in the collision plane for the ^{24}Mg system; different times exhibit the quasiperiodicity of the system.

indicated by the plus on the figure. As the system advances in time, it traces out that part of the energy surface with the drawn contours. The labels on the contours denote the values of the energy above the minimum. This region of the energy surface possesses a broad minimum and the evolution proceeds as if the system is classically bounded in this minimum. Note in this figure the barrier, which "prevents" the system from collapsing into the ground state.

I wish to digress and examine the results of the three-level shell model[14] studied by Koonin and Williams,[15] which will suggest ways in which to interpret these calculations. The model consists of N particles occupying a three-shell model level, with energies ($\pm\varepsilon$,0), and with a two-body interaction between pairs of particles in different levels,

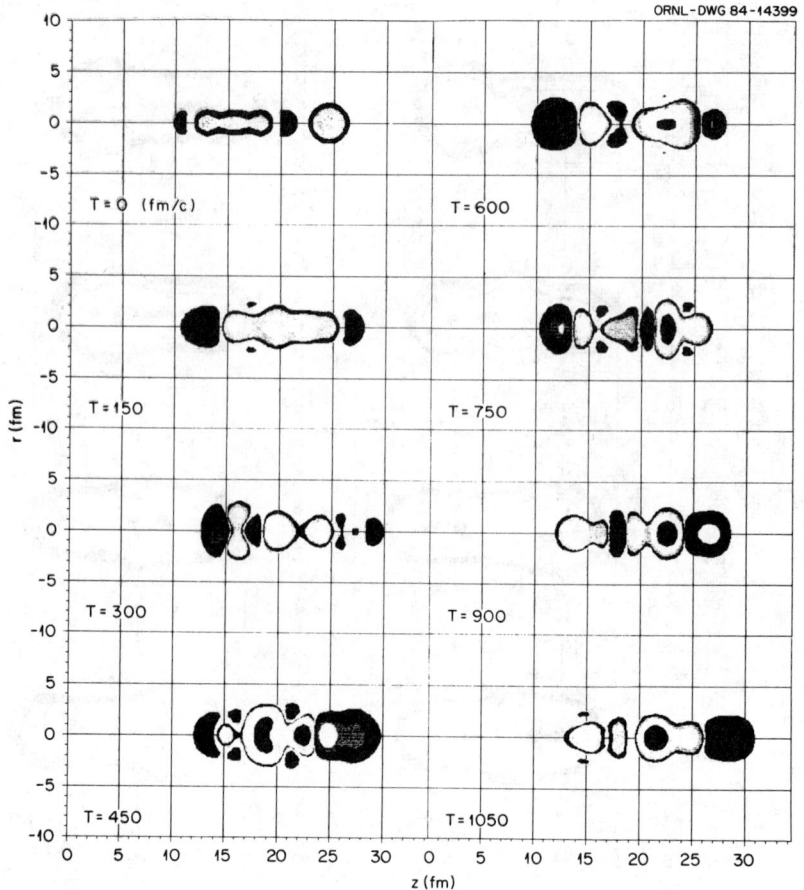

Fig. 6. Same as in Fig. 5 except for the isovector density.

$$H = \sum_k \varepsilon_k \left(\sum_n a^\dagger_{nk} a_{nk} \right) + \sum_{k\ell} V_{k\ell} \left(\sum_n a^\dagger_{nk} a_{n\ell} \right)^2 \quad (5)$$

with

$$\varepsilon_k = \begin{cases} 0 \\ \pm\varepsilon \end{cases} \quad (6)$$

and

$$V_{k\ell} = V(1 - \delta_{k\ell}) \quad (7)$$

This model is completely solvable quantum mechanically, and has a well-defined classical limit in terms of TDHF. The classical

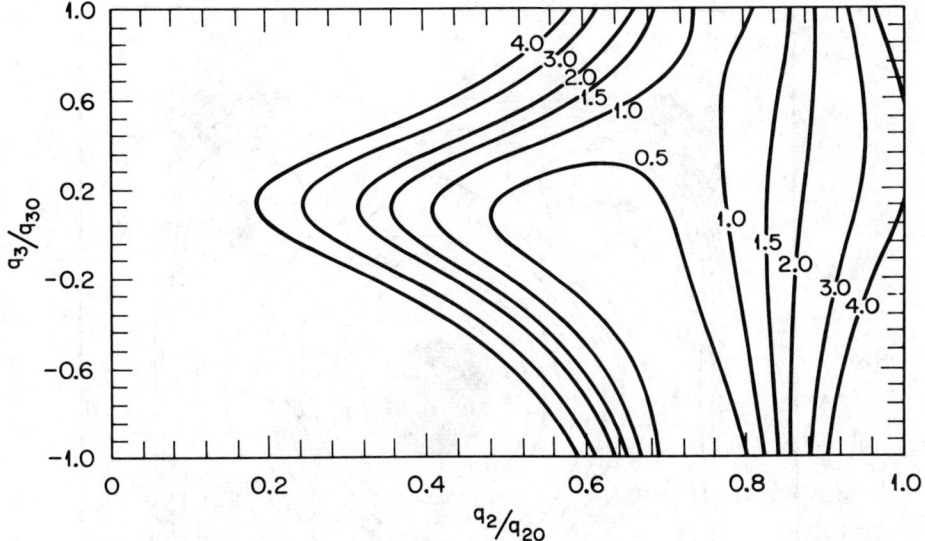

Fig. 7. Dynamical energy surface for ^{24}Mg computed using the Hartree-Fock method. The constants defining the scale are $q_{20} = 56.6$ fm^2 and $q_{30} = -412.6$ fm^3.

solutions have two independent degrees of freedom and, as a function of the interaction strength, exhibit regions of phase space with periodic, nonergotic quasiperiodic, and stochastic motion. Denoting the conjugate coordinates as two-dimensional vectors, q and p, the classical potential energy is obtained as

$$h(\vec{q},\vec{p}=0) = -1 + \vec{q}^2 - \frac{1}{2}q_1 + \frac{\chi}{4}\left[\vec{q}^4 - 2\left(q_1^2 q_2^2 + \vec{q}^2\right)\right], \quad (8)$$

where χ is the dimensionless strength parameter,

$$\chi = \frac{V}{(N-1)\varepsilon}. \quad (9)$$

In the strong coupling limit this potential, as shown in Fig. 8, has one local maximum, at coordinates $\vec{q} = 0$, surrounded by four local minima. The results of solving the TDHF equations with this potential for three different initial solutions are shown in Fig. 9. The three solutions correspond to initial states formed from RPA solutions having the excitation energy indicated in the figure. The first column displays the TDHF trajectory q_2 as a function of q_1, and the last two columns plot the corresponding Poincaré phase-space sections for two different projections, $q_1 = 0$ and $q_2 = 0$. The low-energy solution is periodic and corresponds to the motion of two coupled harmonic oscillators. The intermediate-energy solution is

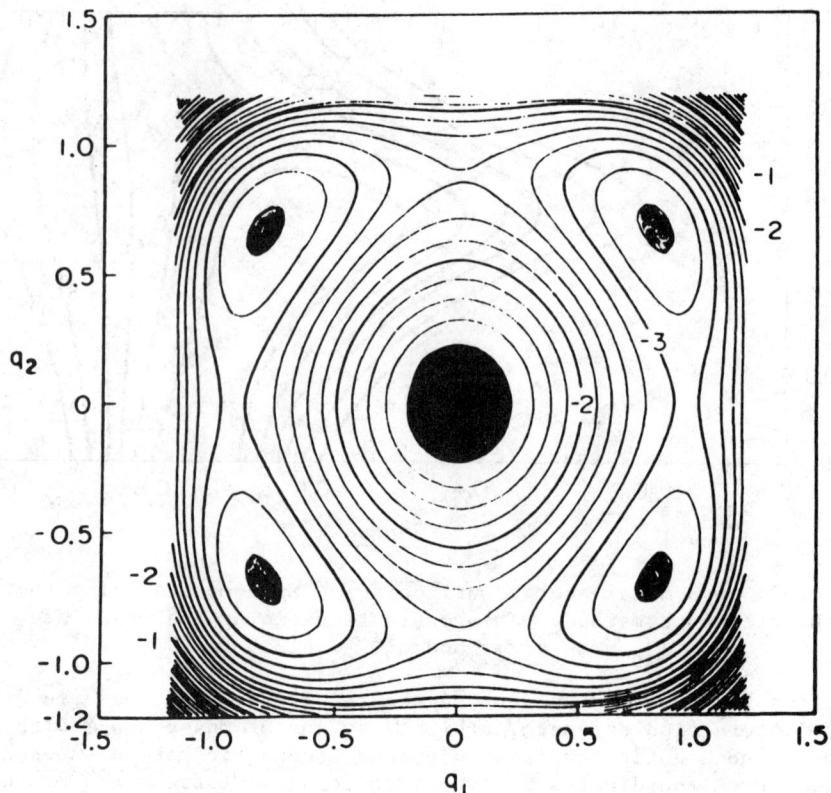

Fig. 8. Classical potential energy surface for the three-level model.

quasiperiodic, and the Poincaré section here is beginning to break up into different phase-space segments, whereas at high energy, the motion becomes stochastic, and begins filling in the region of phase space allowed by considerations of energy and momentum conservation. In passing from this topic, I would like to note that of the three types of motion, the authors were unable to requantize the stochastic motion and found that the corresponding RPA solutions were a reasonable approximation to the quantum-mechanical solutions.

Returning to the numerical experiment, we can introduce the multipole moments of the densities as "descriptive" coordinates, again classical in nature, which will help to diagnose the TDHF motion. In time and frequency space, I define both isoscalar and isovector moments as

$$M_\ell(t) = \int d\vec{r}\, r^\ell Y_{\ell 0}(\hat{r})\, \rho(\vec{r},t)$$
$$M_\ell(\omega) = \int dt\, e^{-i\omega t} M_\ell(t) \qquad (10)$$

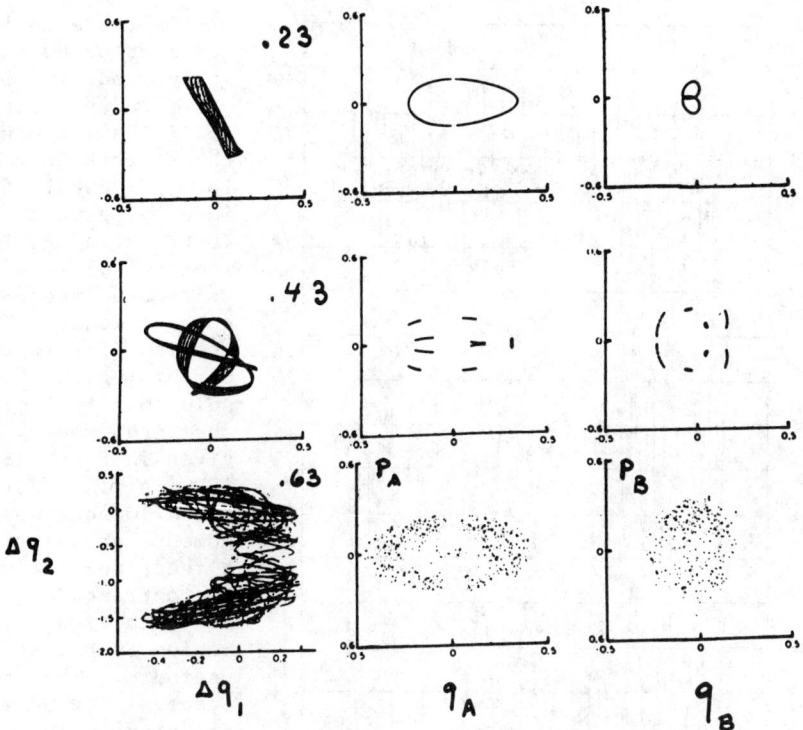

Fig. 9. Trajectory in $\vec{q}$-space and two Poincaré sections for three different TDHF trajectories computed with the three-level model.

and in Fig. 10 the isovector dipole and the isoscalar quadrupole and octupole moments are plotted as a function of time. This represents the full-time evolution corresponding to the densities shown in Figs. 5 and 6, and is nearly 50,000 time steps of the TDHF equations. Here we see clearly the strong quasiperiodic octupole vibrations and also the highly stochastic isovector motion. The quadrupole motion is strongly coupled to the octupole motion, and appears to undergo damping on this time scale. The Poincaré sections for the quadrupole and octupole motion are shown in Fig. 11. These are velocities as a function of the corresponding coordinates for the TDHF trajectory. The octupole coordinates seem to be focused into a band of trajectories which recur, while it is not clear that the quadrupole section is not stochastic and will fill in this region of phase space.

The evidence from the numerical simulation for this classification is not entirely convincing. However, one can appeal to the theorem of Kolmogorov, Arnold, and Moser (KAM),[16] which is sometimes given as follows: "For a Hamiltonian system characterized by an invariant N-torus, e.g., integrable with N frequencies, it remains unchanged under a wide class of nonlinear perturbations of the orbits." In terms of the KAM theorem, the motion we are examining

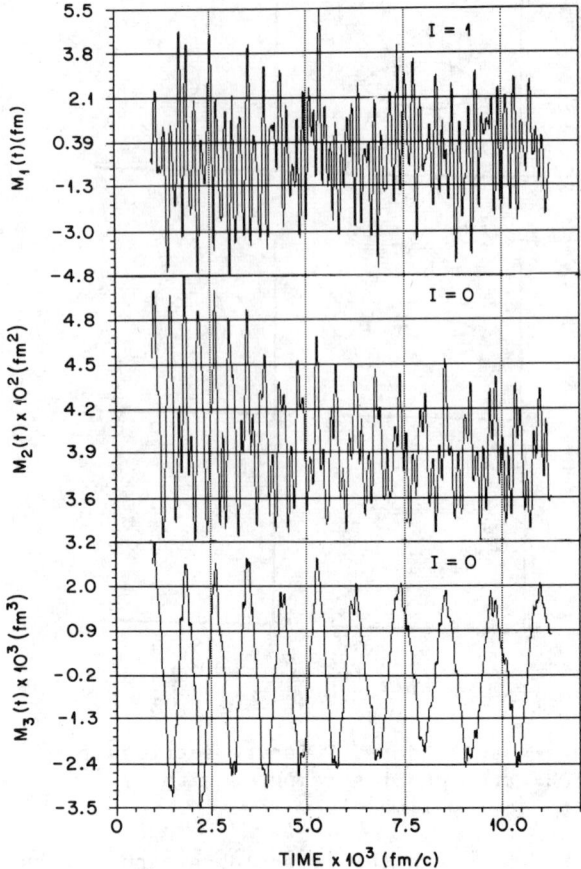

Fig. 10. The time dependence of the isovector dipole and the isoscalar quadrupole and octupole moments for the ^{24}Mg system.

may be close to being quasiperiodic, and the strong coupling to other modes during the collision introduces the stochastic behavior. We can test this idea by perturbing the TDHF orbits and by looking for solutions which are indeed quasiperiodic. The results of slightly perturbing the TDHF solutions by lowering the total energy are given in Figs. 12 and 13. In Fig. 12 the three lowest isoscalar moments are plotted as a function of time. In contrast to the results in Fig. 10, this motion is almost periodic. The frequency spectrum corresponding to this motion is shown in Fig. 13 for the same moments. Note the logarithmic scale on the figure, and the occurrence of harmonics of the principal low frequency, about 1 MeV, vibrational mode, as well as a vibrational mode near 8 MeV.

There are several shortcomings of this approach, and it is useful to mention these briefly. The obvious question concerns the long-time evolution needed to describe the quasiperiodic phenomena. It is usually thought that correlations which are absent from the TDHF approach become increasingly important in time. This is predicated on a variety of experience in fusion, fission, and deep-inelastic reaction studies. However, it clearly is not true for the studies of giant resonances.[9] Here, the correlations are important in building up the decay widths of the resonances but the frequencies can be accurately obtained from long-time TDHF evolution. The second criticism is more serious and raises the issue that there is a continuous spectrum of classical frequencies which emerge from this treatment, whereas in real life there is a discrete spectrum corresponding to

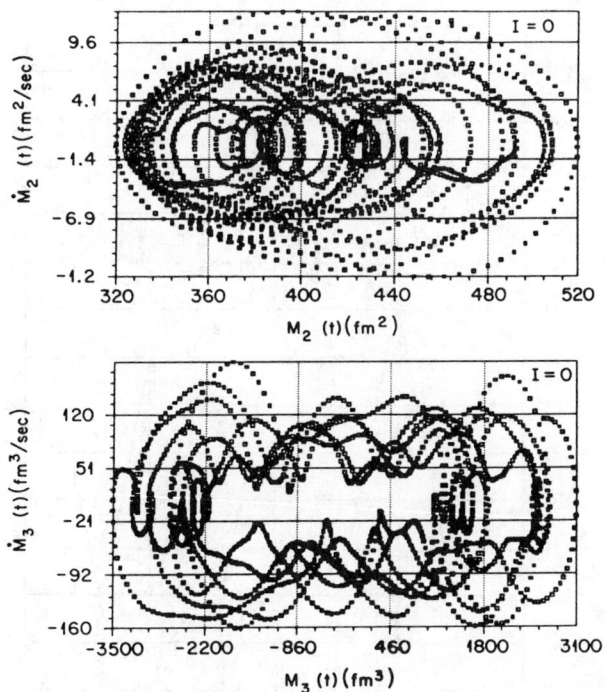

Fig. 11. Poincaré phase-space plots for the isoscalar quadrupole and octupole modes of the ^{24}Mg system.

the level spacing between quantum states. This question becomes important in attempting to calculate transition amplitudes[17] in the many-body system. There have been several approaches to this problem, one of the earliest by Griffin and co-workers[18] is to impose quantization conditions on the TDHF equations thereby yielding discrete frequency, periodic mean-field solutions. Similar ideas have also been developed by Zahed and Baranger,[19] by Klein and Umar,[20] and many others.[6] The above three papers can be understood by the following construction. Replace the initial time-boundary condition that is usually imposed on the TDHF single-particle wave function, ϕ_λ, by the periodicity condition,

$$\phi_\lambda(t+\tau) = e^{-i\Lambda_\lambda t} \phi_\lambda(t) \qquad (11)$$

This development follows that of Zahed and Baranger. The phase Λ_λ is the quasi-energy of the state ψ_λ and leads to the nonlinear equation for ψ_λ and

$$[-i\partial_t - h + \Lambda_\lambda]\psi_\lambda(t) = 0 \qquad (12)$$

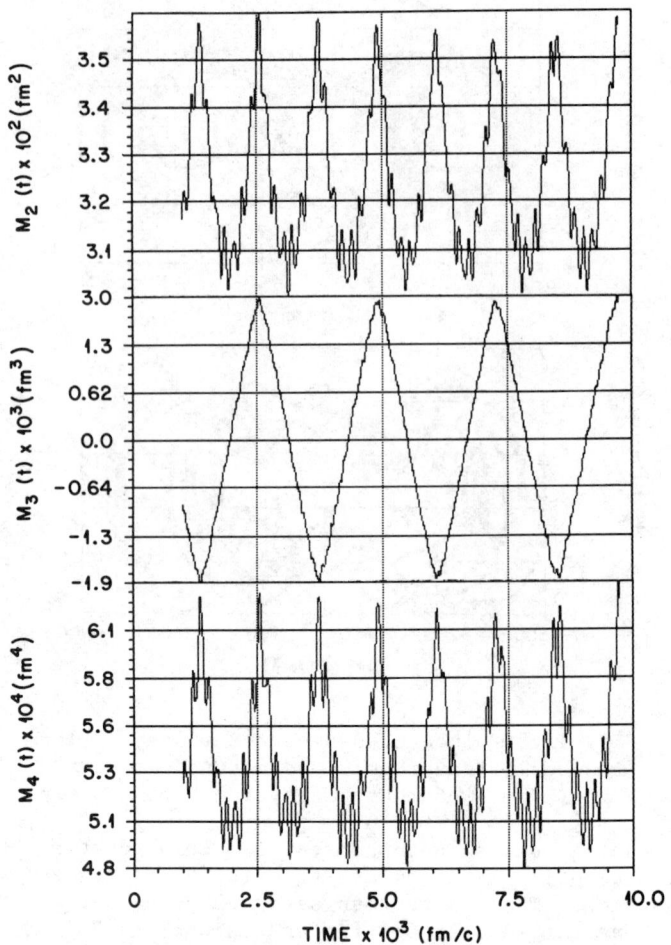

Fig. 12. Time dependence of the isoscalar quadrupole, octupole, and hexadecapole moments for the cooled modes of the ^{24}Mg system.

Under the assumption that solutions of this type of equation exist and can be found, together with the principle of least action, results in a Sommerfeld quantization relation,

$$2\pi n = \int_0^\tau dt \sum_\lambda \langle \psi_\lambda(t) | i\partial_t | \psi_\lambda(t) \rangle \tag{13}$$

Presently, no methods are available to solve systems of equations of this type in the general case. However, within the last two years, a new method has been developed by Baranger and Davies[21] to solve systems of classical equations having this character, and this method may be applicable to the periodic TDHF problem.

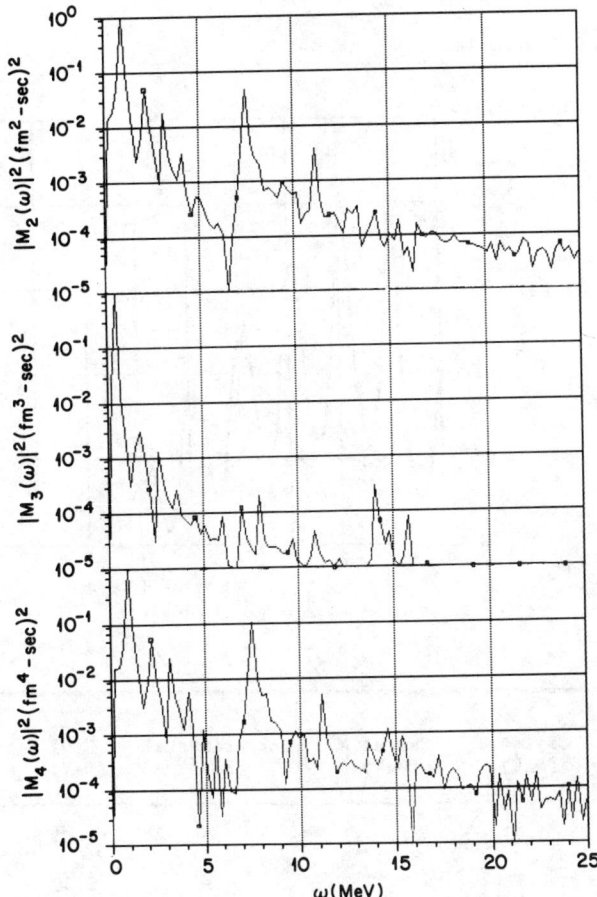

Fig. 13. Frequency dependence of the isoscalar quadrupole, octupole, and hexadecapole moments for the cooled modes of the ^{24}Mg system.

As a final remark, we can follow the suggestion of Koonin and Williams[15] and examine the RPA solutions near the Hartree-Fock minimum in Fig. 7. These are shown in Fig. 14. These are actually linear response solutions expanded about the isomer, and the top part of the figure displays the structure function. The peaks here can be proved to be transitions between a series of vibrational states built on the intrinsic structure of the isomer. The resulting frequencies calculated with two different forces are compared to the experimental data and to those obtained with the U(4) model.[2] Only S-wave states are compared, however, the microscopic calculations do as well as the model.

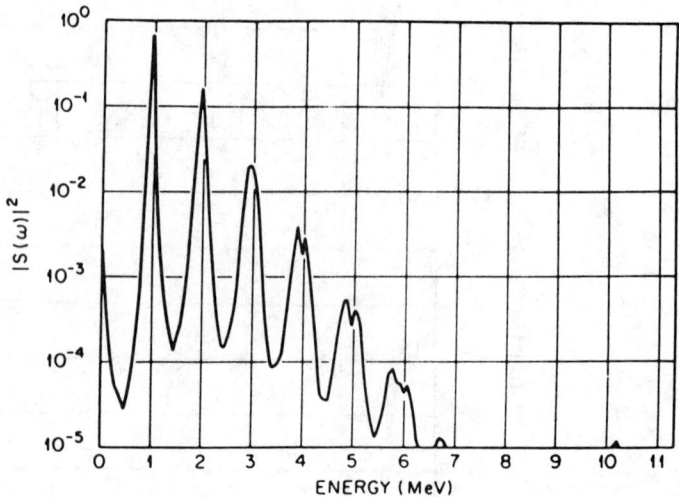

Skyrme I (MeV)	Skyrme M* (MeV)	Experimental (MeV)	U(4) (MeV)
0.98	1.01	–	1.04
1.97	2.12	–	2.58
3.01	3.08	3.17–3.35	3.44
4.00	4.13	4.25	4.44
4.80	5.06	–	5.20
5.75	5.90	5.80	5.84

Fig. 14. Comparison of resonance frequencies in the ^{24}Mg isomer computed using linear response to empirical results obtained with the U(4) model.

REFERENCES

1. A. S. Umar, PhD. thesis, Yale University, 1985 (unpublished);
    A. S. Umar, M. R. Strayer, R. Y. Cusson, P.-G. Reinhard, and
    D. A. Bromley, Phys. Rev. $\underline{C32}$, 175 (1985); A. S. Umar and
    M. R. Strayer, Phys. Lett. $\underline{171B}$, 353 (1986).
2. K. A. Erb and D. A. Bromley, Phys. Rev. $\underline{C23}$, 2781 (1981).
3. R. R. Betts and W.D.M. Rae, Proceedings of the International
    Nuclear Physics Conference, Harrogate, England, Aug. 25-30, 1986,
    Vol. II (in press).
4. W.D.M. Rae, Proceedings of the Xth Symposium on Nuclear Physics
    Oaxtepec, Mexico, January 5-8, 1987 (in press).
5. M. R. Strayer, in Fusion Reactions Below the Coulomb, ed. by
    S. G. Steadman (Springer, New York, 1984).
6. J. W. Negele, Rev. Mod. Phys. $\underline{54}$, 913 (1982).
7. A. K. Kerman and S. E. Koonin, Ann. Phys. (N.Y.) $\underline{100}$, 332 (1976).
8. K.T.R. Davies, K.R.S. Devi, S. E. Koonin, and M. R. Strayer, in
    Treatise on Heavy-Ion Science, Vol. 3, ed. by D. A. Bromley
    (Plenum, New York, 1985).
9. J. Blocki and H. Flocard, Phys. Lett. $\underline{85B}$, 163 (1979).
10. A. K. Kerman, private communication.
11. G. Leander and S. E. Larson, Nucl. Phys. $\underline{A239}$, 93 (1975).
12. S. Marsh and W.D.M. Rae, Phys. Lett. B (in press).
13. N. Takigawa and A. Arima, Nucl. Phys. $\underline{A168}$, 593 (1971).
14. S. Y. Li, A. Klein, and R. M. Dreizler, J. Math. Phys. $\underline{11}$, 975 (1970).
15. R. D. Williams and S. E. Koonin, Nucl. Phys. $\underline{A391}$, 72 (1982).
16. M. Hénon, p. 55 in Chaotic Behavior of Deterministic Systems,
    Les Houches, France, ed. by R.H.G. Helleman and R. Stora (North-
    Holland, New York, 1983).
17. Bela Gazdy and David A. Micha, Phys. Rev. $\underline{A33}$, 4446 (1986).
18. K.-K. Kan, J. J. Griffin, P. C. Lichtner, and M. Dworzecka,
    Nucl. Phys. $\underline{A332}$, 109 (1979).
19. Ismail Zahed and Michel Baranger, Phys. Rev. $\underline{C29}$, 1010 (1984).
20. A. Klein and A. S. Umar, Phys. Rev. $\underline{C34}$, 1965 (1986).
21. M. Baranger and K.T.R. Davies, Ann. Phys. (N.Y.), in press.

"The submitted manuscript has been authored by a contractor of the U.S. Government under contract No. DE-AC05-84OR21400. Accordingly, the U.S. Government retains a nonexclusive, royalty-free license to publish or reproduce the published form of this contribution, or allow others to do so, for U.S. Government purposes."

MICROSCOPIC CLUSTER THEORY IN NUCLEAR PHYSICS

Y. C. Tang
School of Physics, University of Minnesota, Minneapolis, MN 55455

ABSTRACT

The Microscopic cluster theory (MCT), formulated within the framework of the resonating-group method (RGM), has been used over the last forty years to study nuclear bound-state, scattering, and reaction problems from a unified viewpoint. In this theory, cluster correlations, arising as a consequence of the on-the-average attractive nature of the nucleon-nucleon interaction, are explicitly taken into consideration. The formulation is based on a variational procedure, utilizing fully-antisymmetric many-nucleon trial wave functions. Analytical derivations of the kernel functions, representing the nonlocal interactions among the clusters, are facilitated by the use of various generator-coordinate techniques. Multi-configuration calculations have been performed for a number of light and medium-weight systems, yielding generally satisfactory results which shed information concerning nuclear structure and reaction mechanisms. By examining the general features of the kernel functions, detailed studies have been carried out to investigate the main effects of antisymmetrization. Here the most important finding is that, for a system where the nucleon-number difference of the interacting nuclei is small, the effective internuclear potential contains an appreciable parity-dependent component. As a practical application, the resultant wave functions have also been used to study various electromagnetic processes, such as radiative capture and bremsstrahlung.

I. INTRODUCTION

It is generally believed [1] that the explanation of all low-energy nuclear phenomena can, in principle, be achieved by solving the time-independent many-nucleon Schrödinger equation

$$<\delta\psi|H-E_T|\psi> = 0 , \qquad (1)$$

where $\delta\psi$ represents a completely arbitrary variation of ψ in the whole Hilbert space. In reality, one recognizes that this is an unattainable task except for systems containing a very small number of nucleons. For practical calculations, one must necessarily work in an appropriately chosen but restricted model space which is nevertheless extended enough to yield sufficiently accurate results for comparison with experiment.

In the microscopic cluster theory (MCT) [1,2], the choice of a restricted yet flexible model space is accomplished by realizing that, because of the on-the-average attractive nature of the nucleon-nucleon force, there exist in nuclei rather long-ranged correlations which manifest themselves through the formation of nucleon clusters. The intricate properties exhibited by nuclear systems are, therefore, considered to be a consequence of the dynamical interplay among various cluster structures. In this way, one can, as a reasonable approximation, regard a complicated many-nucleon system as behaving like a much simpler system composed of a few clusters of correlated nucleons, thereby greatly reducing the computational requirement. As has been demonstrated in a rather large number of detailed studies [1,2], this cluster theory does successfully explain, quantitatively in many instances, the varied nuclear phenomena in the low-excitation region.

In addition to the feature of accounting explicitly for cluster correlations, the MCT has also the following important characteristics:
(i) It employs totally antisymmetric wave functions and, therefore, the Pauli exclusion principle is fully taken into consideration.
(ii) It treats correctly the motion of the total center of mass.
(iii) It utilizes a nucleon-nucleon potential which explains the essential features of the two-nucleon low-energy scattering data.
(iv) It considers nuclear bound-state, scattering, and reaction problems in a unified manner.
(v) It can be used to study cases where the particles involved in the incoming and outgoing channels are both arbitrary composite nuclei.
(vi) It is based on a variational principle; consequently, the accuracy of the result can be tested and improved by systematically expanding the model or basis-function space employed in the calculation.

The model space of the MCT is spanned by antisymmetrized, translationally-invariant, nonorthogonal wave functions. Because of the choice of such rather complicated basis functions, MCT calculations are frequently tedious to perform, but can be facilitated by the use of various generator-coordinate techniques [2-4], which have been especially developed over the last fifteen years to evaluate the kernel functions representing the nonlocal interactions among the clusters. Additionally, there is one simplifying feature which should be mentioned. As is now well understood [1], the Pauli principle, expressed by the antisymmetrization of the wave function, has the effect of reducing greatly the differences between apparently different nonorthogonal wave functions when the nucleons are close to one another. From this it follows that, at relatively low excitation energies, the number of cluster configurations which one needs for approximate calculations can usually be made so small that MCT calculations become quantitatively feasible.

A brief description of the MCT is given in the next section. In Sec. III, several representative bound-state, scattering, and reaction calculations are described. Section IV is devoted to a general discussion of the effects of antisymmetrization on effective internuclear potentials. The use of MCT wave functions in studying electromagnetic processes, such as radiative capture and bremsstrahlung, is illustrated in Sec. V. Finally, in Sec. VI, concluding remarks are made.

II. BRIEF DESCRIPTION OF THE MICROSCOPIC CLUSTER THEORY

A detailed discussion of the MCT is given in Ref. 2; hence, only a very brief description will be presented here. The MCT trial wave function is chosen to have the form

$$\psi(\tilde{r}_1,\cdots,\tilde{r}_N) = \sum_\tau \int \hat{\Phi}_\tau(\tilde{r}_1,\cdots,\tilde{r}_N;\vec{\beta}_\tau) F_\tau(\vec{\beta}_\tau) d\vec{\beta}_\tau + \sum_\lambda C_\lambda \hat{\eta}_\lambda(\tilde{r}_1,\cdots,\tilde{r}_N), \qquad (2)$$

where $\tilde{r}_i$ represents the space, spin, and isospin coordinates of the ith nucleon (N is the total number of nucleons), $\vec{\beta}_\tau$ denotes a set of continuous parameter or generator coordinates which characterize the basis functions $\hat{\Phi}_\tau$, and λ enumerates the discrete set of square-integrable basis functions $\hat{\eta}_\lambda$. The coefficients $F_\tau(\vec{\beta}_\tau)$ and C_λ represent the continuous and discrete linear variational amplitudes, respectively. The choice of the basis set is quite flexible; the basis functions must be linearly independent, but need not be orthogonal to one another.

The basis set is selected in a way which is most suited to the problem under consideration. In the MCT, the first term on the right side of Eq. (2) is chosen to provide a proper description of nucleon clustering present mainly in the nuclear surface, while the second term, containing the so-called distortion functions $\hat{\eta}_\lambda$, is included to improve the behavior of the wave function in the strong-interaction, compound-nucleus region.

The linear variational amplitudes F_τ and C_λ satisfy a set of coupled equations, obtained by using the projection form of the time-independent Schrödinger equation expressed by Eq. (1). In that equation, E_T denotes the total energy of the system and H is a Galilean-invariant Hamiltonian given by

$$H = \sum_{i=1}^{N} \frac{1}{2M} \vec{p}_i^2 + \sum_{i<j=1}^{N} V_{ij} - T_{cm}, \qquad (3)$$

with T_{cm} being the kinetic-energy operator of the total center of mass and V_{ij} being a nucleon-nucleon potential chosen to fit the two-nucleon scattering data especially in the low-energy region. The function $\delta\psi$ represents the variation, in the chosen model space, of the trial wave function ψ through a completely arbitrary variation of the amplitudes F_τ and C_λ.

The MCT can be formulated either in the framework of the resonating-group method (RGM) or in the framework of the generator-coordinate method (GCM). As has been extensively discussed [2], these two formulations are entirely equivalent. In the RGM formulation, for example, the trial wave function ψ for the simplest case, where one considers a system of two spinless clusters in the no-distortion approximation, has the form

$$\psi(\tilde{r}_1,\cdots,\tilde{r}_N) = \int \hat{\Phi}(\tilde{r}_1,\cdots,\tilde{r}_N;\vec{R}'') F(\vec{R}'') d\vec{R}'', \qquad (4)$$

with the basis function $\hat{\Phi}$ given by

$$\hat{\Phi}(\tilde{r}_1,\cdots,\tilde{r}_N;\vec{R}'') = \mathcal{A}[\phi(A)\phi(B)\delta(\vec{R}-\vec{R}'')Z(\vec{R}_{cm})]. \qquad (5)$$

In the above equation, $\mathcal{A}$ is the antisymmetrization operator, $\phi(A)$ and $\phi(B)$ are translationally-invariant properly-normalized functions describing the internal behavior of clusters A and B, and $Z(\vec{R}_{cm})$ is any normalized function describing the motion of the total center of mass. The interesting point to note here is that, even in this very simple case, the basis functions $\hat{\Phi}$ with different values of $\vec{R}''$ are not mutually orthogonal because of the presence of the antisymmetrization operator.

By substituting Eqs. (4) and (5) into Eq. (1), one obtains the following resonating-group integrodifferential equation for $F(\vec{R}')$:

$$[-\frac{\hbar^2}{2\mu}\nabla^2_{\vec{R}'} + V_D(\vec{R}') - E]F(\vec{R}') + \int K(\vec{R}',\vec{R}'')F(\vec{R}'')d\vec{R}'' = 0, \quad (6)$$

where μ is the reduced mass, E is the relative energy of the two clusters in the c.m. system, V_D is a direct or no-exchange potential, and $K(\vec{R}',\vec{R}'')$ is an energy-dependent kernel function given by

$$K(\vec{R}',\vec{R}'') = H_E(\vec{R}',\vec{R}'') - E_T N_E(\vec{R}',\vec{R}''), \quad (7)$$

with H_E and N_E being the exchange-Hamiltonian and exchange-normalization kernels, respectively; they have the forms

$$H_E(\vec{R}',\vec{R}'') = \langle \phi(A)\phi(B)\delta(\vec{R}-\vec{R}')Z|H| \mathcal{A}''[\hat{\phi}(A)\hat{\phi}(B)\delta(\vec{R}-\vec{R}'')Z]\rangle, \quad (8)$$

$$N_E(\vec{R}',\vec{R}'') = \langle \phi(A)\phi(B)\delta(\vec{R}-\vec{R}')Z| \mathcal{A}''[\hat{\phi}(A)\hat{\phi}(B)\delta(\vec{R}-\vec{R}'')Z]\rangle. \quad (9)$$

In Eqs. (8) and (9), $\hat{\phi}(A)$ and $\hat{\phi}(B)$ are antisymmetric functions obtained by antisymmetrizing the cluster internal functions $\phi(A)$ and $\phi(B)$, and the operator $\mathcal{A}''$ is defined by the equation

$$\mathcal{A}'' = \mathcal{A}' - 1, \quad (10)$$

with $\mathcal{A}'$ being an intercluster antisymmetrization operator which interchanges nucleons in different clusters.

From Eqs. (6) - (10), one notes that, by adopting the resonating-group trial wave function of Eq. (4), the many-nucleon system is approximated as a simple system of two structureless point clusters interacting through an energy-dependent, nonlocal, effective interaction. As is evident from these equations, the nonlocal part arises from the operation of intercluster antisymmetrization. If such an operation is omitted in the calculation, then the effective intercluster potential will just be the direct potential V_D. This latter potential can be easily obtained by a double-folding procedure and is a simple ℓ-independent local potential when a purely central nucleon-nucleon potential containing specifically no Majorana exchange component is employed.

The derivation of the kernel function $K(\vec{R}',\vec{R}'')$ is a central problem in a resonating-group calculation. Generally, this is a rather tedious task, especially when the nucleon number N involved in the system is large. However, there exists now a number of generator-coordinate computational techniques [2-4] which serve to facilitate the derivation. In our work [2], for example, we have made extensive use of the so-called complex-generator-coordinate technique (CGCT) in which the essential idea is to express the trial wave function as a linear superposition of antisymmetrized products of single-particle wave functions. Other techniques have been employed by many groups around the world. As a consequence, even systems containing rather large numbers of nucleons have been successfully treated with the MCT approach for bound-state, scattering and reaction considerations.

III. EXAMPLES OF MCT CALCULATIONS

In this section, the usefulness of the MCT in studying nuclear many-body problems will be described in several examples. These examples are selected mainly for illustrative purposes; however, they do serve to demonstrate the effects of enlarging the model space, the interesting feature of the MCT in treating bound and continuum states in a unified manner, the importance of including cluster specific distortion in the calculation, and so on.

A. Structure of ^{16}O with the microscopic α-cluster model.

For the first example, the investigation of the structure of ^{16}O by Bauhoff et al. [5] will be described. In this investigation, the microscopic α-cluster model, which is a static version of the MCT, is used, with the purpose being to explain the observation of a number of rotational bands in the low-excitation region of this nucleus.

The intrinsic many-nucleon wave functions $\hat{\Phi}$ are constructed as Slater determinants of 1s harmonic-oscillator single-particle orbitals

$$\phi_i(\tilde{r}_i) = (\pi b^2)^{-3/4} \exp[-(\vec{r}_i-\vec{R}_j)^2/2b^2]\chi_i \quad (i=1 \text{ to } N; j=1 \text{ to } N/4), \quad (11)$$

centered around chosen cluster positions $\vec{R}_j$. In the above equation, the indices i and j label the nucleons and the α-cluster centers, respectively, χ_i denotes the spin and isospin part of the wave function, and b is the oscillator width parameter. In accordance with the Pauli principle, all single-nucleon spatial states are occupied by four nucleons. Additionally, we should point out the important fact that a wave function $\hat{\Phi}$ of this type can describe not only states of a shell-model character when all $\vec{R}_j$ are chosen to be equal or close together, but also states with a pronounced cluster structure when the positions $\vec{R}_j$ are well separated. This property makes the model well suited for the study of the low-excitation region of ^{16}O where both types of states are known to be present.

In general, this kind of many-nucleon wave function will not be the eigenstate of either the parity or the angular-momentum operator. Thus, components with good quantum numbers have to be projected out of the intrinsic state. For the parity operator, this is easily achieved by using the linear combinations

$$\hat{\Phi}_{\pm} = \hat{\Phi}(R_j) \pm \hat{\Phi}(-R_j) \quad . \quad (12)$$

The angular-momentum projection can also be straightforwardly carried out [5], but does require lengthy computational periods. Therefore,

in this investigation, the simplification of projecting out good
angular-momentum states only after variation is adopted, although the
parity projection is always performed before variation.

The variational parameters in this model are the cluster positions $\vec{R}_j$ and the oscillator width parameter b. In contrast to the
usual procedure of assuming certain geometrical configurations which
serve to restrict the number of parameters, the important characteristic of this investigation is that all 3x4-6+1=7 variational parameters (the six spurious rigid-body degrees of freedom are subtracted out) are unconstrainedly varied. With the Brink-Boeker [6]
nucleon-nucleon potential B1, extensive searches then yield four
positive-parity and two negative-parity energy minima, corresponding
to ground and excited intrinsic states on which rotational bands may
be built. Upon further projecting out angular-momentum states,
Bauhoff et al. found that the microscopic α-cluster model is, indeed,
capable of explaining most of the experimentally-found T = 0 levels
up to about 15-MeV excitation.

Table I

K = 0 Rotational bands in ^{16}O

J^π	Bent-rhomb band		Linear-chain band	
	Calc. E_x (MeV)	Exp'tal E_x (MeV)	Calc. E_x (MeV)	Exp'tal E_x (MeV)
0^+	5.45	6.05	16.27	
2^+	7.44	6.92	16.67	17.13
4^+	11.69	10.36	17.56	18.02
6^+	17.24	16.28	18.96	19.32
8^+	23.43		20.87	
10^+			23.28	

Two calculated K = 0 positive-parity rotational bands are
particularly worth mentioning. These bands are based on intrinsic
states with bent-rhomb and asymmetrical linear-chain configurations, respectively. In Table I, the calculated and experimental [7]
results for the excitation energy E_x are shown. From this table, it
is seen that the experimentally-found rotational band with bandhead
at 6.05 MeV compares well with the band calculated from the bent-
rhomb configuration. This is quite gratifying for a microscopic
calculation which employs a rather standard effective nucleon-nucleon
interaction and which contains no adjustable parameters. For the
linear-chain band, the agreement between theory and experiment is
also satisfactory. In this case, however, we should mention one
important fact; i.e., the 0^+ member of this band has not yet been
experimentally located. From the viewpoint of obtaining a detailed
understanding of nuclear structure, it would indeed be very interesting to search for this particular level in the excitation-energy
region around 16.7 MeV.

The intrinsic linear-chain configuration is strongly prolate-
deformed; it has a rms radius of 4.80 fm, which is much larger than
the value of 2.68 fm for the intrinsic tetrahedron configuration of

the ground-state band. In Fig. 1, we demonstrate this particular property by showing the density distribution of the chain configuration after projection onto positive-parity states. From this figure, one notes that the α clusters are barely in contact, indicating that, in a shell-model calculation, one will have to take into account an impractically large number of major shells for satisfactory results.

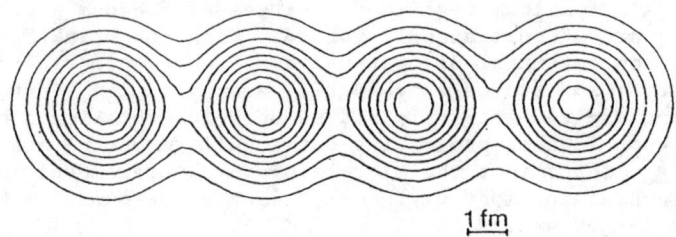

Fig. 1: Density contours for the linear chain configuration in ^{16}O. (From Ref. 5).

B. Multi-configuration study of the seven-nucleon system.

The formulation of this problem is discussed in detail in Ref. 8 and, hence, will not be described here. In this study, the cluster configurations included are the t+α, n+^{6}Li, n+^{6}Li*, and d+^{5}He configurations, with ^{6}Li* being the T = 0 excited state of ^{6}Li having a d+α cluster structure with relative orbital angular momentum equal to 2. Because of the large number of configurations included and the careful choice of cluster internal functions, the resulting calculation turned out to be quite complicated, requiring the formulation of a three-cluster RGM. Therefore, for the sake of reducing computational effort, the decision was made to omit the nonessential features of including the Coulomb and noncentral parts of the nucleon-nucleon potential in the calculation.

Table II

Cluster separation energy $\tilde{E}$ in the ground state of ^{7}Li

Model space	Cluster configurations	$\tilde{E}$ (MeV)
Single configuration 1: SC1	t + α	2.00
Single configuration 2: SC2	n+^{6}Li	-1.06
Double configuration 1: DC1	t+α, n + ^{6}Li*	2.85
Double configuration 2: DC2	n+^{6}Li, n+^{6}Li*	1.59
Double configuration 3: DC3	t+α, d+^{5}He	3.30
Triple configuration 1: TC1	t+α, n+^{6}Li, n+^{6}Li*	3.25
Triple configuration 2: TC2	t+α, n+^{6}Li, d+^{5}He*	3.44
Quadruple configuration: QC	t+α, n+^{6}Li, n+^{6}Li*, d+^{5}He	3.50

In Table II, we show the t+α cluster separation energies $\tilde{E}$ in the ground $(L,S) = (1,1/2)$ state of ^{7}Li, calculated in various model spaces [9]. To obtain these results, we have adopted the Minnesota or MN nucleon-nucleon potential [10] with the exchange-mixture parameter u chosen to be 1 (i.e., a Serber exchange mixture). As is seen from this table, $\tilde{E}$ is equal to 3.50 MeV in the QC case. This value is rather close to the empirical result of 3.17 MeV, obtained by averaging the experimental t+α cluster separation energies in the $3/2^-$ ground and $1/2^-$ first-excited states of ^{7}Li and by making a Coulomb correction of 0.86 MeV.

From Table II, one also finds the following interesting features:
(i) For a reasonable $\tilde{E}$ result, the t+α cluster configuration must be included in the calculation. This finding confirms the common belief that the t+α configuration is the dominant cluster configuration in the ground state of ^{7}Li.
(ii) The values of $\tilde{E}$ in the SC1 and QC calculations differ by an appreciable amount of 1.50 MeV. This indicates that the specific distortion or dynamical polarization of the t and α clusters, effected by the addition of other cluster configurations to enlarge the model space, has a significant influence in the present case.
(iii) The results for $\tilde{E}$ in the DC3 and TC1 calculations are about the same, indicating that, in the ^{6}Li ground state, the d+^{5}He cluster configuration is similar in significance to the n+^{6}Li plus n+^{6}Li* cluster configurations.
(iv) The difference in $\tilde{E}$ values obtained in the TC2 and QC calculations, being only 0.06 MeV, is quite small. This implies that the QC result of 3.50 MeV for $\tilde{E}$ probably cannot be improved much more. Any further calculation with a larger model space employing additional multi-cluster configurations will definitely require a vast increase in computational effort and, in our opinion, would no longer be worthwhile.

To demonstrate the effects of enlarging the model space on continuum states, we show the phase shifts and transmission coefficients calculated with the SC2 and DC2 models in the $(L,S) = (2,3/2)$ case. The results obtained as a function of E_2, the relative energy of the neutron and ^{6}Li in the c.m. system, are shown by the dashed (SC2) and dot-dashed (DC2) curves in Fig. 2, where $\delta^2_{20,20}$ denotes the n+^{6}Li phase shift, and $\eta^2_{02,20}, \eta^2_{22,20}$, and $\eta^2_{42,20}$ represent the transmission coefficients leading from the n+^{6}Li channel to the various n+^{6}Li* channels, characterized by ℓ values (ℓ is the orbital angular momentum between the neutron and the ^{6}Li* cluster) equal to 0, 2, and 4, respectively.

The phase-shift curve in the SC2 case shows that there exists a resonance state at around 10 MeV. This state is, however, so broad that its effects on the n+^{6}Li cross sections are quite limited. The situation is changed in the DC2 case. Now, the phase-shift result exhibits a more rapid rise in the low-energy region and indicates the presence of an additional state at 14.5 MeV. This latter state has predominantly a n+^{6}Li*, $\ell=2$ cluster structure and expresses itself through a dispersion-like resonance behavior in the n+^{6}Li phase-shift curve.

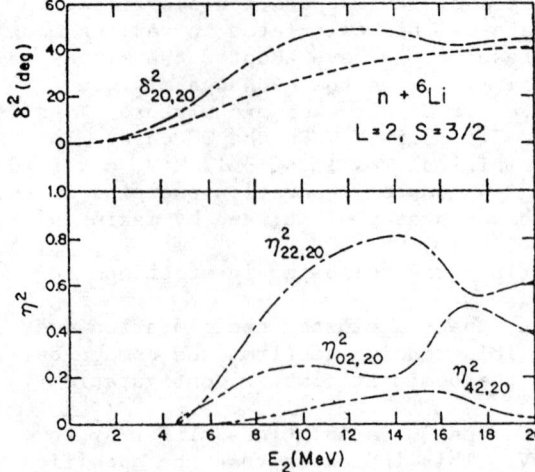

Fig. 2: Calculated phase shifts and transmission coefficients for $(L,S)=(2,3/2)$ in the $n+{}^6Li$ channel. (From Ref. 8).

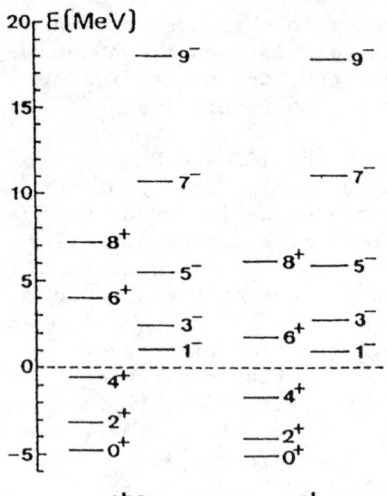

Fig. 3: Comparison of calculated and observed $K^\pi=0^+$ ground-state and 0^- excited rotational bands in ${}^{20}Ne$. (From Ref. 11).

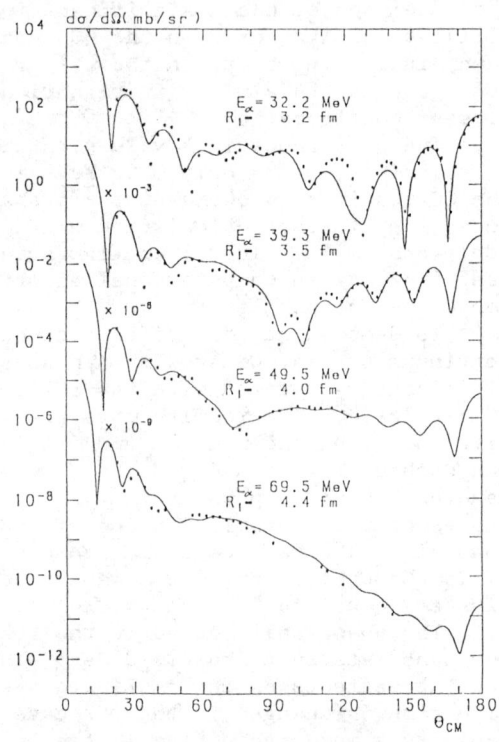

Fig. 4: Comparison of calculated and experimental angular distribution for $\alpha + {}^{16}O$ elastic scattering at four laboratory energies. (Adapted from Ref. 11).

Because of large L and S values in this case, the spin-orbit contribution will be quite appreciable. With this contribution taken into consideration, one expects that the phase shift in the $7/2^+$ state will, in particular, increase rather rapidly in the low-energy region. Thus, it is likely that low-energy n+^{6}Li cross sections may be significantly influenced by the presence of such a state. It would indeed be interesting to carefully analyze the experimental data at excitation energies higher than about 12 MeV to determine whether or not our finding here is borne out.

C. Resonating-group study of the $\alpha+^{16}$O system from bound states to high-energy scattering states.

As an example of utilizing the MCT to study relatively heavy systems, the works of Wada and Horiuchi [11] and Matsuse et al. [12] in investigating the properties of the $\alpha+^{16}$O system will now be described. In these investigations, single-configuration resonating-group calculations are performed, with the internal wave functions of the α and ^{16}O clusters both described by closed-shell configurations in harmonic-oscillator potential wells having a common width parameter. The effective two-nucleon interaction used is the HNY potential [13], but with the triplet-even strength of the medium-range part modified from -546.0 to -524.7 MeV.

Calculated $K^\pi=0^+$ ground-state and 0^- excited rotational bands [12] are compared with experiment in Fig. 3. Quite obviously, the agreement is satisfactory. For the scattering states, the situation is more complicated. Here it is necessary to take reaction channels into account in order to explain reasonably the behavior of the observed scattering cross section. Within the proper framework of the resonating-group method, this would be a difficult problem because, especially at high energies, one will need to include a large number of multi-cluster open channels in the calculation. Thus, the usual, although rather crude, procedure is to merely add into the resonating-group formulation a phenomenological, macroscopic imaginary potential. From the viewpoint of a microscopic investigation, this is clearly undesirable; however, our experience with optical-model studies does indicate that this procedure will, in general, yield rather satisfactory results.

In the calculation of Wada and Horiuchi [11], the operator on the left side of Eq. (6) is modified to include the imaginary-potential term $i\tilde{N}^{1/2}W\tilde{N}^{1/2}$, where $\tilde{N}$ denotes the norm operator whose spatial representation is

$$\tilde{N}(\vec{R}',\vec{R}'') = \delta(\vec{R}'-\vec{R}'') + N_E(\vec{R}',\vec{R}'') \quad . \tag{13}$$

The absorptive potential $W(r)$ is chosen to have the same form as that used by Michel et al. [14] in their optical-model study of the $\alpha+^{16}$O system, i.e.,

$$W(r) = W_0/\{1+\exp[(r-R_I)/2a_I]\}^2 \quad , \tag{14}$$

with $W_0=-25$ MeV and $a_I=0.65$ fm. The radius parameter R_I is an adjustable quantity; its value is adjusted at each energy to obtain a best fit with experiment. The results at various α-particle laboratory energies E_α are compared with measured cross-section data in Fig. 4, where the values of R_I are also indicated.

The agreement between theory and experiment, as shown in Fig. 4, is quite good. Thus, in the $\alpha+^{16}O$ case, one can conclude that the resonating-group method, with a fixed effective nucleon-nucleon potential, can explain quite well the data from bound states to high-energy scattering states. This indicates that the empirically established energy dependence of the real part of the optical potential [14] comes appreciably from the nonlocality of the nucleus-nucleus interaction originating from intercluster antisymmetrization. To gain further confidence in the validity of this statement, one would need to perform additional calculations at higher energies from 20 to 50 MeV/nucleon, when high-quality experimental data, covering a wide angular range, become available.

D. Bound-state, scattering, and reaction calculations in the four-nucleon system.

In this investigation [15], a multi-configuration resonating-group calculation is performed to study in a unified manner the bound-state, scattering, and reaction properties of the four-nucleon system. The cluster configurations considered are the physical configurations p+t, n+h and d+d, as well as the pseudo-inelastic configurations d+d*, d*+d*, d+d**, d*+d**, and d**+d**, with d* and d** being cluster pseudo-states. These latter configurations are included to represent approximately the effects of deuteron breakup processes. Also, for reliability, rather flexible internal wave functions consisting of superpositions of two Gaussian functions for the t and h clusters, and three Gaussian functions for the d cluster and its pseudo-excited states have been adopted in the calculation.

Table III

Calculated energies for the ground and first-excited states of ^{4}He

Model space	Cluster configurations	Ground 0^+ state (MeV)	Excited 0^+ state (MeV)
SC	d+d	-19.21	-3.22
TC1	d+d, d+d*, d*+d*	-21.90	-3.68
Sext.C	d+d, d+d*, d*+d*, d+d**, d*+d**, d**+d**	-23.02	-3.91
DC	p+t, n+h	-25.17	-5.10
TC2	p+t, n+h, d+d	-26.03	-5.98
Quint C.	p+t, n+h, d+d, d+d*, d*+d*	-26.28	-6.10
OC or full	p+t, n+h, d+d, d+d*, d*+d*, d+d**, d*+d**, d**+d**	-26.50	-6.43

With the MN potential and u=1, the results for the energies of the ground and first-excited 0^+ states obtained in a number of model spaces are shown in Table III. Here one notes that, in the OC or

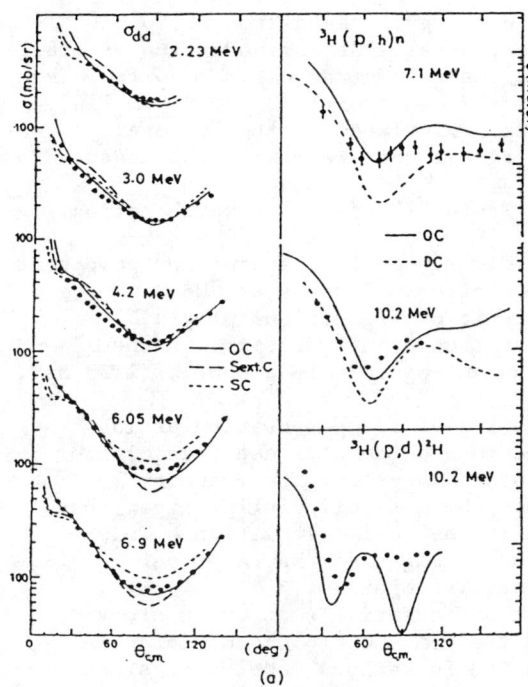

Fig. 5a: Comparison of calculated and experimental cross sections for d+d scattering, and ^{3}H(p,h)n and ^{3}H(p,d)^{2}H reactions. (From Ref. 15).

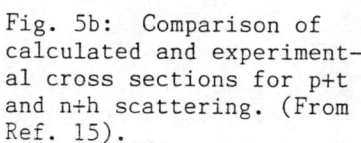

Fig. 5b: Comparison of calculated and experimental cross sections for p+t and n+h scattering. (From Ref. 15).

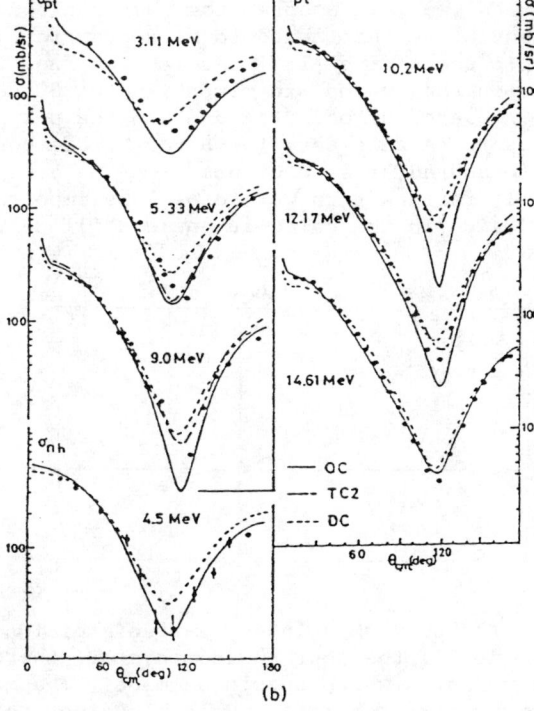

full calculation, the ground-state energy is equal to -26.50 MeV which is rather close to the experimentally determined value of -28.3 MeV. In addition, the calculated charge rms radius is 1.60 fm, also in good agreement with the empirical result of 1.63 ± 0.04 fm [16]. For the first-excited 0^+ state, the calculated excitation energy turns out to be 20.07 MeV, which agrees very well with the measured value of 20.1 MeV [17].

There are other features in Table III which are worth noting. These features are:
(i) The p+t plus n+h cluster configuration is the dominant component in both the ground and the first-excited 0^+ states of ^{4}He.
(ii) The d+d and the pseudo-inelastic configurations play a significant role; the inclusion of these configurations in the calculation lowers the energies by an appreciable amount of 1.33 MeV in both of these states.
(iii) With a trial function consisting of a superposition of 15 spatially symmetric Gaussian functions which does not take cluster correlations into account, one obtains a ground-state energy expectation value of -25.55 MeV which is significantly larger than the value of -26.50 MeV obtained in the full calculation. This indicates that, even for a nucleus as light as ^{4}He, a proper consideration of clustering effects is important.

Calculated d+d, p+t, and n+h differential scattering cross sections, n+h polarizations, and t(p,h)n and t(p,d)d differential reaction cross sections are computed in various model spaces and compared with experiment in Figs. 5 and 6. To obtain the calculated results, we have adopted the same nucleon-nucleon spin-orbit potential as that found to be appropriate in the neighboring system of p+α scattering [18]. As is seen from these figures, the agreement between theory and experiment in the OC case is quite satisfactory. The presence of both the d+d and the p+t plus n+h configurations seems to be necessary for a reasonable description of the behavior of the measured cross sections. On the other hand, the pseudo-inelastic configurations turn out to be less important, although their inclusion in the calculation is still desirable in some situations.

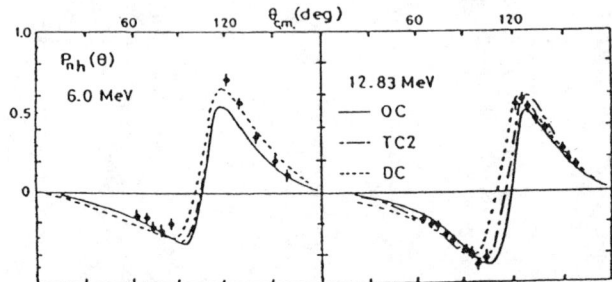

Fig. 6: Comparison of calculated and experimental polarizations for n+h scattering (From Ref. 15).

Although this investigation explains reasonably well the behavior of the four-nucleon system, there are still some improvements which should be made. The most important of these is to adopt a more realistic nucleon-nucleon potential in which the tensor part is included. Another improvement is to better represent the breakup effects by introducing more deuteron pseudo-states, including also those in singlet configurations. Quite clearly, these improvements would require a large increase in computational effort; however, we do believe that the resulting calculation will definitely be feasible at the present stage of development.

IV. EFFECTS OF ANTISYMMETRIZATION IN NUCLEAR SCATTERING

The MCT can be used not only to study the detailed properties in specific cases, but also to learn the general characteristics of nuclear systems. This is carefully discussed in Ref. 1. In this talk, I shall concentrate on only one special topic, namely, the effects of antisymmetrization on the effective internuclear potential.

A. Features of exchange kernel functions.

Consider the general case of A+B scattering, where the nuclei A and B consist of N_A and N_B ($N_A > N_B$) nucleons, respectively, and are described by translationally-invariant shell-model functions in harmonic-oscillator wells having width parameters α_A and α_B. As is seen from Eq. (6), the effective interaction between these two nuclei has a direct part V_D and a nonlocal part represented by the exchange-kernel function $K(\vec{R}',\vec{R}'')$. By studying the structure of $K(\vec{R}',\vec{R}'')$, one finds that it is composed of a sum of nucleon-exchange terms, i.e.,

$$K(\vec{R}',\vec{R}'') = \sum_x \sum_q K_{xq}(\vec{R}',\vec{R}'') , \qquad (15)$$

where

$$K_{xq}(\vec{R}',\vec{R}'') = P_{xq}(\vec{R}',\vec{R}'')\exp(-A_{xq}\vec{R}'^2 - C_{xq}\vec{R}'\cdot\vec{R}'' - B_{xq}\vec{R}''^2) + \text{h.c.}, \qquad (16)$$

with P_{xq} being a polynomial in $\vec{R}'^2$, $\vec{R}'\cdot\vec{R}''$, and $\vec{R}''^2$. In Eq. (15), the index x denotes the number of nucleons interchanged between A and B ($1 \leq x \leq N_B$) and the index q denotes the interaction type which can be conveniently classified by the diagrammatic representation shown in Fig. 7 (see Ref. 19 for a detailed explanation). For each value of x from 1 to (N_B-1), there are five interaction types which are denoted by the index q = a,b,c,d, and e, while for x = N_B (core exchange) there are only three interaction types, namely, q = a,c, and d.

From Eq. (16), it is seen that each nucleon-exchange term, represented by the function K_{xq}, consists of an exponential factor multiplied by a polynomial factor. The exponential factors, depending on x and the nucleon numbers of the nuclei involved, collectively determine the general features of antisymmetrization, while the polynomial factors, which depend additionally on the dynamical structures of the interacting clusters, contain information concerning more specific features, such as blocking and clustering effects. At present, we have a clear understanding about the roles played by the exponential factors, but only some limited knowledge about the structures of the polynomials [20,21]. However, numerical

resonating-group investigations [20,22] in specific nuclear systems
did show that, by simply examining the exponential factors, one can
already learn many essential characteristics of the effects of
antisymmetrization.

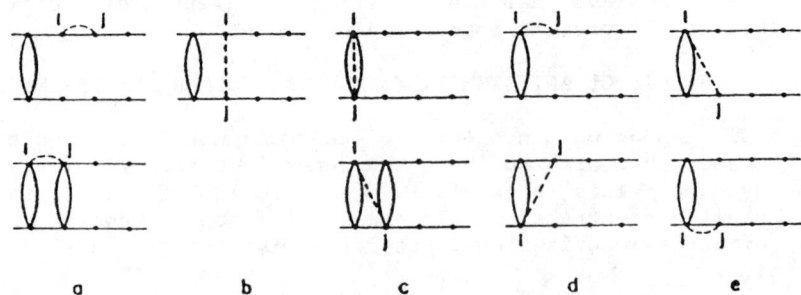

Fig. 7: Diagrammatical representation of nucleon-exchange terms.
(From Ref. 19).

The general expressions [19] of A_{xq}, B_{xq}, and C_{xq}, for any A+B
system, can be derived by utilzing the complex generator-coordinate
technique. As is expected, the quantities A_{xq} and B_{xq} are always
positive, but the quantity C_{xq} can acquire either a positive or a
negative value. This latter fact can be advantageously employed to
further classify the nucleon-exchange terms into two classes:
(i) Class-A terms with $C_{xq}<0$. For these terms, the Born scattering
amplitudes are forward-peaked and can be exactly reproduced by
equivalent local, energy-dependent potentials having a Wigner
character.
(ii) Class-B terms with $C_{xq}>0$. For these terms, the Born scattering
amplitudes are backward peaked and can be exactly reproduced by
equivalent local, energy-dependent potentials having a Majorana
character.

The classification into class-A and class-B exchange terms is
particularly useful at energies higher than about 25 MeV/nucleon
where rather sharp resonance levels no longer exist. At these
energies, the cross-section angular-distribution curve normally
exhibits a distinct V-shape with its tip occurring at an angle θ_m
which is a measure of the relative importance of class-B exchange
terms. When these terms make important contributions, θ_m will be
relatively small (i.e., close to about 90°) and vice versa. For $\theta < \theta_m$, the cross-section curve has a decreasing trend and the
contribution comes essentially from the direct and class-A terms,
while for $\theta > \theta_m$, it has an increasing trend and the contribution
comes essentially from the class-B terms. Therefore, by just
visually inspecting the experimental angular-distribution curves at
such energies, one can already learn whether there are significant
class-B contributions or not, although such an inspection will not
yield much information concerning the importance of class-A terms in
comparison with the direct term.

As an illustration, the V-shaped curve in ^{3}He + α scattering at
an energy of 35.1 MeV/nucleon is shown in Fig. 8, where the solid

circles represent experimental data taken from Ref. 23 and the solid curve represents the result of a resonating-group calculation [18] containing a phenomenological imaginary potential. From this figure, one notes that θ_m is equal to about 90°, indicating that class-B exchange terms are very important. The situation is similar in the case of $\alpha + {}^6\text{Li}$ scattering [24] at 41.5 MeV/nucleon, although the experimentally determined value of θ_m has a somewhat larger value of about 105°. On the other hand, one finds that the behavior of the experimental angular distribution for $p + {}^{16}\text{O}$ scattering [25] at 40 MeV/nucleon is rather different. Here the differential cross section continues its decreasing trend to about 130° and then increases only weakly in the larger-angle region. This suggests strongly that the class-B terms are fairly unimportant and, unlike the ${}^3\text{He} + \alpha$ and $\alpha + {}^6\text{Li}$ cases, exchange effects are not a dominant factor in determining the cross-section behavior at large angles beyond 130°.

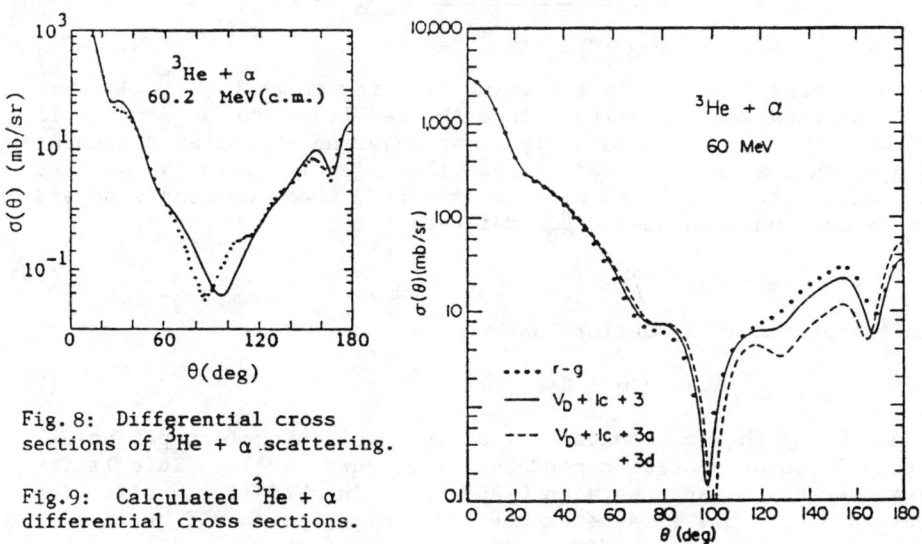

Fig. 8: Differential cross sections of ${}^3\text{He} + \alpha$ scattering.

Fig. 9: Calculated ${}^3\text{He} + \alpha$ differential cross sections.

B. Characteristic ranges and characteristic wave numbers of equivalent local potentials.

The simplest procedure to determine the relative importance of various nucleon-exchange terms is to analyze the kernel functions K_{xq} in the Born approximation. First, one computes the Born scattering amplitude corresponding to each exchange term, and then construct an equivalent local, energy-dependent "exchange" potential $\tilde{V}_{xq}$ which yields exactly the same Born amplitude as the exchange term under consideration. This equivalent potential, which can have either a Wigner or a Majorana (parity-dependent) character depending upon whether the corresponding exchange term belongs to class A or class B, has the form

$$\tilde{V}_{xq} = \tilde{P}_{xq} \exp[-(k/k_{xq})^2] \exp[-(R/R_{xq})^2](1 \text{ or } P^R) \quad , \tag{17}$$

where k denotes the asymptotic wave number, $\tilde{P}_{xq}$ is a polynomial in k^2 and R^2, and P^R is a Majorana space-exchange operator interchanging the position coordinates of the clusters which are now treated as structureless point particles. it is characterized by two quantities, a characteristic range R_{xq} and a characteristic wave number k_{xq}. The information which one seeks about antisymmetrization effects can then be extracted by carefully studying the properties of these characteristic quantities.

Using the formulas given in Ref. 26, one easily obtains general expressions for R_{xq} and k_{xq}. These expressions are

$$R_{xq} = \left[\frac{|x-x_q|}{J_q - (x-x_q)^2} \frac{4}{\alpha} \right]^{1/2}, \tag{18}$$

$$k_{xq} = \left\{ \frac{J_q - (x-x_q)^2}{[\mu_0 - |x-x_q|]^2 + K_q} \mu_0 \alpha \right\}^{1/2}. \tag{19}$$

In these equations, x_q is the value of x for which C_{xq} equals zero. It is an important quantity, since the resultant exchange potential $\tilde{V}_{xq}$ will turn out to have a Wigner or Majorana character depending upon whether x is smaller or larger than x_q. In Table IV, we list the constants x_q, J_q, and K_q. As is noted, these constants depend on the reduced nucleon number μ_0, defined as

$$\mu_0 = N_A N_B / (N_A + N_B), \tag{20}$$

and the parameter λ, defined as

$$\lambda = \kappa/(\alpha + 2\kappa), \tag{21}$$

with κ being the range parameter of the nucleon-nucleon potential with a Gaussian spatial dependence (i.e., $\exp(-\kappa r^2)$). This latter parameter has a range between 0 and 1/2. In a realistic situation where $\kappa \simeq \alpha$, λ has a value around 1/3. Also, it should be mentioned that, for simplicity in discussion, the constants x_q, J_q, and K_q are computed under the assumption of equal cluster width parameters, i.e., $\alpha_A = \alpha_B = \alpha$ (the general features of antisymmetrization effects are similar in the unequal-width-parameter case; see Ref. 27). In this special case, the exponential factors in type-d and type-e terms are then the same, and these two types can be grouped together in our consideration.

Table IV

Expressions for x_q, J_q, and K_q

Exchange type q	x_q	J_q	K_q
a	μ_0	μ_0^2	0
b	$\mu_0 - \lambda$	$(\mu_0 + \lambda)^2$	$-\lambda^2$
c	$\mu_0 + \lambda$	$(\mu_0 + \lambda)^2$	$-\lambda^2$
d	μ_0	$\mu_0(\mu_0 + \lambda)$	0

By examining the expressions for R_{xq} and k_{xq}, one finds that (i) for $x < x_q$, R_{xq} and k_{xq} decrease monotonically with increasing x and have largest values when $x = 1$, and (ii) for $x > x_q$, R_{xq} and k_{xq} increase monotonically with increasing x and have largest values when $x = N_B$ (q = a,c,d). Since it has been shown [28] that, because of the existence of Pauli forbidden or almost-forbidden states, the effective internuclear potential is necessarily deep in the interior region, the local wave number has large values for small values of R even at low energies; therefore, it is expected that equivalent potentials with large characteristic ranges and characteristic wave numbers will generally make dominant contributions. In other words, this study indicates that, among all exchange terms, the one-exchange terms ($x = 1$) have the largest influence for $x < x_q$ and the core-exchange terms ($x=N_B$) have the largest influence for $x > x_q$. Especially at relatively high energies and in situations where absorptions are strong and, hence, grazing collisions are dominant, one anticipates that this conclusion will be valid.

The above discussion suggests that it may be useful to define two alternative characteristic quantities, i.e., the characteristic energy E_{xq} defined as

$$E_{xq} = \frac{\hbar^2}{2M\mu_0} k_{xq}^2 , \qquad (22)$$

with M being the nucleon mass, and the characteristic weight ζ_{xq} defined as

$$\zeta_{xq} = (k_{xq} R_{xq})^2 . \qquad (23)$$

This latter quantity (i.e., ζ_{xq}) may be considered as providing a qualitative measure of the relative importance of the corresponding nucleon-exchange term.

The usefulness in studying the features of the characteristic quantities is illustrated in the case of $^3_2$He + α scattering. In this case, α is chosen to be equal to 0.44 fm^{-2}, and the nucleon-nucleon potential of Eq. (8) in Ref.29 with $\kappa = 0.46$ fm^{-2} and a Serber exchange mixture is adopted. Also, for clarity in discussion, we omit all Coulomb effects by letting the charge of the proton be infinitesimally small.

Various values of the characteristic quantities are given in Table V. Here one notes that the type 1c or knock-on exchange term among class-A terms and the 3-exchange terms among the class-B terms have especially large values for the characteristic energy and the characteristic weight. This may be taken as an indication that these particular terms must be included in the calculation in order to yield good results.

That this is indeed so is demonstrated in Fig. 9, where we show a cross-section comparison at 60 MeV, an energy which is considerably higher than the characeristic energies of many of the exchange terms. Here it is seen that the differential cross sections obtained with the resonating-group calculation which includes all exchange terms (solid circles) and with a calculation including only the direct potential, the type-1c term and the 3-exchange terms (solid curve) are quite similar at all angles. In fact, even if we omit further the type-3c term which has a large characteristic energy but a comparatively smaller characteristic weight, the cross-section behavior (dashed curve) is still reasonably satisfactory and becomes worse only in the backward angular region.

Table V

Characteristic quantities in the ^{3}He + α case

x	Exchange type q	R_{xq} (fm)	k_{xq} (fm^{-1})	E_{xq} (MeV)	ζ_{xq}	Class
1	a	1.64	1.35	22	4.89	A
	b	0.92	1.35	22	1.54	A
	c	1.76	2.69	88	22.30	A
	d	1.47	1.51	27	4.89	A
2	a	0.95	1.03	13	0.96	B
	b	1.22	1.64	33	3.98	B
	c	0.34	1.10	14	0.14	A
	d	0.87	1.13	15	0.96	B
3	a	3.02	2.30	64	48.00	B
	c	1.61	2.30	64	13.72	B
	d	2.50	2.77	93	48.00	B

C. One-exchange and core-exchange potentials.

In the one-exchange case, the type-c term has the largest characteristic weight and, hence, is the dominant term. The characteristic range and characteristic wave number of the corresponding exchange potential $\tilde{V}_{1c}$ are

$$R_{1c} = \left[\frac{4(\mu_0 + \lambda - 1)}{2\mu_0 + 2\lambda - 1} \frac{1}{\alpha} \right]^{1/2}, \tag{24}$$

$$k_{1c} = \left[\frac{2\mu_0 + 2\lambda - 1}{1 - 2\lambda} \mu_0 \alpha \right]^{1/2}. \tag{25}$$

Based on these expressions, one finds that $\tilde{V}_{1c}$ has the following features:
(i) R_{1c} has a value similar to, but smaller than, the characteristic range [27] of the direct potential V_D. Using additionally the information that the highest powers of R^2 in the polynomial factors of $\tilde{V}_{1c}$ and V_D are the same, one can conclude that $\tilde{V}_{1c}$ is only somewhat shorter-ranged than the direct potential.
(ii) The characteristic energy per nucleon E_{1c}/μ_0 depends very weakly on the nucleon numbers N_A and N_B, and has a large value around 50 MeV/nucleon in all scattering systems.

The above findings indicate that the one-exchange type-c (knock-on) term has generally an important influence in all scattering systems and over a wide energy range. Therefore, this particular term must always be taken into account in a practical calculation. Additionally, we should point out that the characteristic energy per nucleon of the one-exchange type-a,b, or d equivalent potential is around 15 MeV/nucleon for any system, which is considerably smaller than the value of about 50 MeV/nucleon for E_{1c}/μ_0. This implies that, at energies higher than about 30 MeV/nucleon, the type-1c term may be the only class-A term which needs to be included. On the other hand, it should be emphasized that, when the energy is low (see Table V), other one-exchange terms, especially the type-1a and type-1d terms, must also be considered. This is illustrated in Fig. 10, where we show the equivalent local potentials in the $\alpha + {}^{40}Ca$ case at a rather low energy of 10 MeV/nucleon obtained by Horiuchi [28] with the WKB approximation. Here it is seen that the one-exchange and full-exchange potentials are indeed rather similar, but the knock-on exchange potential alone is not a sufficiently good representation of the one-exchange contributions.

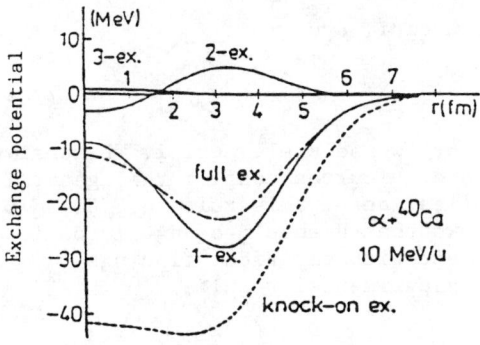

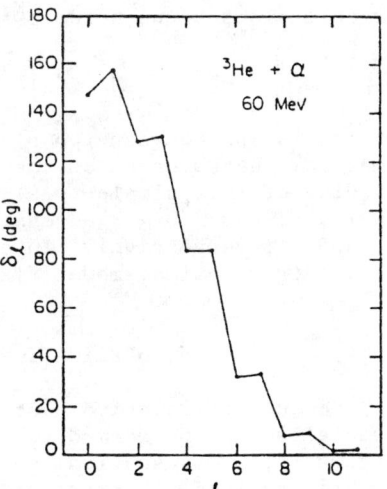

Fig. 10: Exchange potentials in the $\alpha + {}^{40}Ca$ case. (From ref. 28)

Fig. 11: Calculated ${}^3He + \alpha$ phase shifts at 60 MeV.

In the core-exchange case, type-a and type-d terms make the largest contribution. For these terms, the corresponding exchange potentials $\tilde{V}_{ca}$ and $\tilde{V}_{cd}$ have the following characteristic ranges:

$$R_{ca} = \left(\frac{4}{N_A - N_B} \frac{1}{\alpha} \right)^{1/2}, \qquad (26)$$

$$R_{cd} = \left[\frac{4}{(N_A - N_B) + (N_A/N_B)\lambda} \frac{1}{\alpha} \right]^{1/2}. \qquad (27)$$

The characteristic weights for these two potentials turn out to be the same and are given by

$$\zeta_{ca} = \zeta_{cd} = \frac{4N_A N_B}{(N_A - N_B)^2} \ . \tag{28}$$

From the above equation, one finds clearly that core-exchange terms are generally important only when the nucleon-number difference $(N_A - N_B)$ of the interacting nuclei is rather small. This is illustrated in Fig. 11, where we show that, for ^{3}He + α scattering at 60 MeV, the phase shift exhibits a distinct zigzag behavior, indicating that there exists an appreciable parity-dependent component which favors the odd-state internuclear interaction in this particular case.

By studying the diagonal elements of the parity-projected GCM norm kernel, Baye [30] has shown that the parity Π of the Pauli-favored states (i.e., states with the stronger interaction) is given by

$$\Pi = (-1)^{N_A} \tag{29}$$

for collisions between two p-shell nuclei, and

$$\Pi = (-1)^{N_B} \tag{30}$$

for collisions between two s-shell or two sd-shell nuclei. Extensive numerical checks based on the GCM energy curves confirm the general validity of this simple rule. Application of this rule to ^{12}C + ^{13}C, ^{18}O + ^{19}F, and ^{16}O + ^{20}Ne cases shows that Π should be equal to -1, +1, and +1, respectively, in agreement with empirical findings [31,32] from optical-model fits of experimental results.

V. APPLICATIONS OF MCT WAVE FUNCTIONS

Because of relative ease in computation, MCT calculations have especially been performed to study the properties of very light systems. In this section, we shall describe several investigations in which various electromagnetic quantities are computed in these systems for comparison with experiment.

A. Electric quadrupole moment of ^{6}Li.

As an example of applying the MCT wave function to study the electromagnetic property of a nucleus, we discuss the work of Mertelmeier and Hofmann [33] in evaluating the electric quadrupole moment Q of ^{6}Li. In this work, a relatively simple resonating-group calculation, consisting of a single d+α cluster configuration, is performed. The effective nucleon-nucleon potential used is derived from a modified version of the rather realistic two-nucleon interaction obtained previously by Eikemeier and Hackenbroich [34]; it contains not only central, spin-orbit and Coulomb parts, but also a tensor component.

For an explanation of the experimental value [35] of -0.0644 fm^2 for the ground-state quadrupole moment, it was found that a tensor force must not only be included but also be treated in a consistent manner. If the tensor force is omitted from the calculation, the quadrupole moment will necessarily be equal to zero. With this force included, Q is equal to -1 fm^2 when a d-state component is added only into the relative-motion wave function of the d and α clusters. This value is much larger in magnitude than the experimental result. On the other hand, Q becomes positive when a d-state component in the internal function of the d cluster alone is incorporated in the formulation. In fact, Mertelmeier and Hofmann found that a satisfactory result can only be obtained when d-state components are present in both the d+α relative-motion function and the d-cluster internal function. The resultant value of Q turns out to be equal to -0.076 fm^2, which agrees very well with experiment.

The resonating-group wave function used in this calculation has rather limited flexibility; the spatial structure of the d cluster is described by a single Gaussian function and the d+α relative-motion function consists of a superposition of only two Gaussian functions. Thus, the excellent agreement between theory and experiment may be somewhat fortuitous and a more refined calculation should be performed. However, this investigation does call to attention the necessity of treating consistently the tensor interaction. Many existing resonating-group calculations (see, e.g., Ref. 36), which include such an interaction, do not seem to pay proper consideration to this important criterion and should perhaps be further examined to ascertain the reliability of the obtained results.

B. ^{3}He(α,γ)^{7}Be and ^{3}H(α,γ)^{7}Li reactions at astrophysical energies.

Because of its important role in stellar evolution and, in particular, the solar-neutrino problem, the radiative-capture reaction ^{3}He(α,γ)^{7}Be has received extensive attention, both theoretically and experimentally (see Ref. 37 and references contained therein). In this subsection, we describe the calculation of the capture cross sections by Kajino and Arima [38], using wave functions obtained from a MCT investigation.

In this calculation, the internal wave functions of the clusters are assumed to be harmonic-oscillator shell-model functions with the highest spatial symmetry. Different width parameters are used for the various clusters, in order to ensure that the variational stability conditions are satisfied. Results obtained for the ^{3}He(α,γ)^{7}Be and ^{3}H(α,γ)^{7}Li reactions, with a modified Hasegawa-Nagata nucleon-nucleon potential [39], are shown in Fig. 12. Here one notes that the agreement between calculation and experiment is quite satisfactory. The cross-section factors at zero energy S(0), important for astrophysical considerations, are determined to be 0.51 and 0.10 keV-b for these two reactions, respectively.

If the solar-neutrino problem were to be resolved by the ^{3}He(α,γ)^{7}Be reaction alone, the value of S(0) would have to be as small as about 0.12 keV-b. This required value is much smaller than the presently recognized value, from both theoretical studies and experimental measurements, of around (or somewhat larger than) 0.50 keV-b. Thus, one may confidently state that the solar-neutrino problem does not have its origin in this capture reaction and its resolution will need to be sought elsewhere.

The S(0) value of 0.10 keV-b for the ^{3}H(α,γ)^{7}Li reaction is larger than that previously estimated [40]. This is useful information to have, since this capture reaction plays an essential role in determining the primordial abundance of the element ^{7}Li in big-bang nucleosynthesis.

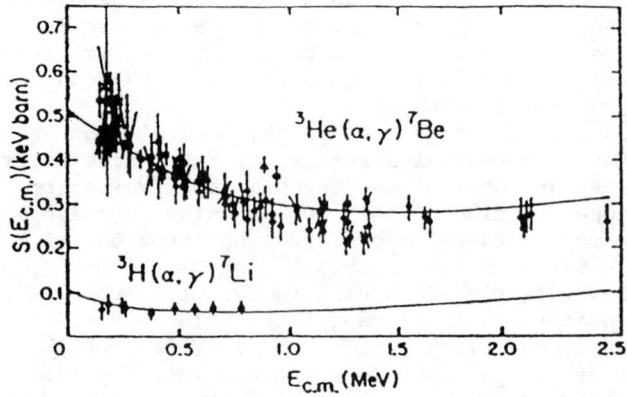

Fig. 12: Values of the cross-section factor S(E). (Adapted from Ref. 38).

C. Microscopic description of $\alpha+\alpha$ bremsstrahlung.

Wave functions obtained with the MCT can also be used to study transitions between continuum states. Here we illustrate this by describing a calculation of Baye and Descouvemont [41] on $\alpha+\alpha$ bremsstrahlung.

The MCT wave function used has a rather simple structure; it consists of a single $\alpha+\alpha$ cluster configuration, with the spatial part of each cluster described by a space-symmetric Gaussian function. The nucleon-nucleon interaction employed is the V1 potential of Volkov [42]. Results for the laboratory cross section obtained in the so-called Harvard geometry (i.e., coplanar geometry with $\theta_1=\theta_2$, where θ_1 and θ_2 define the final directions of the two α particles in the laboratory system) are plotted as a function of E_i, the initial relative energy of the α particles in the c.m. system, and compared with experimental data [43] in Fig. 13. From this figure, one finds that there is a fair agreement between calculation and experiment. The major discrepancy occurs in the E_i region below about 6.5 MeV (the apparent structure in the 37° data above 6.5 MeV may have no physical reality and should be further checked experimentally). Baye and Descouvemont attribute this to the fact that their scattering calculation does not reproduce too well the $\ell=2$ phase shifts near the resonance at about 3-MeV excitation. A later potential-model study of this problem by Langanke [44] does seem to support this suggestion. In this respect, we should mention that it is rather simple to obtain a much more satisfactory description of the $\alpha+\alpha$ phase-shift behavior in the $\ell=2$ resonance region. All what one needs to do is to introduce specific distortion effects of the α clusters into the MCT formulation. In fact, such an improved calculation [45] exists at the present moment, and it would indeed be interesting to compute again the bremsstrahlung cross sections with the MCT wave functions from this calculation.

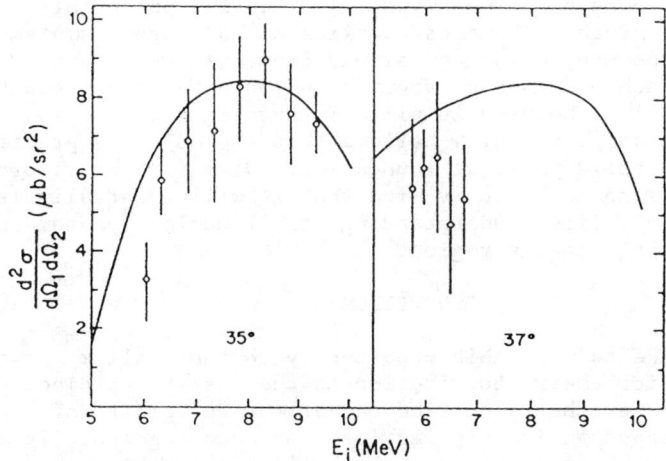

Fig. 13: Comparison of calculated and experimental laboratory cross sections in the Harvard geometry with $\theta_1 = \theta_2 = 35°$ and $37°$. (Adapted from Ref. 41).

One interesting finding from the work of Baye and Descouvemont concerns the accuracy of the so-called external bremsstrahlung approximation [41], based on the first term (i.e., the E_γ^{-1} term, with E_γ being the photon energy) of the Feshbach-Yennie expansion [46]. The results of this investigation show that this approximation yields reasonable cross-section values for α+α bremsstrahlung only when the photon energy is smaller than about 3 MeV and when the α+α scattering angle in the c.m. system is close to $90°$.

VI. CONCLUSION

In this talk, I have briefly discussed the formulation of the microscopic cluster theory (MCT), and illustrate its utility in studying nuclear bound-state, scattering, and reaction problems, and in explaining the features of various types of electromagnetic processes. Additionally, a general discussion is presented to show the importance of nucleon-exchange effects in determining the characteristics of the effective internuclear potential.
    Results from MCT calculations have been universally encouraging. However, except for very light systems, usually only single-configuration studies with rather simplified cluster internal functions have been carried out. It is clear that, for further advancing the MCT, much more effort has to be devoted to heavier

nuclear systems by performing high-quality multi-configuration investigations utilizing realistic cluster functions.

Because of difficulties in treating the strong repulsive core present in a realistic two-nucleon interaction, MCT calculations have always been performed with effective nucleon-nucleon potentials. In this respect, there exists the problem that different research groups have tended to adopt different types of effective potentials. This is unfortunate, since it becomes complicated to extract systematic features from the many existing calculations. It would indeed be desirable to reach a consensus about a common effective nucleon-nucleon potential to be used in all MCT investigations.

In conclusion, I strongly believe that the MCT is a practical theory built on solid physical foundation. With further vigorous effort, there is no doubt in my mind that it will eventually lead us into achieving a unified understanding of all nuclear phenomena in the nonrelativistic energy region.

ACKNOWLEDGMENT

I would like to take this opportunity to thank all my coworkers and colleagues for their contribution to the research project of microscopic cluster theory carried out at the University of Minnesota. Without their help, advice, and encouragement, it would be doubtful that I can maintain my enthusiasm over this project for the past twenty years.

REFERENCES

1. K. Wildermuth and Y. C. Tang, A Unified Theory of the Nucleus (Vieweg, Braunschweig, Germany, 1977).
2. Y. C. Tang, Microscopic Description of the Nuclear Cluster Theory, in Lecture Notes in Physics, Vol. 145 (Springer-Verlag, Berlin, 1981).
3. H. Horiuchi, Prog. Theor. Phys. (Suppl.) $\underline{62}$, 90 (1977).
4. M. Hanck, Nucl. Phys. $\underline{A439}$, 1 (1985), and references contained therein.
5. W. Bauhoff, H. Schultheis, and R. Schultheis, Phys. Rev. $\underline{C29}$, 1046 (1984).
6. D. M. Brink and E. Boeker, Nucl. Phys. $\underline{A91}$, 1 (1967).
7. F. Ajzenberg-Selove, Nucl. Phys. $\underline{A460}$, 1 (1986).
8. Y. Fujiwara and Y. C. Tang, Phys. Rev. $\underline{C31}$, 342 (1985) and University of Minnesota Report UM-RGM2 (1984).
9. Y. Fujiwara and Y. C. Tang, Phys. Rev. $\underline{C32}$, 1428 (1985).
10. D. R. Thompson, M. LeMere, and Y. C. Tang, Nucl. Phys. $\underline{A286}$, 53 (1977).
11. T. Wada and H. Horiuchi, preprint (1987).
12. T. Matsuse, M. Kamimura, and Y. Fukushima, Prog. Theor. Phys. $\underline{53}$, 706 (1975).
13 Y. Yamamoto, Prog. Theor. Phys. $\underline{52}$, 471 (1974); A. Hasegawa and S. Nagata, Prog. Theor. Phys. $\underline{45}$, 1786 (1971).
14. F. Michel, J. Albinski, P. Belery, Th. Delbar, Gh. Grégoire, B. Tasiaux, and G. Reidemeister, Phys. Rev. $\underline{C28}$, 1904 (1983).
15. H. Kanada, T. Kaneko, and Y. C. Tang, Phys. Rev. $\underline{C34}$, 22 (1986).

16. H. Frank, O. Haas, and H. Prange, Phys. Lett. **19**, 391, 719 (1965).
17. S. Fiarman and W. E. Meyerhof, Nucl. Phys. **A206**, 1 (1973).
18. R. D. Furber, R. E. Brown, G. L. Peterson, D. R. Thompson, and Y. C. Tang, Phys. Rev. **C25**, 23 (1982).
19. M. LeMere, D. J. Stubeda, H. Horiuchi, and Y. C. Tang, Nucl. Phys. **A320**, 445 (1979).
20. K. Aoki and H. Horiuchi, Prog. Theor. Phys. **69**, 857 (1983).
21. D. Baye, Nucl. Phys. **A460**, 581 (1986).
22. M. LeMere, Y. Fujiwara, Y. C. Tang, and Q. K. K. Liu, Phys. Rev. **C26**, 1847 (1982).
23. P. G. Roos, A. Nadasen, P. E. Frisbee, N. S. Chant, T. A. Carey, M. T. Collins, and B. T. Leeman, Phys. Rev. **C21**, 799 (1980).
24. D. Bachelier, M. Bernas, J. L. Boyard, H. L. Harney, J. C. Jourdain, P. Radvanyi, M. Roy-Stephan, and R. DeVries, Nucl. Phys. **A195**, 361 (1972).
25. J. M. Cameron, J. R. Richardson, W. T. H. van Oers, and J. W. Verba, Phys. Rev. **167**, 908 (1968).
26. G. W. Greenlees and Y. C. Tang, Phys. Lett. **34B**, 359 (1971).
27. Y. C. Tang, in Proc. of the Intern. Symp. on Clustering Phenomena in Nuclei, Tübingen, Germany, 1981 (Attempto Verlag Tübingen), p. 24.
28. H. Horiuchi, in Proc. of the 4th Intern. Conf. on Clustering Aspects of Nuclear Structure and Nuclear Reactions, Chester, United Kingdom, 1984 (D. Reidel Publishing Co., 1985), p. 35.
29. D. R. Thompson and Y. C. Tang, Phys. Rev. **C4**, 306 (1971).
30. D. Baye, Nucl. Phys. **A460**, 581 (1986).
31. S. Ohkubo, Phys. Rev. **C25**, 2498 (1982).
32. Y. Kondō, B. A. Robson, R. Smith, and H. H. Wolter, Phys. Lett. **162B**, 39 (1985).
33. T. Mertelmeier and H. M. Hofmann, Nucl. Phys. **A459**, 387 (1986).
34. H. Eikemeier and H. H. Hackenbroich, Nucl. Phys. **A169**, 407 (1971).
35. F. Ajzenberg-Selove, Nucl. Phys. **A413**, 1 (1984).
36. M. Hanck, Y. C. Tang and D. Baye, Nucl. Phys. **A419**, 308 (1984).
37. T. Kajino, Nucl. Phys. **A460**, 559 (1986).
38. T. Kajino and A. Arima, Phys. Rev. Lett. **52**, 739 (1984).
39. F. Tanabe, A. Tohsaki, and R. Tamagaki, Prog. Theor. Phys. **53**, 677 (1975).
40. W. A. Fowler, G. R. Caughlan, and B. A. Zimmerman, Ann. Rev. Astron. Astrophys. **13**, 69 (1975).
41. D. Baye and P. Descouvemont, Nucl. Phys. **A443**, 302 (1985).
42. A. B. Volkov, Nucl. Phys. **74**, 33 (1965).
43. U. Peyer, J. Hall, R. Müller, M. Suter, and W. Wölfli, Phys. Lett. **41B**, 151 (1972); B. Frois, J. Birchall, C. R. Lamontagne, U. von Moellendorf, R. Roy, and R. J. Slobodrian, Phys. Rev. **C8**, 2132 (1973).
44. K. Langanke, Phys. Lett. **174B**, 27 (1986).
45. Y. C. Tang, M. LeMere, and D. R. Thompson, Phys. Rep. **47C**, 167 (1978).
46. H. Feshbach and D. R. Yennie, Nucl. Phys. **37**, 150 (1962).

SIMULATIONS OF STRUCTURE, DYNAMICS AND TRANSFORMATIONS IN FINITE AGGREGATES

Uzi Landman, R. N. Barnett, C. L. Cleveland and Jia Luo
School of Physics, Georgia Institute of Technology
Atlanta, Georgia 30332

Dafna Scharf and Joshua Jortner
School of Chemistry, Tel-Aviv University,
69978 Tel-Aviv, Israel

ABSTRACT

Classical and quantum molecular dynamics simulation methods for investigations of the energetics, structure and dynamics of small clusters are opening new avenues for detailed microscopic studies of these systems. We review the simulation techniques and demonstrate their versatility, and richness of information which they afford, via studies of isomerization and melting transformations in small clusters, modes of electron localization in clusters and the dynamics of fragmentation and energy pathways following an electronic excitation of rare-gas clusters.

INTRODUCTION

The nature, properties and behavior of physical systems depend upon the identity of the constituents and the nature of interactions between them and upon the degree of aggregation and the ambient conditions. Investigations of the energy level structure, elementary excitations, morphology (shape or crystallographical structure), phase transformations and dynamics of finite systems, and their dependence upon the degree and form of aggregation, are common endeavors in the physical sciences spanning a wide spectrum of interaction forms and strengths, spatial dimensions, and temporal scales. Moreover, the theoretical tools which are developed for and employed in the exploration of such diverse fields and phenomena as in elementary particles, nuclear physics, condensed matter and in the molecular and atomic sciences share a high degree of common features, as even a brief scan of papers in these fields indicates. Thus, for example, the use of symmetry groups (discrete and continuous), techniques for calculations of energy level spectra (Hartree-Fock based methods, configuration interaction, correlated wave functions etc.), many-body techniques for calculations of dispersion relations and elementary excitations, statistical mechanics formulations and analysis of energy level structure, and of reactive events and fragmentation phenomena, calculations of phase-diagrams and phase-transformations, methods of discretization on lattices, the use of newtonian dynamics, stochastic dynamics, formal scattering methods and time-dependent Hartree-Fock techniques for studies of reactions, and the use of path-integral based techniques for calculations of equilibrium finite temperature properties and time dependent processes are all prevalent, with rather small variations across disciplinary boundaries.

Finite systems, beyond the very small end of the size spectrum present an immense theoretical challenge since the number of particles in these systems renders the use of molecular science techniques rather cumbersome (or impractical) while their finiteness prohibits the employment of condensed matter methodology based on translational symmetry, and complicates, due to size defects, the adaptation of the analytical framework and techniques of statistical mechanics which are formulated on the premise of the thermodynamic limit. Computer simulations, using either Monte Carlo (MC) or Molecular Dynamics (MD) techniques alleviate certain of the major difficulties which hamper other theoretical approaches, thus opening new avenues for investigations of physical systems,[1-3] finite ones in particular.[4,5] These methods allow simulations at finite temperatures of equilibrium as well as non-equilibrium (i.e., time dependent) phenomena. For equilibrium studies both the MC and MD methods sample in an efficient manner, the phase-space of the system. These methods are most useful even for the "modest" objective of the determination of the ground-state configuration of a small cluster at $T = 0$, since with these methods the search is not restricted to only a few points (representing highly symmetric structures) on the potential energy surface of the system as is the case in most other techniques. Since the number of local minima is expected to increase exponentially with N, the use of these simulation method offers a decided advantage.[5] The MD techniques allows in addition studies at finite temperatures of dynamical processes in equilibrium and non-equilibrium situations.

Guided by the above considerations we review in this paper Molecular Dynamics Simulations of classical and quantum phenomena in finite systems, illustrating the methods and the richness of microscopic information revealed via these techniques. In the second section we outline briefly the basic elements of classical and quantum path-integral molecular dynamics simulations. Studies of the equilibrium behavior and phase transformations in small clusters are discussed in the third section. The attachment of an electron to small cluster systems and the new modes of electron localization exhibited by these systems is described in the fourth section. Finally, the dynamics of energy transfer and cluster fragmentation following excitation is discussed in the fifth section.

CLASSICAL AND QUANTUM MD SIMULATIONS

Classical MD[1-3]

The classical Molecular-Dynamics (MD) method consists of a numerical generation of the phase-space trajectories for a system of N particles, interacting via a potential function $V(\vec{r}_1,\ldots,\vec{r}_N)$, where $\vec{r}_i$ is the coordinate of particle i. In case of an extended system periodic boundary conditions (pbc's) are imposed. In addition, formulations have been developed which allow simulations of various equilibrium ensembles, such as constant volume microcanonical (E,Ω,N), constant pressure (HPN), or constant temperature macrocanonical (T,Ω,N) ensembles. Since our interest here is in finite systems, no pbc's are employed.

The starting point of a MD simulation is a well-defined microscopic description of the physical system, in terms of a Hamiltonian or a Lagrangian from which the equations of motions are derived. Thus, given a Hamiltonian

$$\mathcal{H} = \frac{1}{2}\sum_i \vec{P}_i \cdot \vec{P}_i/m_i + V(\vec{r}_1,\ldots,\vec{r}_N) \;, \tag{1}$$

the equations of motion are

$$\frac{d\vec{r}_i}{dt} = \frac{\partial \mathcal{H}}{\partial \vec{P}_i} = \frac{\vec{P}_i}{m} \tag{2a}$$

$$\frac{d\vec{P}_i}{dt} = \frac{\partial \mathcal{H}}{\partial \vec{r}_i} = -\frac{\partial V}{\partial \vec{r}_i}\;, \quad (i = 1,\ldots,N). \tag{2b}$$

numerical integration of the equation of motions allows investigations of the dynamics of equilibrium as well as non-equilibrium properties.

In equilibrium studies the properties of the system under investigation are calculated as averages along the dynamical trajectory of the system, instead of the customarily used ensemble averages. The trajectory average $\langle A \rangle$ of the property $A(\vec{r}^N(t),\vec{P}^N(t))$, where $\vec{r}^N$ and $\vec{P}^N(t)$ stand for the collection of position and momentum vectors of the N particles, is defined as

$$\langle A \rangle = \lim_{t\to\infty} \frac{1}{t}\int_0^t A(\vec{r}^N(t'),\vec{P}^N(t'))dt'. \tag{3}$$

The property A can be an equal-time characteristic of the system, such as the average potential energy $\langle V \rangle$ or the average pair (or higher order) distribution function, or a time-correlation function, such as the velocity-velocity correlation function, $\langle \vec{v}(o)\cdot\vec{v}(\tau)\rangle$. When $A \equiv K = \sum_i \vec{P}_i \cdot \vec{P}_i/2m$ and using the equipartition theorem, the temperature of the microcanonical ensemble is obtained from

$$\langle K \rangle = \frac{3}{2} NkT \tag{4}$$

where k is Boltzmann's constant.

For detailed descriptions of the MD method and of the numerical algorithms used we refer to recent reviews.[1-3]

Quantum MD[7-14]

Recent developments of simulation methods of quantum systems rest upon the path-integral formulation[7] of quantum statistical mechanics. Consider an electron interacting with a set of N particles via a potential V_e. The Hamiltonian for the system can be written as

$$H = K + V + K_e + V_e , \qquad (5)$$

where K is the kinetic energy operator of the N-particle cluster with masses M_I,

$$K = -\sum_{I=1}^{N} \frac{\hbar^2}{2M_I} \nabla_I^2 , \qquad (6)$$

and V is the interionic potential energy operator of the N-particle cluster, which assuming pair-interactions can be written as

$$V = \frac{1}{2} \sum_{I=1}^{N} \sum_{J=1}^{N}{}' \Phi_{IJ}(R_{IJ}) , \qquad (7)$$

K_e is the kinetic energy of the electron (mass m)

$$K_e = -\frac{\hbar^2}{2m} \nabla^2 , \qquad (8)$$

and V_e is the operator for the potential energy of the electron which consists of a sum of the electron-particle interaction potentials

$$V_e = \sum_{I=1}^{N} \Phi_{eI}(\vec{r}-\vec{R}_I) . \qquad (9)$$

The partition function, Z, of the system is given by

$$Z = \text{Tr} \exp(-\beta H) , \qquad (10)$$

where $\beta = 1/kT$ is the inverse temperature. An approximate expression for the partition function can be obtained through the use of Trotter's formula[15] to replace the sum of operators in the exponent by a product of P terms,

$$Z_P = \text{Tr} \left[\exp\left(-\frac{\beta K}{P}\right) \exp\left(-\frac{\beta V}{P}\right) \exp\left(-\frac{\beta K_e}{P}\right) \exp\left(-\frac{\beta V_e}{P}\right) \right]^P, \quad (11)$$

and

$$Z = \lim_{P \to \infty} Z_P.$$

Using the expression for the free particle propagator in the coordinate representation, Z_P can be written as

$$Z_P = \left(\frac{mP}{2\pi\hbar^2\beta}\right)^{3P/2} \prod_I^N \left(\frac{M_I P}{2\pi\hbar^2\beta}\right)^{3P/2} \int \exp[-\beta(V_{eff}^e + V_{eff}^I)] d\tau \quad (12)$$

where the effective potential for the electron is

$$V_{eff}^e = \sum_{i=1}^{P} \left[\frac{Pm}{2\hbar^2\beta^2} (\vec{r}_i - \vec{r}_{i+1})^2 + \frac{V_e(r)}{P} \right], \quad (13)$$

and it is understood here that $\vec{r}_{(P+1)} \equiv \vec{r}_1$. The effective potential for the ions is given by

$$V_{eff}^I = \sum_i^P \left[\sum_I^N \frac{PM_I}{2\hbar^2\beta^2} (\vec{R}_{iI} - \vec{R}_{(i+1)I})^2 + \frac{V}{P} \right], \quad (14)$$

and the integration in Eq. (12) is taken over the volume

$$d\tau = d^3R_1 \ldots d^3R \, d^3r_1 \ldots d^3r_P.$$

When $M_I \gg m$, the thermal wave length of the ions $(\beta\hbar^2/M_I)^{1/2}$ is much smaller than that of the electron and the Gaussian functions corresponding to the heavy atoms reduce to δ-functions. The resulting expression for the partition function is

$$Z_P = \prod_{I=1}^{N} \left[\frac{M_I P}{2\pi\beta\hbar^2}\right]^{3P/2} \left[\frac{mP}{2\pi\beta\hbar^2}\right]^{3P/2} \int \exp\{-\beta V_{eff}\} d\tau , \quad (15)$$

where

$$V_{eff} = V_{eff}^e + V. \quad (16)$$

This result reflects the frequently used approximation that in a system consisting of a light particle and interacting with heavier ones, the heavy particles can be treated using classical mechanics whereas a quantum treatment is required for the light particle.

Equation (13) and (15) establish an approximate isomorphism between the quantum problem and a corresponding classical one.[8] In the isomorphic classical problem, the electron is mapped onto a closed chain (necklace) of P pseudo-classical particles (beads) with nearest-neighbor harmonic interactions whose strength is determined by the mass (m), the inverse temperature (β) and the number of beads (P). The average energy of the system can then be evaluated from the relation

$$\langle E \rangle = -\frac{\partial}{\partial\beta} \ell n Z = \frac{3N}{2\beta} + \langle V \rangle + K_e + \frac{1}{P} \langle \sum_{i=1}^{P} V_e(\vec{r}_i) \rangle \quad (17)$$

where the pointed brackets indicate averages over the Boltzmann distribution as defined in Eq. (12). The electron kinetic energy estimator is

$$K_e = \frac{3P}{2\beta} - \frac{Pm}{2\hbar^2\beta^2} \langle \sum_{i=1}^{P} (\vec{r}_i - \vec{r}_{i+1})^2 \rangle . \quad (18)$$

An alternative expression for K_e can be obtained[13] in the form

$$K_e = \frac{3}{2\beta} + \frac{1}{2P} \sum_{i=1}^{P} \langle \frac{\partial V_e(\vec{r}_i)}{\partial \vec{r}_i} \cdot (\vec{r}_i - \vec{r}_p) \rangle . \quad (19)$$

The first term on the r.h.s. of Eq. (19) is the free particle contribution, K_{free}, and the second term K_{int} is the contribution from the interaction, with the classical particles

In the Quantum Path-Integral Molecular Dynamics (QUPID) method statistical sampling over the Boltzmann distribution is replaced by the equivalent[10] averaging over the phase-space trajectories generated via classical MD by the Hamiltonian

$$H = \sum_{i=1}^{P} \frac{m^* \dot{\vec{r}}_i^2}{2} + \sum_{i=1}^{N} \frac{M_I \dot{\vec{R}}_i^2}{2} + \sum_{i=1}^{P} \left[\frac{Pm}{2\hbar^2 \beta} (\vec{r}_i - \vec{r}_{i+1})^2 + \frac{1}{P} V_e \right] + V , \qquad (20)$$

where the masses m^* and M_I can be chosen to be arbitrary, since in classical systems configurational averages do not depend on the masses appearing in the kinetic energy term. In our calculations we choose $m^* = 1$ a.m.u. and M_I to be the ionic masses. This choice is made such that the internal frequencies of the quantum necklace $\omega \sim [mP/m^* \beta^2 \hbar^2]^{1/2}$ will match the frequencies of the nuclear motion.

As seen from our discussion the above formulation allows simulations of finite-temperature equilibrium properties of quantum systems. Extensions to real time evolution using Monte-Carlo and Molecular Dynamics methods are currently in progress.[16]

ISOMERIZATION AND MELTING OF SMALL CLUSTERS

The nature of phase transformations in finite systems is a subject of considerable conceptual and practical interest.[5,17-24] While the monotonic decrease in the melting temperature with decreasing particle size has been long observed,[21] satisfactory theoretical understanding of the melting transition, in bulk as well as small cluster systems, is lacking. Since thermodynamics is founded on the premise of the thermodynamic limit, the applicability of a thermodynamic description to finite systems is in question. Indeed for small systems the magnitude of fluctuations precludes a sharp definition of a phase transformation.[23,24] In fact, for finite systems which are large enough so that the total entropy change per particle accompanying a first-order transition is of the same order of magnitude as in the infinite system and taking the mean square fluctuation in temperature $\langle(\Delta T)^2\rangle$ as the intrinsic temperature uncertainty in the system, i.e., $\langle(\Delta T)^2\rangle = (\Delta T_{tr})^2$, one obtains[23]

$$\frac{\Delta T_{tr}}{T_{tr}} = \frac{1}{N\sigma} \qquad (21)$$

where ΔT_{tr} is the width of the finite peak in the curve of specific heat vs. temperature around the peak at the "transition temperature", T_{tr}. σ is the latent entropy per particle, measured in units of Boltzmann's constant, i.e., $\sigma = L/T_{tr}$, where L is the latent heat per particle.

In addition, the construction of theoretical models of melting of small clusters is further complicated by the intrinsic inhomogeneity of the system (exterior, surface, versus interion atoms), and by the spatial and system size dependencies of the spectrum of motional degrees of freedom.[22]

In this section we use classical MD simulations to study the energetics, structure and dynamics of alkali-halide clusters of variable sizes [$(NaCl)_n$, for n = 4, 16 and 108] over a wide temperature range[25] and investigate using quantum simulations electron induced isomerization in the $(NaCl)_4$ system.[26] Our simulations show that the very nature of the phase transformation and the underlying physical processes involved, depend on the size of the system. We conclude that at the lower end of the size spectrum, the phase space of the system is characterized by a small number of stable configurations (solid isomers) between which the system transforms in a diffusionless manner, with temperature dependent rates and branching rations. As the system size is increased the number of accessible conformers increases leading to a hierarchial kinetics of isomerization events which exhibits itself as a broadening of the transition region. For these small clusters coexistence is between solid isomers rather than inter-phase (solid-liquid) coexistence. The latter develops for clusters of sufficient size, characterized by a dense spectrum of accessible states, separated by thermally surmountable barriers.[27] Under these circumstances conventional melting is observed as a sharp transition, the separation of time-scales for inter-well and intra-well dynamics ceases,[18,19] and true solid-liquid coexistence is found. The attachment of an electron to the small $(NaCl)_4$ cluster is found to cause structural modifications, the appearance of structural isomers, which do not have a parentage in the neutral cluster. In addition we find a lowering of the transition temperatures between the various structural forms as compared to the corresponding neutral cluster.

Classical

We have performed constant energy MD simulations for $(NaCl)_n$ clusters, with interionic potentials consisting of a point Coulomb term and a Born-Mayer repulsion.[28,29] At each temperature the system was equilibrated for a prolonged period (typically 10^{-10} sec) after which data was collected (typically for $\sim 10^{-9}$ sec). The integration time step was chosen as 3×10^{-15} sec to assure energy conservation.

In Fig. 1 the caloric curves (total energy, U, versus kinetic energy, KE) for $NaCl)_n$ with n = 4, 16 and 108 (a, b and c, respectively) are shown. (The kinetic temperature is defined as kT = 2U/3(N-2), since the clusters are setup initially with zero total

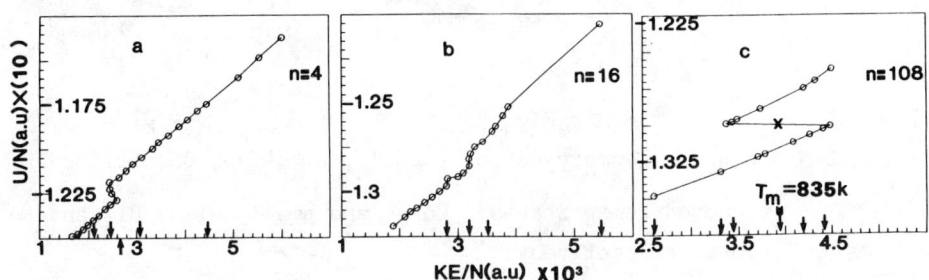

Fig. 1. Caloric curves for n = 4,16,108. Arrows denote temperatures in increasing order: (a) 589, 698, 740, 874, 1249K; (b) 636, 726, 806K, 1684K; (c) 555, 705, 730, 890, 938. Note change in character of transition upon increase in size.

210

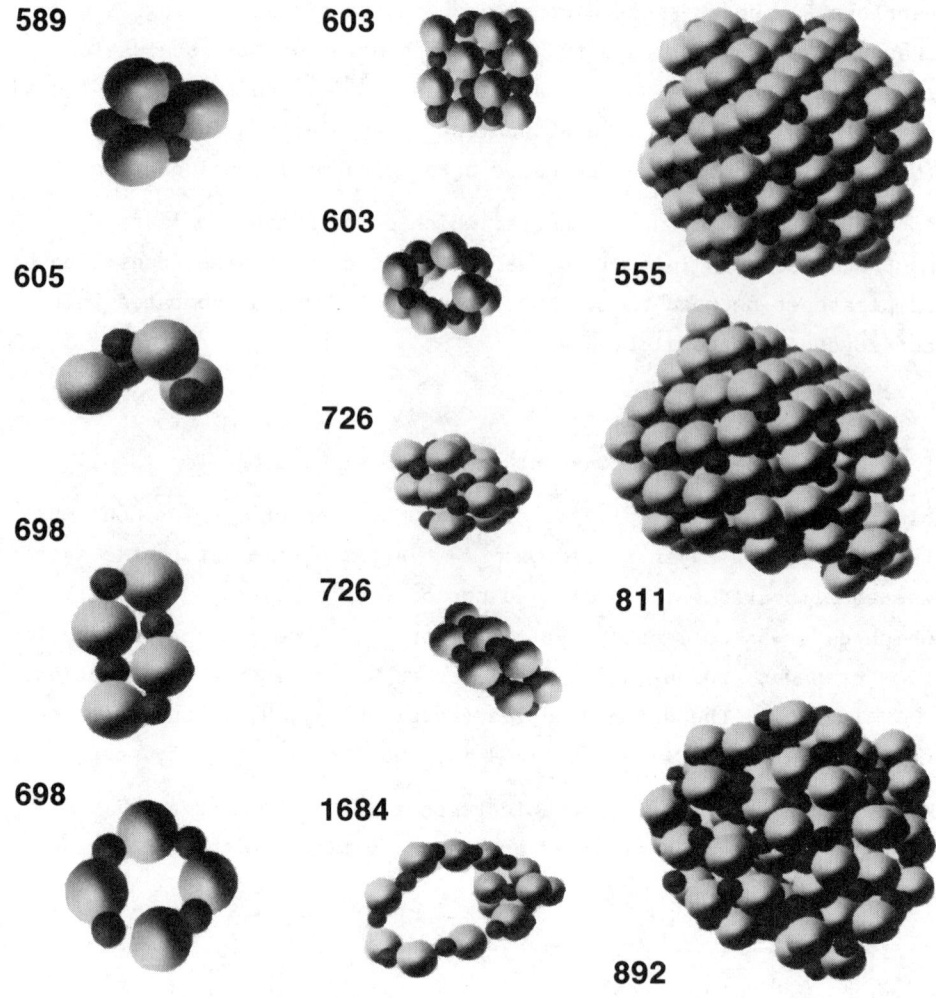

Fig. 2 Fig. 3 Fig. 4

Figs. 2-4. Cluster geometries for $(NaCl)_n$, n = 4,16,108, at the denoted temperatures. Large and small balls, Cl^- and Na^+ ions, respectively.

angular and linear momenta. Throughout, energy is expressed in atomic units 1a.u. = 2 Rydbergs). For the n = 4 cluster the break in the curve corresponds to isomerization between a cube and ring conformations (see Fig. 2). The development of a hierarchy of structural transformations is demonstrated for n = 16, in Fig. 1b. The lower temperature break corresponds to diffusionless solid to solid isomerization while the higher temperaure break is accompanied by the development of disorder and diffusion characteristics (see Figs. 3,6). The caloric curve shown in Fig. 1c for n = 108 cluster is markedly different from those of the smaller systems. We observe in it a "Van der Waals loop phenomenon" with the true melting point (solid-liquid coexistence, see Fig. 4) at 835K. While gradual increase in the total energy leads to the lower (solid) branch of the caloric curve, terminating ultimately and resulting in the liquid branch, the coexistence point (marked by x) is achieved by providing the system via scaling of the momenta, when on the solid branch at 835K, with sufficient energy to reach the horizontal line (melting of part of the system was observed to be initiated at the surface, progressing inwards). The structural "phase diagram" of the smaller clusters (n = 4 and 16) is seen to be characterized by branches corresponding to stable isomers, separated by regions of isomer-coexistence. Associated with the latter is a time scale dictated by the residence times at the different isomers. A sample of the time evolution of the kinetic energy of the n = 4 cluster, at T = 698 (coexistence region in Fig. 1a.), exhibiting a transformation between the ring and cube isomers (see Fig. 2) is shown in Fig. 5a. This record was obtained by short time averaging of KE over sets of 100 time-steps. The distribution of short-time averages of the kinetic energy (for that point on the caloric curve), shown in Fig. 5b, is trimodal, with each of the peaks corresponding to a distinct structure as noted (transition state (L), cube and ring). Note that the choice of the averaging time span is of importance in obtaining the resolution which affords such structural identification. Away from the coexistence region the distribution becomes, monotonously, unimodal. Further structural characterization of the systems is provided by the function $n(r)$, related to the pair correlation function, and defined as the average number of atoms whose distance from an atom chosen as origin is between r and $r + \Delta r$. Additional information is provided by the coordination number function $C(R) = \int_0^R n(r')dr'$. These are shown

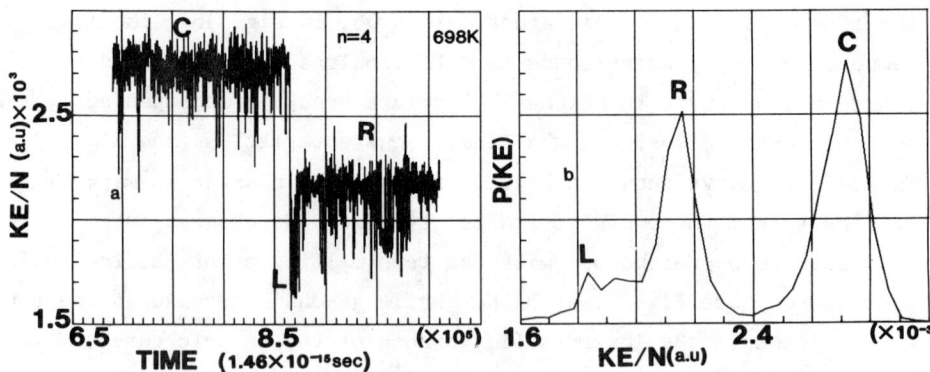

Fig. 5(a). Sample of KE vs. time for (NaCl) showing isomerization from cube (C) to ring (R), through a ladder (L) transition state, conformations. (b) Short-time average, KE distribution, P(KE).

in Fig. 6 for the n = 4, 16 and 108 clusters, respectively. It is evident that the sharp isomerization for the smallest cluster is accompanied by a sudden decrease from 3 to 2 in the first neighbor coordination while for the intermediate size cluster the decrease in coordination (from ~ 4 at the low temperature) across the transition region is gradual. Finally for the large (n = 108) cluster the melting is characterized by a sharp drop in the first neighbor coordination (from ~ 4.5 to ~ 3). The structural transformations are also clearly seen by following the evolution of features in n(r) as a function of temperature. In particular, except for the n = 4 cluster, upon increase in EK the first peak grows broader and decreases in magnitude and further coordination shells lose their structure (more pronounced for the n = 108 cluster). Note also that this characteristic trend, which corresponds to liquification, does not occur for the intermediate cluster (n = 16) till past the first, isomerization, break in the caloric curve.

A probe of the dynamical nature of the transition is provided by the density of states (power spectrum) calculated as the fourier transform of the velocity autocorrelation function.[30] These are given in Fig. 7 for n = 4, 16 and 108, respectively. First, we notice the

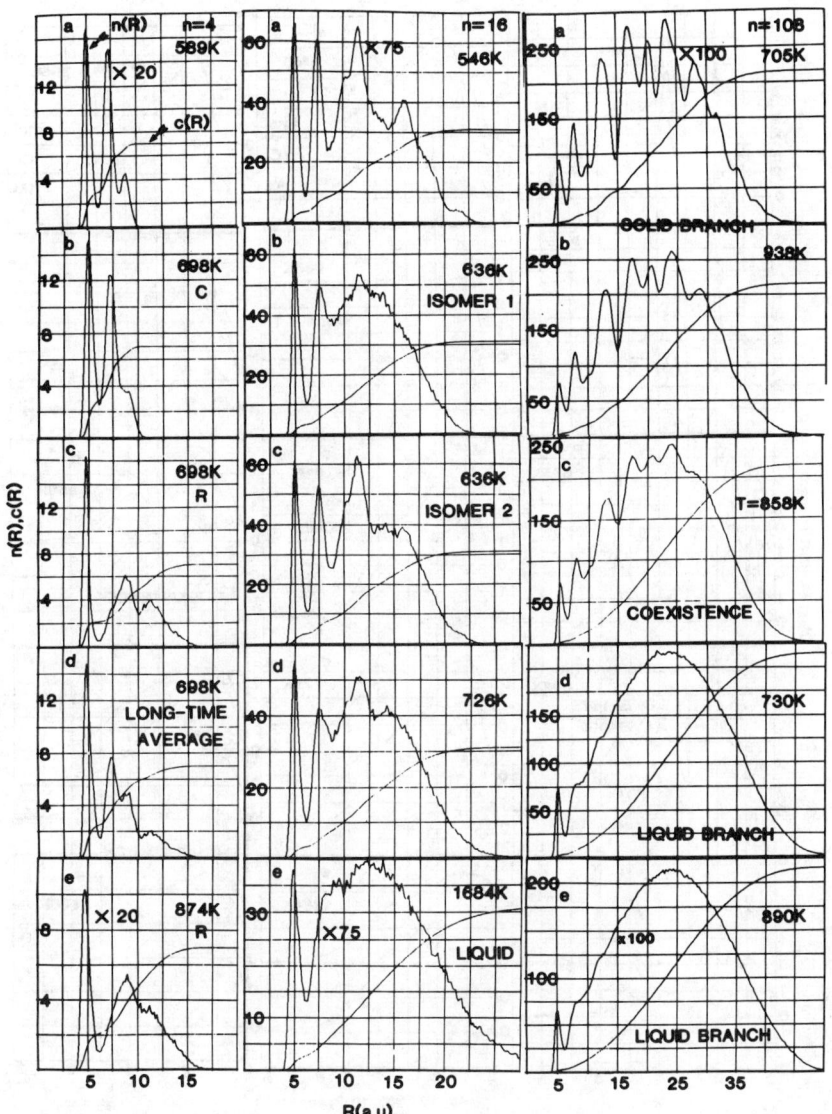

Fig. 6. Particle pair distribution, n(R), and coordination number C(R), functions for n = 4, 16, and 108 at the indicated temperatures. Note dependencies on size and on temperature.

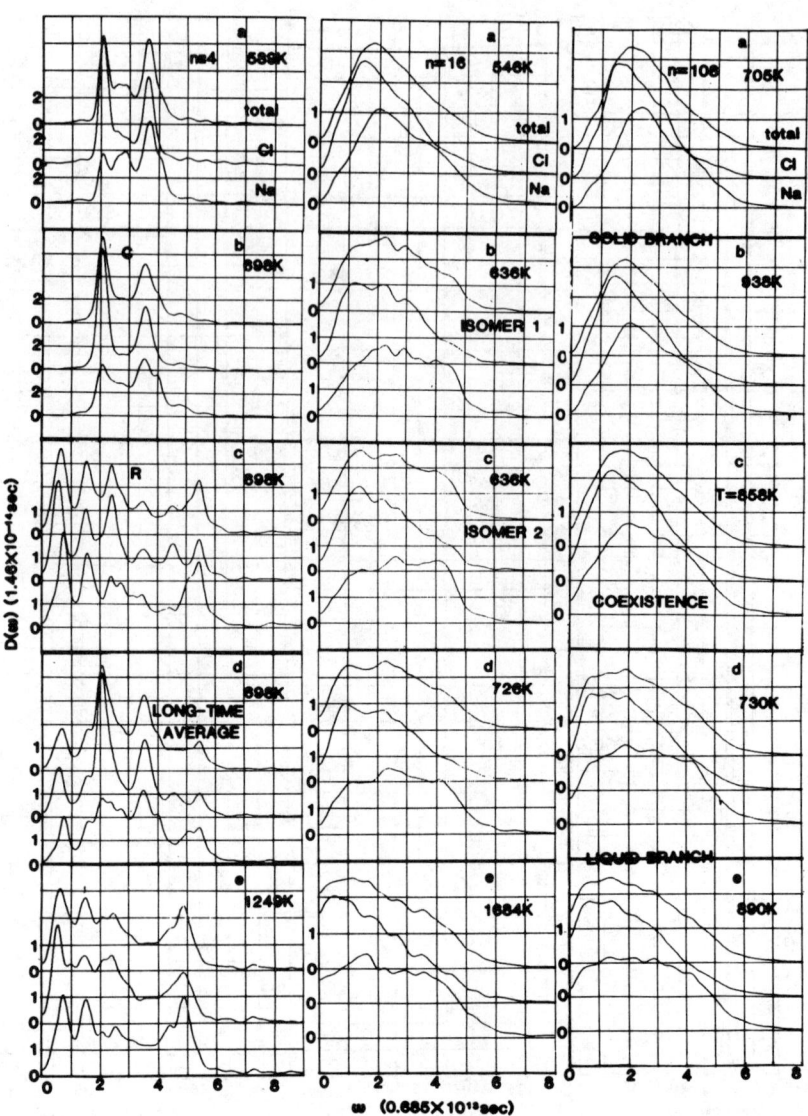

Fig. 7. Densities of states for n = 4, 16 and 108 at the denoted temperatures. Note shift to lower frequencies with increase in temperature, and onset of diffusion for n = 16 and 108 at higher temperatures.

trend of stabilization at higher temperature of the structural form (solid isomers for the smaller clusters and liquid for the large one) which possesses a higher density of low frequency modes. This can be understood in terms of the vibrational contribution to the free energy,[29] which the systems attempts to minimize. Secondly, for the n = 4 cluster we do not find any indication of soft mode or onset of diffusion (see value near $\omega = 0$ in $I(\omega)$ plots). For the intermediate size cluster, however, we do observe what may be construed as soft mode development and onset of diffusion at temperatures <u>above the first, isomerization, break</u> in the caloric curve (see Fig. 1b), thus corroborating our distinction between isomerization and melting for intermediate size clusters. True diffusion character is observed for the n = 108 cluster both at coexistence and on the liquid branch.

It is interesting to note that similar isomerization and phase-transformation patterns were obtained recently in simulations of rare-gas clusters.[31] In the case of rare gases the interaction potentials are much softer resulting in a less distinct character, and a higher number of the structural isomers than those exhibited by the ionic systems.

<u>Quantum</u>

An added perspective on the interesting problem of configurational changes in clusters is provided via temperature dependent studies of cluster isomerization induced by non-reactive electron attachment.[14,26] Studies which we performed on electron attachment to the small, neutral $(NaCl)_4$ cluster over a broad temperature range ($50°K-1200°K$) employed the QUPID method described in Section 2 with the interionic interactions given, as in the classical simulation, by the coulomb and Born-Mayer potentials and the electron interactions with the ions (Eq. 9) were given by:

(a) a Coulomb repulsive interaction from a closed shell anion of radius R_A, complemented by the electron induced polarization (α_A) interaction, operative at distances larger than R_A, and further corrected for the "self energy" of the induced dipole

$$\phi_{eI} = \begin{cases} e^2/r - \dfrac{e^2\alpha_A}{2r^4} ; & r > R_A \\ \\ e^2/r ; & r \le R_A, \end{cases} \quad (22)$$

(b) a local model pseudo-potential for the electron-cation interaction

$$\phi_{eI} = \begin{cases} -e^2/R_c ; & r \le R_c \\ \\ -e^2/r ; & r \ge R_c, \end{cases} \quad (23)$$

where R_c is a cutoff radius, $R_c = 3.24a_o$.[10,32-34] In these investigations we used constant-temperature simulations with a velocity form of the Verlet algorithm.[35] The number of "beads", P, was varied from P = 1000 at 50°K, 520 at other temperatures and 250 at 1000°K.

Equilibrium configurations at various temperatures are shown in Fig. 8. At low temperature (T ≤ 500K) the electron distribution is extended with an equal probability charge distribution around all the four Na^+ ions, and the ionic configuration is similar to that of the low-T neutral cluster. At the intermediate temperature domain, 500K ≤ T ≤ 750K, a configurational change of the negative cluster is exhibited (see Fig. 8b). This is a distorted planar configuration and is similar to the stable structure of the related classical $Na_4Cl_5^-$ cluster, which we found in simulations performed in the range 50K < T < 1100K. At the high-T domain, 750K < T < 1200K, a coexistence of several isomers is found. At ~ 750K a boat-like configuration (Fig. 8c) dominates (in this range a bent chain is also found) and at T ~ 1000K an elongated chain in coexistence with a planar ring (8d and 8e) are found. We observe a tendency toward spatial localization of the electron upon increase in T. Furthermore, the presence of the excess electron induces two types of configurational modifications which are either quantitatively or qualitatively different from those in the neutral cluster:

(i) The localized excess electron can play the role of a pseudonegative ion, with appreciable kinetic enregy, which is

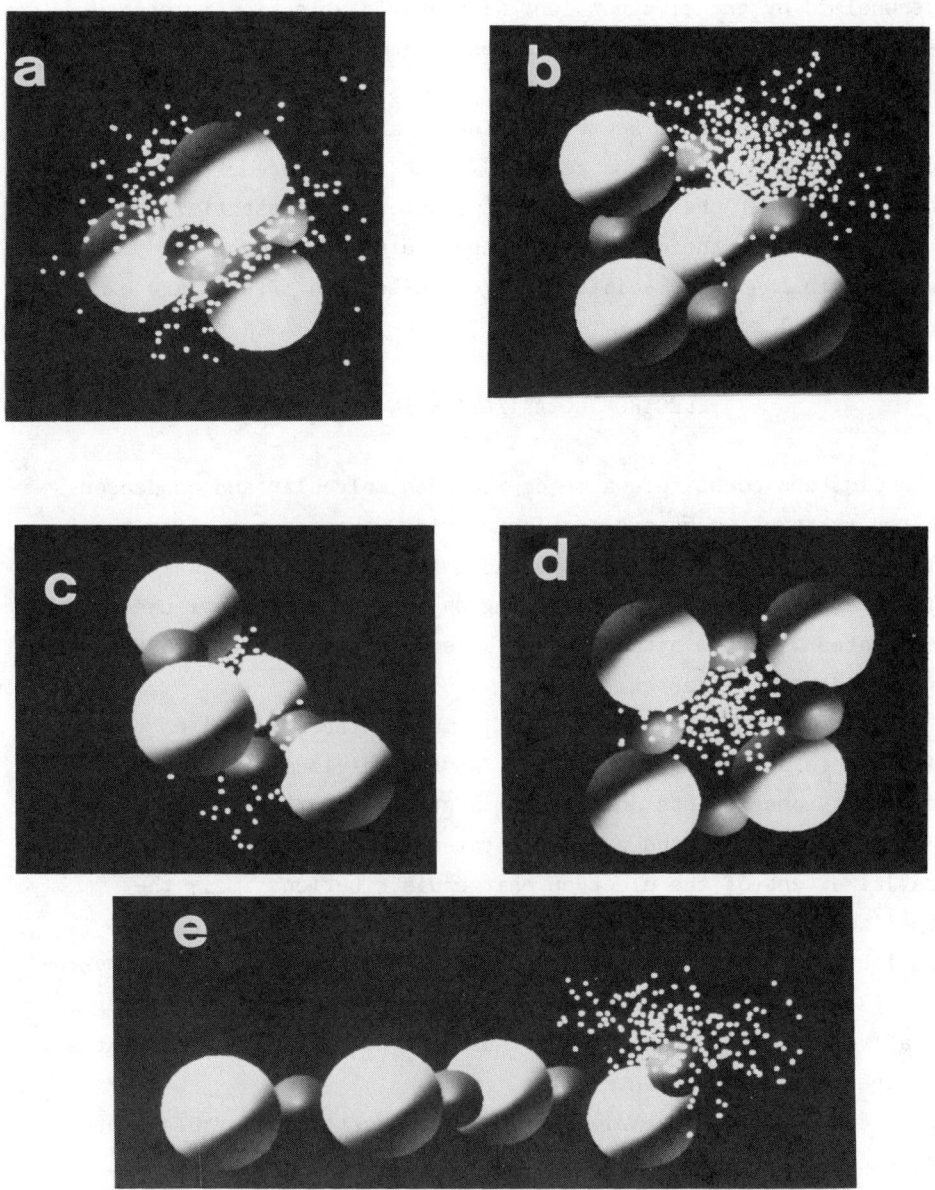

Fig. 8. Cluster configurations for e-Na_4Cl_4. (a) 50K, (b) 575K, (c) 750K, (d,e) 1000K. Large balls represent Cl^- anions and small dark balls represent Na^+. Small dots represents the electron distribution.

overwhelmed by the electron ion, with appreciable kinetic energy, leading to new nuclear configurations which have no counterpart in the neutral cluster.

(ii) The partial neutralization of a single positively charged ion by the excess electron results in the appearance of the high-T configuration of the neutral parent cluster at substantially lower temperatures for the negatively charged cluster. The first effect is observed in the intermediate T range, while both effects are exhibited in the high-T domain.

ELECTRON LOCALIZATION IN CLUSTERS

Clusters constitute a bridge between molecular and condensed matter systems, whose exploration provide ways and means for the elucidation of quantum, dynamic and chemical size effects in large finite systems.[36-38] The following physical and chemical phenomena associated with the attachment of an excess electron to clusters are of considerable interest:

(1) The parentage problem. In many interesting cases the excess electron is attached to a cluster, whose individual atomic or molecular constituents do not form a stable negative ion.

(2) Localized and extended states of the excess electron. The spatial extent of the electron charge distribution, i.e., the localization length, relative to the cluster size, provides a distinction between localized and extended states in a finite system.

(3) Bulk and surface states of the excess electron. The relative energetic stability of these two distinct types of states in different clusters is intriguing.

(4) Cluster reorganization. The energetically stable state of the electron attached to the cluster may involve a nuclear reorganization, where electron attachment is accompanied by a large cluster reorganization energy.

(5) Cluster isomerization induced by electron attachment. Electron attachment to a cluster may result in a configurational change resulting in a close-energy isomer.

We have utilized the quantum path integral molecular dynamics method (Section 2) to explore the compositional, structural and size dependence of electron localization in ionic and hydrogen bonded clusters, which provide novel information regarding the electron localization modes in large finite systems, resulting in novel phenomena of the formation of energetically stable surface states of excess electron in clusters.

Ionic Clusters

To investigate the modes of electron localization in ionic clusters we have performed[14] QUPID simulations of electron interaction with $[Na_{14}Cl_{13}]^+$ and $[Na_{14}Cl_{12}]^{++}$, at 300K. The structure of the bare clusters is cubic[29] (with small distortions) and the doubly-charged one contains an internal anion vacancy. Both the interionic-interactions and the electron-ion potentials were the same as those described in the previous section (with $\alpha_A = 0$ in Eq. 22). The number of "beads" used in these simulations was 399, to obtain convergent discretization. Employing an integration step of $\Delta t = 1.03 \times 10^{-15}$ sec, long equilibration runs were performed ($1-2\times10^4$ Δt). The reported results were obtained via averaging over 8×10^3 Δt following equilibration.

In Figs. 9a and 9b we present our results for the equilibrium electron charge distribution (small dots), and for the nuclear configuration of the clusters. The energetics of these systems may be summarised in terms of the adiabatic electron binding energy, EABE, of the cluster.

$$EABE = E_B^e + E_c, \qquad (24)$$

which is obtained by summing the electron binding energy

$$E_B^e = K_{int} + P^{-1} \sum_{i=1}^{P} \langle V_e(\vec{r}_i) \rangle \qquad (25)$$

and the cluster reorganization energy

$$E_c = \langle V_{AHC} \rangle - \langle V_{AHC} \rangle_0, \qquad (26)$$

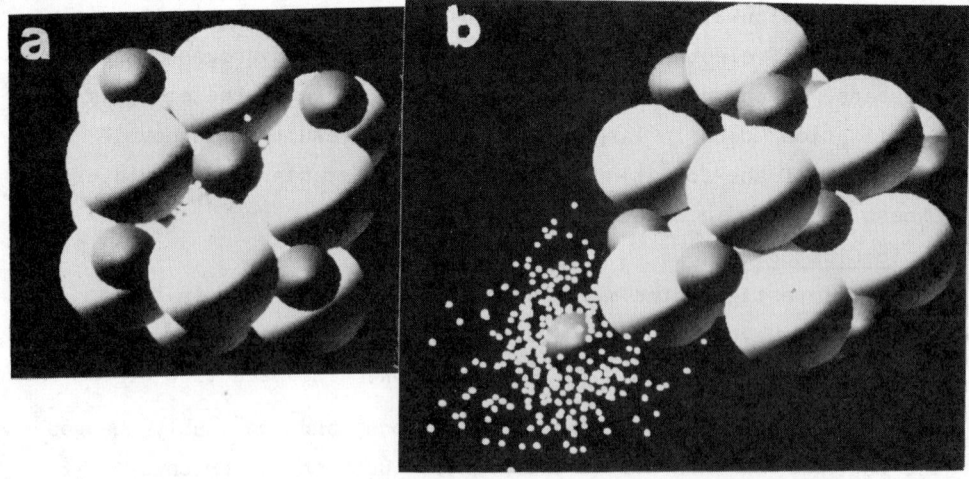

Fig. 9. Cluster configurations for (a) e-$[Na_{14}Cl_{12}]^{++}$ and (b) e-$[Na_{14}Cl_{13}]^{+}$, exhibiting internal and external electron localization, respectively.

where $\langle V_{AHC}\rangle_0$ is the potential energy of the "bare" AHC in the absence of the electron. Our results concerning the energetics of electron attachment are given in Table I.

As is apparent from Fig. 9a the $[Na_{14}Cl_{12}]^{++}$ cluster stabilizes an internally localized electron state, with the ionic configuration similar to that found in a bulk F^- center defect, thus establishing the dominance of short-range interactions for internal electron localization.

A drastically different localization mode is obtained in the $[Na_{14}Cl_{13}]^{+}$ system (Fig. 9b), where a novel surface state is exhibited. We refer to this state as a cluster surface localized state. Such surface states were considered for macroscopic alkali halide crystals by Tamm about fifty years ago, but were never experimentally documented. It is of interest to note that the total energy of e-$[Na_{14}Cl_{12}]^{++}$ is close to that of the $[Na_{14}Cl_{13}]^{+}$, whereupon the electron (internal) binding energy in the cluster is similar to that of a negative ion. Finally, we note that the magnitude of the adiabatic binding energies for these clusters are considerably higher

Table I. Energetics and electron charge distribution for bulk and surface states of electrons attached to sodium cloride clusters at 300K, P = 399. Components of radius of gyration of the charge distribution $R_g^{x,y,z}$ in units of a_o, and energies in Hartrees.

CLUSTER	$e\text{-}[Na_{14} C\ell_{12}]^{++}$	$e\text{-}[Na_{14} C\ell_{13}]^{+}$
E_B^e	-6.77	-4.34
E_c	2.17	2.87
EABE	-4.6	-1.47
$(R_g^x)^2$	4.6	5.4
$(R_g^y)^2$	6.8	7.0
$(R_g^z)^2$	6.3	9.8

than those found for the neutral clusters discussed in the previous section, were EABE = (0.8±0.2) eV.

Water Clusters

The existence of the solvated electron was experimentally demonstrated in 1863 for liquid ammonia[39] and in 1962 for water.[40] The localization of an excess electron in the bulk of a polar fluid originates from the combination of long-range and short-range attractive interactions,[41] and is accompanied by a large local molecular reorganization. Nonreactive electron localization in water clusters was experimentally documented to originate either from electron binding during the cluster nucleation process,[42,43] or by electron attachment to preexisting clusters.[44] The occurrence of a weakly bound state in $(H_2O)_2^-$ (vertical electron binding energy -3 meV for the equilibrium state[4,45] and -13 to -27 meV for a persistent metastable state),[4] characterized by a diffuse excess electron charge distribution (radius of gyration of ~$36a_o$)[4,45] can be understood on the basis of QUPID calculations, to originate from weak electron-dipole interactions. On the other hand, the experimental observation of stable $(H_2O)_2^-$ (n > 11) clusters,[42-44] which are characterized[43] by a large vertical electron binding energy, i.e., -0.7 eV (for n = 11) to -1.2 eV (for n = 20), poses a challenging theoretical problem. Quantum mechanical calculations[46,47] for $(H_2O)_6^-$ and $(H_2O)_8^-$ reveal that the adiabatic electron binding energy of these, and presumably also larger, water clusters will be positive, precluding the existence of such stable excess electron clusters, in contrast with experiment.[42-44] These theoretical studies followed faithfully the conventional wisdom in the field of solvated electron theory,[48] invoking the implicit assumption that the excess electron state in $(H_2O)_n^-$ constitutes an interior localization mode. QUPID calculations are ideally suited to explore alternative localization modes of the excess electron in water clusters.

A key issue in modeling the system is the choice of interaction potentials. Fortunately, for small water clusters, interaction potential functions which provide a satisfactory description for a range of properties are available.[49-50] We have used the RWK2-M model[50] for the intra and inter-water interactions. Less is known about the electron-water interaction.[51] We have constructed[4] a pseudopotential which consists of Coulomb, polarization, exclusion, and exchange contributions:

$$V(\vec{r}_e, \vec{R}_o, \vec{R}_1, \vec{R}_2) = V_{coul} + V_p + V_e + V_x . \quad (27a)$$

The position of the oxygen and hydrogen nuclei of the water molecule are given by $(\vec{R}_o, \vec{R}_1, \vec{R}_2)$ and $\vec{r}_e$ is the position of the electron.

The Coulomb interaction is

$$V_{coul}(\vec{r}_e,\vec{R}_o,\vec{R}_1,\vec{R}_2) = -\sum_{j=1}^{3} q_j e/\max(|\vec{r}_e-\vec{R}_j|,R_{cc}) \qquad (27b)$$

where $\vec{R}_3 = \vec{R}_o + (\vec{R}_1+\vec{R}_2-2\vec{R}_o)\delta$ is the position of the negative point charge of the RWK2-M model; q_1 and q_2 = 0.6e, q_3 = -1.2e and δ = 0.2218756. The value of q_j and δ were chosen[49-50] to give a good representation of the dipole and quadrupole moments of the water monomer. The cut-off radius, R_{cc}, was taken to be 0.5 a_o, and the results are insensitive to the precise value of R_{cc}.

The polarization interaction is given by

$$V_p(\vec{r},\vec{R}_o) = -0.5\alpha e^2/(|\vec{r}_e-\vec{R}_o|^2+R_p^2)^2 \quad , \qquad (27c)$$

where α = 9.7446 a.u. is the spherical polarizability of the water molecule. The form of V_p and the value of R_p = 1.6 a_o were chosen to fit approximately the adiabatic polarization potential as calculated by Douglass et al.,[51] for an approach of the electrons along the H-O-H bisector (see Table 7 and Fig. 2 in ref. 51).

The exclusion, V_e, and exchange, V_x, contributions both require the electron density, $\rho(\vec{r},\vec{R}_o,\vec{R}_1,\vec{R}_2)$, of the water molecule. We find that a reasonable fit to the calculated electron density,[52] in the regions of importance, is provided by

$$\rho(\vec{r};\vec{R}_o,\vec{R}_1,\vec{R}_2) = 8\,a_o^{-3}\,e^{-3|\vec{r}-\vec{R}_o|/a_o} + a_o^{-3}\sum_{j=1}^{2} e^{-3|\vec{r}-\vec{R}_j|/a_o}. \qquad (27d)$$

A contour plot of $\log_{10}\rho$ for $\vec{r}$ in the plane of the molecule is shown on the right of Fig. 10.

The repulsion, due to the exclusion principle, is modeled as a "local kinetic energy" term,[53]

$$V_e(\vec{r}_e,\vec{R}_o,\vec{R}_1,\vec{R}_2) = 0.5\,e^2 a_o\,(3\pi^2\rho)^{2/3} \quad . \qquad (27e)$$

The exchange interaction is modeled via the local exchange approximation,[54]

$$V_x(\vec{r}_e, \vec{R}_o, \vec{R}_1, \vec{R}_2) = -\alpha_x e^2 a_o (3\pi^2 \rho)^{1/3}/\pi . \qquad (27f)$$

The parameter α_x was taken to be $\alpha_x = 0.3$ in order to obtain good agreement between our simulation results and the SCF results of Rao and Kestner[46] for $(H_2O)_8^-$ at fixed octahedral configuration of the water molecule.

A contour plot of the total electron-water pseudopotential is shown on the left in Fig. 10 (in the plane of the molecule, with the oxygen located at the origin). In our QUPID simulation the water molecules are treated classically (there is no evidence of an isotope effect on solvation) while the quantum electron "necklace" contains a number of beads (P) which depends on the temperature (T). We have found that, as a rule of thumb, adequate discretization is obtained if $P k_B T \geq e^2 a_o$.

Equipped with these potentials we have embarked upon an investigation of the energetics and geometry of $(H_2O)_n^-$ (n = 8-18) clusters. In correspondence with the alternative experimental preparation methods[42-44] we invoked two initial conditions. (i) First condensing the water molecules around a classical negatively charged particle with a radius of $5a_o$, and subsequently replacing the classical particle with the electron necklace. (ii) Placing a compact distribution of beads next to an equilibrated neutral cluster. For the smaller clusters $n \leq 12$ a surface state develops rapidly, regardless of the initial setup of the calculation, while for n = 18 (i) and (ii) yield an "internal" and "surface" state, respectively.

In Table II we summarize the energetic data for the electron vertical binding energy, EVBE = $PE(e-(H_2O)_n) + K_{int}$ where PE is the averaged interaction potential energy between the electron and the water molecules, the cluster reorganization energy, E_c, and the electron adiabatic binding energy, EABE = EVBE + E_c. The neutral cluster reference states were obtained by simulated annealing. The lowest energy configuration for each cluster size was then used to

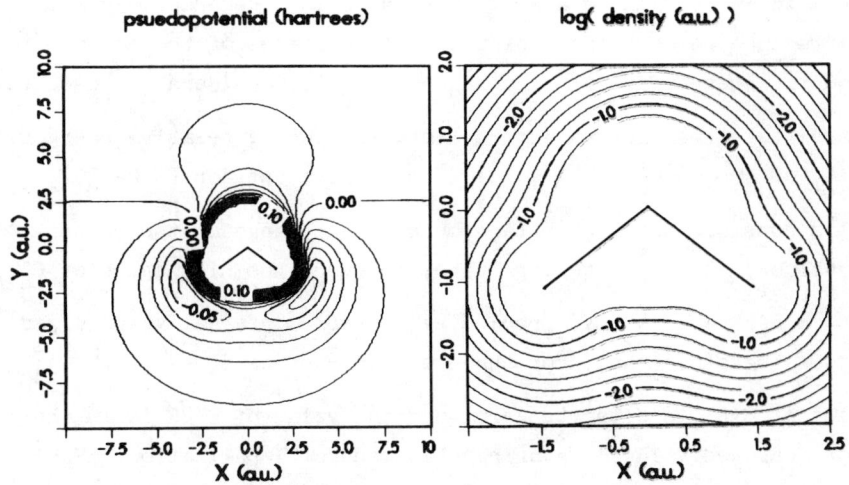

Figure 10: Contours of electron-water interaction, left, and the electron density ($Log_{10}\rho(r)$), right, (see Eq. 27d) in the plane containing the nuclei. The oxygen is located at the origin.

calculate E_c (the difference between the molecular potential energies of the negatively charged and neutral clusters). The energetic stability of the negatively charged cluster with respect to the equilibrium neutral cluster plus free electron is inferred from the magnitude and sign EABE, (negative values corresponding to a stable bound state). The bead distributions for the excess electron (Fig. 11) are characterized by the radius of gyration, $R_g^2 = \frac{1}{2P^2} \langle \sum_{i,j} (\vec{r}_i - \vec{r}_j)^2 \rangle$, and the degree of localization by the complex time correlation function[55] $R(t-t') = \langle |\vec{r}(t)-\vec{r}(t')|^2 \rangle^{1/2}$ for $(t-t') \in (0,\beta\hbar)$, yielding the correlation length $R(\beta\hbar/2)$, (Table I), which for a free particle is denoted by $R_f = \sqrt{3\lambda_T}/2$, where λ_T is the thermal wavelength of the particle. All calculations were at constant temperature with the velocity form of the Verlet integration algorithm.[35,56]

From these results we assert that there is a remarkable quantitative difference between internal and surface states of the excess electron in water clusters. The value of E_c is considerably lower for a surface state than for an internal state insuring relative energetic stability of the former (Table II). As is apparent for $(H_2O)_{18}^-$, |EVBE| is considerably higher (and outside the range of the experimental values) for the interior state, however, the high value of E_c results in EABE = 0.245 eV, precluding a stable internally localized state. On the other hand, for the electron surface state of $(H_2O)_{18}^-$ the value of EABE is close to zero (and the value of EVBE is in the range of measured values) favoring this mode of localization. For $(H_2O)_{12}^-$ (Table II and Fig. 11), only a surface state is found. Finally, for $(H_2O)_8^-$, a very small electron binding energy is found and the state is characterized by a diffuse charge distribution (Fig. 11 and Table II).

The next issue which we address is that of the transition from surface to internal and eventually bulk localization. In Table II we include results for an internally, stable, localized state in $(H_2O)_{32}^-$ at $79°K$. Furthermore we also show results for $(H_2O)_{32}^-$ and $(H_2O)_{64}^-$ at $300°K$. These results show a trend of increase in the magnitudes of the vertical and adiabatic electron binding energies. It is also seen that the convergence to the bulk limit is apparently very slow in the cluster size, demonstrating the importance of long-range interactions.

From these results we conclude that: (1) The electron localization mode in $(H_2O)_n^-$ (for $n \leq 18$) clusters involves the formation of a surface state. (2) The onset of electron localization in a tightly bound state in $(H_2O)_n^-$ clusters is exhibited for $n > 8$, in accord with experiment[42-44] ($n > 11$). (3) The vertical electron binding energy for the cluster-electron-surface-state in $(H_2O)_{12}^-$ and $(H_2O)_{18}^-$ (Table II) are in adequate agreement with the experimental photo-electron spectroscopic data.[43] Additionally, |EVBE| rises sharply in the range n = 8-12. For the $(H_2O)_8^-$ cluster the diffuse

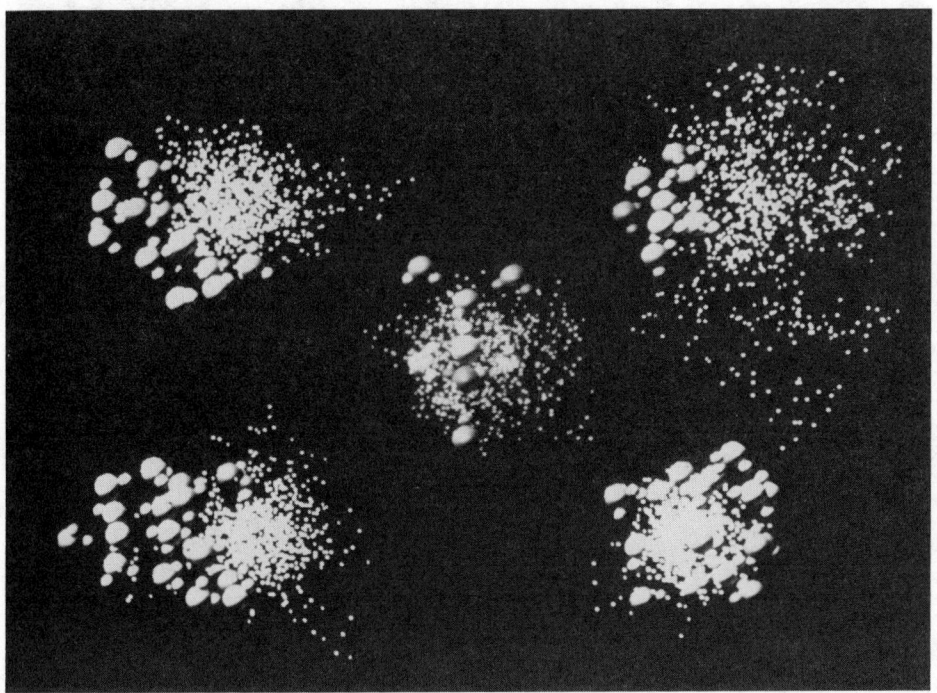

Fig. 11. Cluster configurations of $(H_2O)_n^-$, via quantum path-integral molecular dynamics simulations. Balls, large and small, correspond to oxygen and hydrogen, respectively. The dots represent the electron (bead) distributions. Shown at the center is $(H_2O)_8^-$, for a static molecular configuration as in Ref. 46. From top right and going counterclockwise: (i) diffuse surface state of $(H_2O)_8^-$; (ii) surface state of $(H_2O)_{12}^-$; (iii) surface state of $(H_2O)_{18}^-$ and (iv) internal state of $(H_2O)_{18}^-$.

nature of the excess electron distribution could lead to a large collision induced electron detachment cross-section, which may account for the absence of n < 11 clusters from the experimental spectra.[42-44]
(4) A transition to internal localization occurs for 18 < n < 32.
(5) Convergence to the bulk limit is slow, demonstrating the important role of long-range interaction. Considering the complexity of the system, statistical uncertainties and those implicit in the model

Table II. Energetics and excess electron charge distribution for $(H_2O)_n^-$ clusters. All calculations, except where indicated, were at T = 79°K, using P = 4096 beads for the electron necklace. Energies in eV, radius of gyration, R_g, of the electron charge distribution in a_o units.

Cluster	$(H_2O)_8^-$	$(H_2O)_{12}^-$	$(H_2O)_{18}^-$	$(H_2O)_{18}^-$	$(H_2O)_{32}^-$	$(H_2O)_{32}^{-*}$	$(H_2O)_{64}^{-*}$
	Diffuse	Surface	Surface	Internal	Internal	Internal	Internal
EVBE	−0.190	−0.97	−1.31	−1.96	−2.40	−2.47	−3.20
E_c	0.136	0.871	1.333	2.204	2.38	2.43	2.70
EABE	−0.054	−0.136	0.023	0.245	−0.023	−0.037	−0.48
R_g	10.6	6.1	5.5	4.1	4.0	3.8	3.8
$\dfrac{R(\beta\hbar/2)}{R_f}$	0.28	0.15	0.14	0.11	0.10	0.2	0.19

*Calculated at 300°K

interaction potentials, we are encouraged by these results which provide a consistent energetic and structural picture of electron localization in small water clusters.[57]

DYNAMICS OF RELAXATION AND FRAGMENTATION PROCESSES

The processes of energy acquisition, storage and disposal in clusters are of considerable interest for the elucidation of dynamic reactive and non-reactive processes in finite systems, whose energy spectrum for electronic nuclear, and other particle excitations can be varied continuously by changing the cluster size. These processes, and the simulation methods developed for probing them, are of interest to research in areas where the characteristic energies, interaction strength and spatial and temporal scales may be vastly different. These include studies of the mechanisms of nuclear fragmentation[58] in high energy proton-nucleus[59a] and heavy ion collisions[59b] and reactive and nonreactive vibrational energy redistribution, reactive dissociation or vibrational predissociation in atomic and molecular clusters.[60-64] The goals of these studies is to discover the modes of excitation produced via collisions, optical photoselective vibrational excitation or electronic excitation, their foundation on the microscopic processes and their mechanisms of decay. In particular the ability to vary the cluster size, control the initial conditions (e.g., initial state of the cluster, relative kinetic energy between collision partners) and measure the final products (e.g., mass, energy and angular distribution) allow systematic studies of the kinetics and dynamics of microscopic cooperative phenomena and relaxation processes occuring in small, quantum and classical, systems that are prepared, via the excitation, in a far from equilibrium state.

In this section we review our studies[61,65] of the dynamics of energy relaxation and fragmentation processes in neutral rare-gas clusters following an electronic excitation. In charged clusters, the vibrational excitation resulting in both nonreactive and reactive relaxation, can originate from ionization followed by hole trapping in inert-gas clusters[60] and from electron attachment to alkali halide clusters.[14] In neutral clusters, it was found, using MD simulations,[62] that the dissociation of Ar_n (n = 4-6) clusters can be accounted for in terms of the statistical theory of unimolecular reactions, which implies the occurrence of vibrational energy randomization in small clusters. In our studies we focus on the interesting problem of the dynamical consquences of exciton trapping in rare-gas clusters (RGCs). Extensive information is currently available regarding exciton trapping in solid and liquid inert gases.[63] Exciton trapping in the heavy rare-gas solids, i.e., Ar, Kr, and Xe, exhibits two-centre localization, resulting in the formation of electronically excited, diatomic inert-gas excimer molecules. Electronic excitation of a RGC, R_n, is expected to result in an exciton state, which subsequently becomes trapped by self localization. Although the details of the energetics and spatial charge distribution of excitons in finite RGCs have not yet been explored, some information can be drawn from the analogy with the lowest electronic excitations in solid and liquid rare gases.[63] The two lowest, dipole allowed, electronic excitations in RGCs can adequately be described in terms of tightly bound, Frenkel-type excitations with a parentage in the $^1S_1 \to {}^3P_1$ and $^1S_0 \to {}^1P_1$ atomic excitations which are modified by large nonorthogonality corrections.[63] The process of exciton trapping in the heavy RGCs of Ar, Kr and Xe involves the formation of the diatomic excimer molecule R_2^*, which is characterized by a substantial binding energy at a highly vibrational state. Energy exchange between the R_2^* excimer and the cluster in which it is embedded involves two processes.

(1) Short-range repulsive interactions between the expanded, Rydberg-type excited state of the excimer and the other cluster atoms result in a dilation of the local structure around the excimer, leading to energy flow into the cluster.

(2) Vibrational relaxation of the excimer induces vibrational energy flow into the cluster.

The vibrational energy released into the cluster by processes (1) and (2) may result in vibrational predissociation.

We have explored the dynamic implications of exciton trapping in RGCs by conducting classical molecular dynamics (MD) calculations on electronically excited states of such clusters. As model systems we have chosen, Ar_n (with n = 13,55) and mixed $Xe_m Ar_{n-m}$ (m = 1,2 and n = 13,55) clusters. The ground states of the RGCs were described by Lennard-Jones pair potentials $V(r) = 4\epsilon[(\sigma/r)^{12} - (\sigma/r)^6]$ with the well-depth and distance parameters appropriate for Ar - Ar[66] (121K, 3.4Å), Ar - Xe[67] (177.6K, 3.65Å) and Xe - Xe[68] (222.3K, 4.10Å), respectively. In the electronically excited state the excimer potential is represented by a Morse curve, $V(r) = D_e\{\exp[-2B(r/R_e - 1)] - 2\exp[-B(r/R_e - 1)]\}$ with the parameters D_e, R_e and B taken as 9125.3K, 2.319Å and 5.12 for Ar_2^{*}[69] and 11609K, 2.99Å and 4.228 for Xe_2^*.

An important consequence of the electronic excitation involves the drastic modification of the interaction between the excimer and the ground-state atoms. On the basis of the analysis of Xe^*-Ar interactions[67] (for which a description in terms of a 6-12 LJ potential with ϵ = 92.8K and σ = 4.13Å is adequate) the Ar^*-Ar potential for each of the constituents of the excimer has been described in terms of a Lennard-Jones potential with the parameters ϵ^* and σ^*. We have taken for the energy parameter $\epsilon^* = \epsilon$, while the distance scale ratio $\bar{\sigma} = \sigma^*/\sigma$ has been chosen in the range $\bar{\sigma} =$ 1.0-1.2. The appropriate Ar^*-Ar interaction is characterized by[67] $\bar{\sigma} = 1.10$-1.15, reflecting the enhancement of short-range repulsive interactions in the electronically excited Rydberg-type state.

Following equilibration of the ground state system, electronic excitation was achieved (at the time $t = 0$) by the instantaneous switching-on of the excimer potential between a pair of nearest neighbor atoms and of the potentials between excimer and the ground-state atoms comprising the rest of the cluster (using a fifth-order predictor-corrector method, the integration time step in the ground state was 1.6×10^{-14} sec and in the excited state 1.6×10^{-16} sec).

We consider first the dynamics of excited homonuclear clusters, Ar_{13}^* and Ar_{55}^*. In figures 12a and 12b we show an overview of the dynamics of the nuclear motion following the electronic excitations in Ar_{13}^* which is expressed in terms of the interatomic distances. The excimer exhibits a large amplitude motion in a highly excited vibrational state, whill all the other interatomic distances increase, indicating the initiation of the escape of the ground-state cluster atoms. Insight into the energy flow from the excimer into the cluster is obtained from the time dependence of the kinetic energy (KE), the potential energy (PE) and the total energy of the excimer (fig. 13). The strong oscillations in the PE and KE clearly indicate the persistence of the vibrational excitation of the excimer over a long time scale. Further detailed information concerning the implications of this energy flow on the cluster dissociation was inferred by considering the composition and the energetics of the "main fragment", i.e., the fragment which consists of the excimer together with ground-state atoms, with all the nearest-neighbor separations being smaller than 3σ, beyond which all interatomic interactions are negligibly small. The total energy within the main fragment was partitioned into two separate contributions: (i) The energy of the

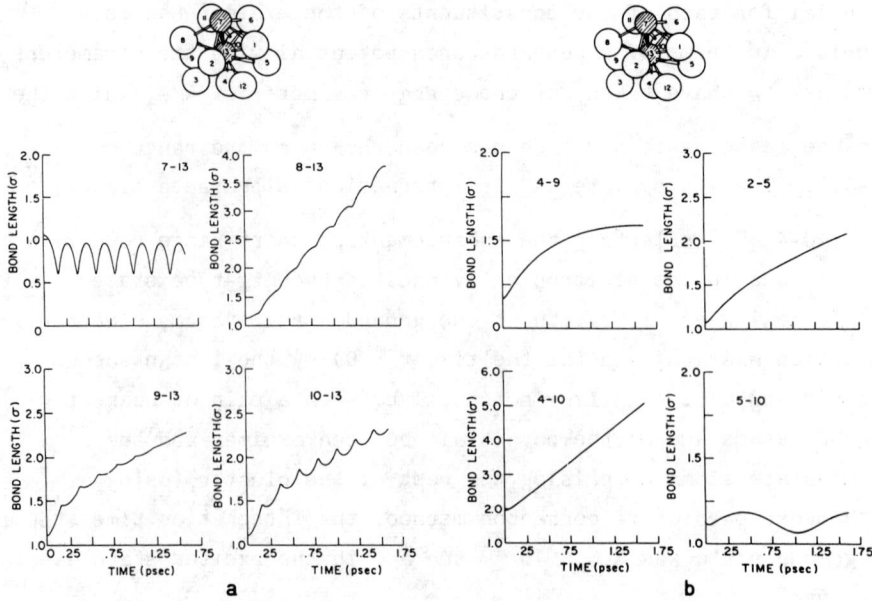

Fig. 12. Time dependence of interatomic distances in the electronically excited Ar_{13} cluster. The insert shows the ground-state equilibrium structure at 24K. The labelling of atoms is indicated. The dashed atoms 7 and 13 form the excimer. Distances are in units of σ_{Ar}. The Ar^*-Ar interaction is characterized by $\bar{\sigma} = 1.20$. (a) Interatomic distances within the excimer and between the excimer atoms and some ground-state atoms. (b) Interatomic potentials between ground-state atoms.

"reaction centre" which consists of the excimer PE and KE together with half of the sum of the potential energy of the Ar^*-Ar interactions, and (ii) the energy of the "bath subsystem", which involves the KE of the ground-state Ar atoms, the potential energies of the Ar-Ar interactions and half of the sum of the potential energies of the Ar^*-Ar interactions. The time evolution of the various contributions to the total energy (fig. 14) portray the energy

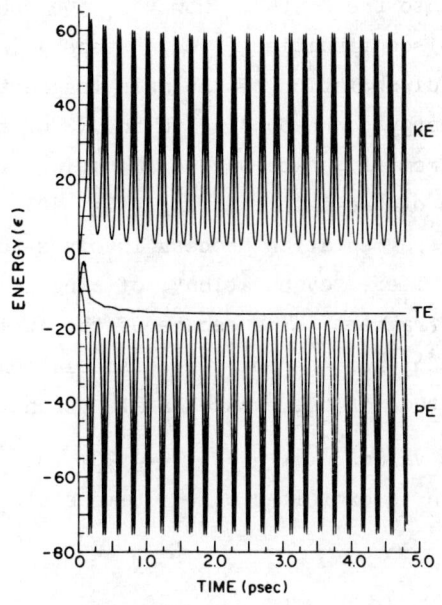

Fig. 13. The dependence of the potential energy (PE), the kinetic energy (KE), and the total energy (TE) of the bare excimer in Ar_{13}^* ($\bar{\sigma} = 1.20$). Energies are given in units of ϵ.

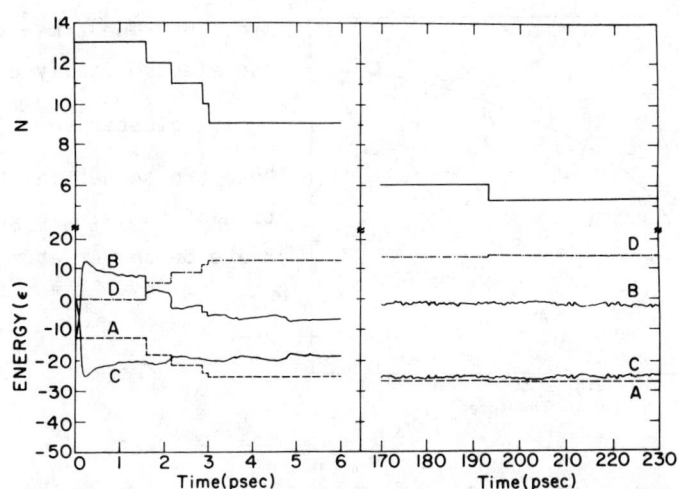

Fig. 14. Time evolution of fragmentation dynamics of the electronically excited $(Ar)_{13}$ cluster ($\bar{\sigma} = 1.15$). (A) Total energy of the main fragment. (B) Total energy of the "bath subsystem". (C) Total energy of the "reaction centre". (D) Kinetic energy of the dissociated atoms. (E) Number of atoms in the main fragment. The steps in curves (A), (D), and (E) mark the stepwise dissociation of individual Ar atoms from the main fragment.

flow from the excitation centre into the "bath". However, the energy
per atom does not equilibrate. Discontinuities (i.e., "steps") in the
energy plots of fig. 14 mark the dissociation of the main fragment,
with the decrease of the total energy corresponding to the KE of the
ground state atoms dissociating from it. A cursory examination of the
time evolution of the composition of the main fragment (fig. 14)
clearly indicates that the major fragmentation process involves the
sequential stepwise dissociation, i.e., "evaporation", of single
ground-state atoms from the main fragment. The escape of the excimer
from the main fragment has not been encountered. The dissociation
process is dominated by the magnitude of the excited-state potential
scale parameter $\bar{\sigma}$. For realistic values[67] of $\bar{\sigma}$ = 1.10-1.20, the
threshold for cluster dissociation is exhibited on the time scale of ~
2-20 psec (fig. 15).

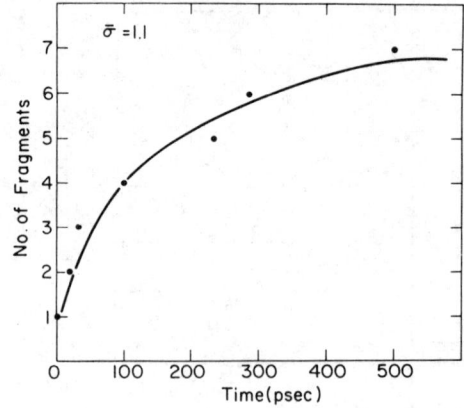

Fig. 15. The time evolution of
the electronically excited
$(Ar)_{13}$ cluster ($\bar{\sigma}$ = 1.10).
Note the sequential "evapora-
tion of the single ground-
state of the Ar atoms.

From these MD results the following picture concerning vibrational energy flow and reactive dynamics of the Ar_{13} cluster emerges.
The temporal persistence of the vibrational excitation of the excimer
(fig. 13) and of the "reaction centre" (fig. 14) corresponds to a
"mode selective" excitation of the excimer, with vibrational energy
redistribution within the cluster being precluded by two effects.
First, the difference in the characteristic frequencies of the (high
frequency) dimer motion and the (low frequency) motion of the cluster.

Second, the local dilation of the cluster structure around the excimer, which is induced by the short-range excimer-cluster repulsive interactions. The vibrational energy flow from the dimer into the cluster (figs. 13 and 14) consists of two stages:

(A) Ultrafast vibrational energy transfer due to repulsion, which occurs on the time scale of ~ 200 fsec (figs. 13 and 14). This energy transfer process is dominated by the magnitude of the scaleparameter $\bar{\sigma}$.

(B) "Slow" energy transfer on the time scale of tens of picoseconds (for $\bar{\sigma}$ = 1.2) and up to hundreds of psec (for $\bar{\sigma}$ = 1.0) due to vibrational relaxation of the excimer. The dynamics of the cluster induced by these energy transfer processes involves reactive vibrational predissociation, as is apparent from figs. 14 and 15. This state of affairs is, of course, drastically different from that encountered in infinite systems, where a nonreactive process prevails when the photon modes of the system are excited. A cursory examination of the dissociative dynamics (fig. 15) of the Ar_{13} cluster following excimer formation indicated that two reactive processes prevail.

(i) A fast stepwise "evaporation" of Ar atoms is exhibited on the time scale of ~10 psec. This process is induced by the energy transfer process (A). Subsequently, an additional reactive process appears (fig. 15), which involves,

(ii) Slower vibrational predissociation of Ar atoms on the time scale $\geq$ 10 psec. This dissociative process is induced by both energy transfer processes (A) and (B).

It is imperative to note that the short-time "explosion" of the electronically excited cluster is induced by energy transfer due to short-range repulsive interactions. When these interactions are switched off by taking $\bar{\sigma}$ = 1.0, only mechanism (B) is operative for vibrational energy flow into the cluster and the cluster dissociative process, which again occurs by stepwise "evaporation" occurring on the time scale of 30-1000 psec. The appropriate excited-state repulsive physical parameter characterizing excimer-cluster interactions in RGCs is $\bar{\sigma}$ = 1.1-1.2, and we expect the occurrence of energy flow predissociation induced by excited repulsive interactions to occur on the time scale ~ 10 psec.

In order to investigate the dependence of the nature and rate of energy redistribution processes following excitation, on the degree of aggregation, we have performed similar situations for an Ar_{55}^* cluster, with the excimer located at either the center or at the outer shell of the cluster. These simulations revealed that while the ultrafast vibrational energy flow from the vibrationally-excited excimer to the rest of the cluster is operative, no fragmentations are observed, i.e., nonreactive vibrational energy redistribution, characteristic to infinite condensed matter systems, prevails. This of course is a direct consequence of the large number of densly spaced available "bath" modes. We note however that following the ultrafast energy flow which occurred on the 0.1-0.3 psec time scale, the subsequent vibrational relaxation rate is rather slow, due to the frequency mismatch and local dilation effects.

We conclude this section with a few observations on mixed clusters. We have performed extensive studies of $Ar_{n-m}Xe_m^*$ cluster, with n = 13, 55 and m = 1, 2, with the Xe atoms located at either the center or on the outer shell. In the case of $Ar_{12}Xe^*$, with the Xe^* centrally located, the excited atom escaped from the cluster on a time scale of 50±20 psec (see Fig. 16). In view of the fast escape, we predict that the radiative decay will be characteristic of the free excited atom rather than from the cluster-shifted species. In the case of $A_nXe_2^*$ we observed a complete dissociation of the cluster within 200 psec following the excitation. Curiously, we observed no such phenomena for $Ar_{54}Xe^*$ or $Ar_{53}Xe_2^*$ (either centrally or peripherally locating the excitations) demonstrating the effect of the degree of aggregation. In these systems the effectiveness of the cluster mediated vibrational relaxation dominates, aided by characteristics of the interaction potentials and the heavy mass of the impurity (the defect modes are embedded within the vibrational manifold of the host cluster). For the case of the $Ar_{54}Xe^*$ system the

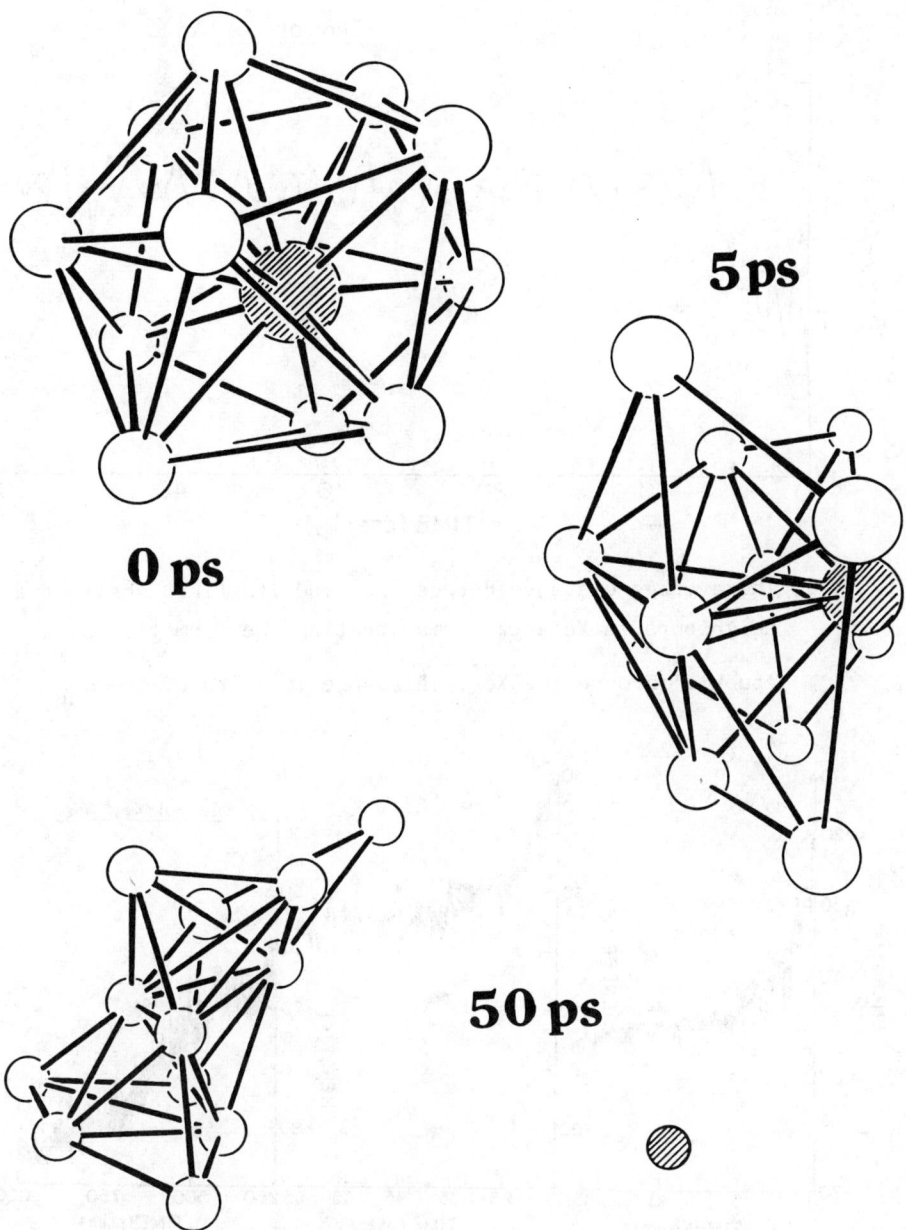

Fig. 16. Cluster configurations for a Xe^*Ar_{12} cluster at t = 0, 5 and 50 psec, showing the escape of the Xe^* atom. Empty spheres represent Ar atoms and the dashed sphere the Xe^*.

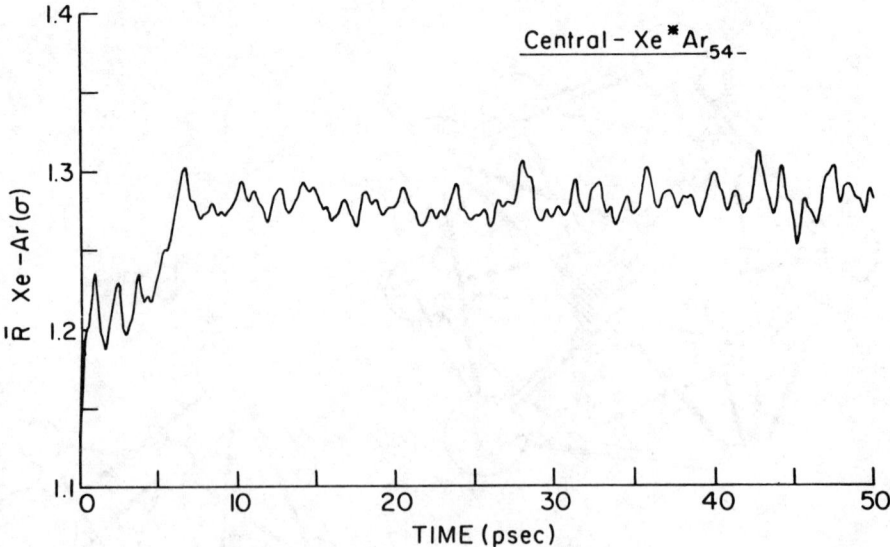

Fig. 17. The average distance between Xe^* and its first shell of Ar neighbours in Xe^*Ar_{54}, demonstrating the formation of a 'bubble' around the Xe^*. Distance in units of the σ_{Ar}.

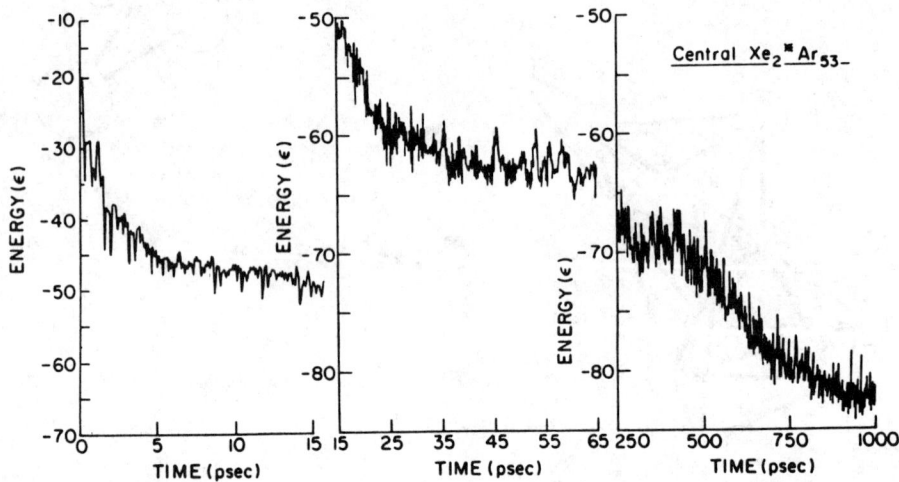

Fig. 18. The total internal energy of the excimer for Xe_2^* centrally located in a $Xe_2^*Ar_{53}$ cluster.

initial process after excitation involved an ultrafast energy transfer to the Ar "bath" which occurred on a time scale of (250 ± 100) fsec, accompanied by "bubble" formation around the Xe^*. This 'bubble' formation, or dilation process, can be followed by calculating the average distance between the Xe^* and its first coordination shell of Ar atoms as a function of time shown in Fig. 17.

For the $Ar_{53}Xe_2^*$ clusters (with the Xe_2^* located centrally or on the surface of the cluster) we observe again an initial ultrafast energy transfer occurring on the 350 fsec time-scale followed by a monotonous relaxation of the excess vibrational energy (see Fig. 18), and a continuous heating of the Ar- bath, resulting in a structural isomerization and eventual evaporations of one to four Ar atoms at very long times. In the final state, for both initial locations of the excimer, the Xe_2^* is found on the surface of the cluster. (Figure 19).

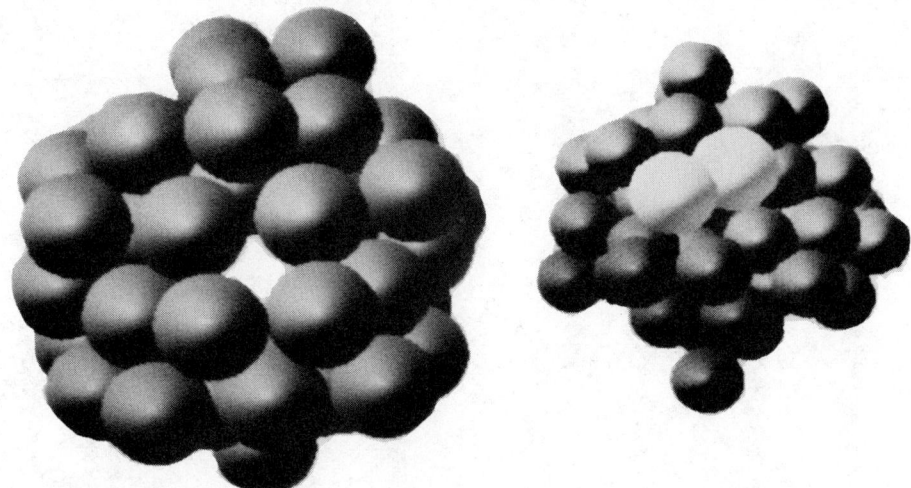

Fig. 19. Cluster configurations for $Xe_2^* Ar_{53}$, for an initially centrally located Xe_2^* at $t = 0^+$ (immediately past the excitation), on the left, and after 1 nsec, on the right. The light atoms are the Xe_2^* and the dark ones are the Argon atoms.

Acknowledgement: We gratefully acknowledge helpful collaboration with N. R. Kestner on water clusters, and the invaluable assistance of V. Mallette in the preparation of the figures and A. Ralston in the preparation of the manuscript. This work is supported by the U. S. Department of Energy under grant No. FG05-86ER45234 (to UL) and by the US-Israel Binational Science Foundation.

References

1. F. F. Abraham, Adv. in Phys. <u>35</u>, 1 (1986); J. Vac. Sci. Technol. <u>B2</u>, 534 (1984).
2. D. W. Heermann, Computer Simulation Methods, (Springer, Berlin, Heidelberg, 1986).
3. U. Landman, R. N. Barnett, C. L. Cleveland, W. D. Luedtke, M. W. Ribarsky, D. Scharf and J. Jortner, Mat. Res. Soc. Symp. Proc. <u>63</u>, 273 (1985).
4. U. Landman, C. L. Cleveland, R. N. Barnett, D. Scharf and J. Jortner, J. Phys. Chem., 1987 (in press).
5. R. S. Berry, T. L. Beck, H. L. Davis and J. Jellinek, Adv. Chem. Phys., 1987 (in press).
6. R. Car, M. Parrinello and W. Anderoni in Microclusters, eds. S. Sugano, Y. Nishina and S. Ohnishi, (Springer, Berlin, 1987), p. 134; R. Car and M. Parrinello, Phys. Rev. Lett. <u>55</u>, 2471 (1985).
7. R. P. Feynman and A. R. Hibbs, Quantum Mechanics and Path Integrals (McGraw-Hill, New York, 1965).
8. D. Chandler and P. G. Wolynes, J. Chem. Phys. <u>79</u>, 4078 (1981).
9. D. Chandler, J. Phys. Chem. <u>88</u>, 3400 (1984).
10. M. Parrinello and A. Rahman, J. Chem. Phys. <u>80</u>, 860 (1984).
11. B. De Raedt, H. Sprik and H. L. Klein, J. Chem. Phys. <u>80</u>, 5719 (1984).
12. M. F. Herman, E. J. Bruskin and B. J. Berne, J. Chem. Phys. <u>76</u>, 5150 (1982).
13. B. J. Berne and D. Thirumalai, in Ann. Rev. Phys. Chem. (to appear, 1986).
14. U. Landman, D. Scharf, and J. Jortner, Phys. Rev. Lett. <u>54</u>, 1860 (1985).
15. L. S. Schulman, Techniques and Applications of Path Integrals (Wiley, New York, 1981).
16. See article by W. H. Miller, and references therein, in this volume.
17. T. L. Hill, Thermodynamics of Small Systems, (Benjamin, N. Y., 1963).
18. G. Natanson, F. Amar and R. S. Berry, J. Chem. Phys. <u>78</u>, 399 (1983).

19. J. Jellinek, T. L. Beck and R. S. Berry, J. Chem. Phys. 84, 2783 (1986); see also references to earlier work here and in ref. 2.
20. R. W. Hockney and J. W. Eastwood, Computer Simulation Using Particles, (McGraw-Hill, N. Y., 1981), pp. 488-498.
21. P. R. Couchman and W. A. Jessor, Nature 269, 481 (1977).
22. P. R. Couchman and C. L. Ryan, Phil. Mag. A37, 369 (1978); R. Balian and C. Bloch, Ann. Phys. N.Y. 60, 401 (1970).
23. Y. Imry, Phys. Rev. B21, 2042 (1980).
24. Y. Imry and D. J. Bergman, Phys. Rev. A3, 1416 (1971).
25. J. Luo, U. Landman and J. Jortner, in The Physics and Chemistry of Small Clusters, (Plenum, New York, 1987).
26. D. Scharf, U. Landman and J. Jortner, J. Chem. Phys., 1987 (in press).
27. This is reminiscent of the picture advanced by F. M. Stillinger and T. A. Weber for a liquid, see: Kinam A5, 159 (1981); Phys. Rev. A25, 978 (1972).
28. F. G. Fumi and M. P. Tosi, J. Phys. Chem. Solids 25, 31 (1964).
29. T. P. Martin, Phys. Rep. 95, 167 (1983).
30. D. A. McQuarrie, Statistical Mechanics, (Harper and Row, N. Y., 1976), Chap. 21, 22.
31. See references 5, 18, 19 and the article by R. S. Berry in this volume.
32. R. W. Shaw, Phys. Rev. 174, 769 (1968).
33. J. V. Abarenkov and V. Heine, Phil. Mag. 12, 529 (1965).
34. D. Scharf, J. Jortner and U. Landman, Chem. Phys. Letts. 130, 504 (1986).
35. J. R. Fox and H. C. Anderson, J. Phys. Chem. 88, 4019 (1984).
36. J. Jortner, Ber. Bunsenges, Physik. Chem. 88, 188 (1984), and other papers.
37. T. D. Mark and A. W. Castleman, Jr. in Adv. Atomic and Mol. Phys. (D. R. Bates and B. Bederson, Eds.), Vol. 20 (1984).
38. See papers in J. Phys. Chem. 88 (1984).
39. W. Weyl, Ann. Phys. 197, 601 (1963).
40. E. J. Hart and J. W. Boag, J. Am. Chem. Soc. 84, 4090 (1962).
41. J. Jortner, J. Chem. Phys. 30, 839 (1959); D. A. Copeland, N. R. Kestner and J. Jortner, J. Chem. Phys. 53, 1189 (1970).

42. H. Haberland, H. G. Schindler and D. R. Worsnop, Ber. Bunsenges, Chem. 88, 3903 (1984); J. Chem. Phys. 81, 3742 (1984).
43. J. V. Coe, D. R. Worsnop and K. H. Bowen (J. Chem. Phys., 1987).
44. M. Knapp, O. Echt, D. Kreisle and E. Recknagel, J. Chem. Phys. 85, 636 (1986); J. Phys. Chem. (preprint, 1986).
45. A. Wallquist, D. Thirumalai and B. J. Berne, J. Chem. Phys. 85, 1583 (1986).
46. B. K. Rao and N. R. Kestner, J. Chem. Phys. 80, 1587 (1984) and references therein.
47. N. R. Kestner and J. Jortner, J. Phys. Chem. 88, 3818 (1984); M. D. Newton, J. Chem. Phys. 58, 5833 (1973).
48. J. Jortner, J. Chem. Phys. 30, 839 (1959); D. A. Copeland, N. R. Kestner and J. Jortner, J. Chem. Phys. 53, 1189 (1970).
49. J. R. Reimers, R. O. Watts and M. L. Klein, Chem. Phys. 64, 95 (1982).
50. J. R. Reimers and R. D. Watts, ibid. 85, 83 (1984). This paper as well as the description of the potential in ref. 49 (Eq. 13 and Table I) contain several ambiguities and typographical errors. When corrected we reproduce their results.
51. C. H. Douglass, Jr., D. A. Weil, P. A. Charlier, R. A. Eades, D. G. Truhlar and D. A. Dixon, Chemical Applications of Atomic and Molecular Electrostatic Potentials, Eds., P. Politzer and D. G. Truhlar, (Plenum, New York, 1981), p. 173.
52. C. W. Kerr and M. Karplus in Water, F. Franks, ed., (Plenum, New York, 1972), p. 21; M. W. Ribarsky, W. D. Luedtke and U. Landman, Phys. Rev. B32, 1430 (1985).
53. J. N. Bardsley, case studies in Atomic Physics 4, 299 (1974); G. G. Kleiman and U. Landman, Phys. Rev. B8, 5484 (1973).
54. See e.g., D. G. Truhlar, in Chemical Applications of Atomic and Molecular Electrostatic Potentials, Eds., P. Politzer and D. G. Truhlar, (Plenum, New York, 1981), p. 123.
55. A. L. Nichols, D. Chandler, V. Singh and D. M. Richardson, J. Chem. Phys. 81, 5109 (1984).
56. In the QUPID method the averaged results do not depend on the dynamic masses used to generate the classical trajectories. We used a mass of 1 amu for the classical particles and 0.025 amu

for the beads. The integration time step was 2.625×10^{-16} sec. Prior to averaging the systems evolved till no discernable trend was observed. Averaging was then performed typically over 2×10^4 time steps.

57. R. N. Barnett, U. Landman, C. L. Cleveland and J. Jortner (to be published).
58. A. Vicentini, G. Jacucci and V. P. Pandharipande, Phys. Rev. C31, 1783 (1985); R. J. Lenk and V. R. Pandharipande, Phys. Rev. C34, 177 (1986).
59a. J. E. Finn et al., Phys. Rev. Lett. 49, 1321 (1982); R. W. Minnisch et al., Phys. Lett. 118B, 458 (1982).
59b. See a comprehensive review by W. U. Schroder and J. R. Huizenga in Treatise on Heavy-Ion Science, Ed., D. A. Bromley, (Plenum, New York, 1984), Vol. 2, p. 115, and other articles in that volume.
60. H. Haberland, Surf. Sci. 156, 305 (1985); J. J. Saenz, J. M. Soler and N. Garcia, ibid. 156, 121 (1985).
61. D. Scharf, J. Jortner and U. Landman, Chem. Phys. Letts. 126, 495 (1986).
62. J. W. Brady and J. D. Doll, J. Chem. Phys. 73, 2767 (1980).
63. J. Jortner, E. E. Koch and N. Schwentner in: Photo-Physics and PhotoChemistry in the Vacuum UV, S. P. McGlynn et al., Eds., (Reidel, Dordrecht, 1985), p. 515.
64. J. Jortner, Ber. Bunsenges. Physik. Chem. 88, 188 (1984).
65. D. Scharf, U. Landman and J. Jortner, J. Chem. Phys. (to be published).
66. W. D. Kristensen, E. J. Jensen and M. J. Cotterill, J. Chem. Phys. 60, 4161 (1974).
67. I. Messing, B. Raz and J. Jortner, J. Chem. Phys. 66, 2239 (1977).
68. A. E. Sherwood and J. M. Prasunitz, J. Chem. Phys. 41, 429 (1964).
69. K. T. Gillen, R. P. Saxon, D. C. Lorentz, G. E. Ice and R. E. Olson, J. Chem. Phys. 64, 1925 (1975).

REACTIVE FLUX CORRELATION FUNCTIONS AND MONTE CARLO EVALUATION OF REAL TIME PATH INTEGRALS

William H. Miller
Department of Chemistry, University of California,
and Materials and Molecular Research Division,
Lawrence Berkeley Laboratory, Berkeley, California 94720

ABSTRACT

The reactive flux correlation function provides a "direct" way to express the rate constant for a chemical reaction, i.e., it avoids having to solve first the complete state-to-state reactive scattering problem (and then discard the state-to-state information). If the reaction dynamics is "direct", only the short time (of order $\hbar\beta$) quantum dynamics is necessary to determine the rate. Longer time ($t \gg \hbar\beta$) dynamics is required if the dynamics is not "direct", and a new methodology is described for the Monte Carlo evaluation of real time path integrals for time up to two orders of magnitude longer than $\hbar\beta/2$.

INTRODUCTION

A potentially powerful way for computing thermally averaged rate constants for chemical reactions is *via* use of the quantum mechanical flux-flux autocorrelation function (aka, the reactive flux correlation function),[1]

$$C_f(t) = \text{tr} [F e^{-\beta_c^* H} F e^{-\beta_c H}], \qquad (1)$$

where H is the Hamiltonian operator for the system, $\beta_c = \beta/2 + it/\hbar$ (t being the time and $\beta = (k_B T)^{-1}$), F is the flux operator

$$F = \frac{i}{\hbar} [H, h(s)]. \qquad (2)$$

© American Institute of Physics 1987

and tr denotes a quantum mechanical trace. Here h(s) is the usual step function

$$h(s) = \begin{cases} 1, & s>0 \\ 0, & s<0 \end{cases},$$

where s is the coordinate (the reaction coordinate) perpendicular to the "dividing surface" which separates reactants (s<0) from products (s>0). (For a standard cartesian Hamiltonian F is given by

$$F = \tfrac{1}{2}[\delta(s)\frac{p_s}{m} + \frac{p_s}{m}\delta(s)], \qquad (3)$$

where p_s is the momentum operation conjugate to s.) The thermal rate constant for a bimolecular reaction, for example, is given[1] simply by the integral of $C_f(t)$,

$$k(T) = Q_R^{-1} \int_0^\infty dt\, C_f(t), \qquad (4)$$

where Q_R is the partition function (per unit volume) of the non-interacting reactants. The purpose of this paper is to review the nature and properties of reactive flux correlation functions and to describe a new approach for calculating them via Monte Carlo path integral methods.

The reason this expression for the rate constant is so useful is that it provides a "direct" way for calculating the rate, i.e., it avoids having to solve first the complete state-to-state reactive scattering problem and then discarding all the state-to-state information by summing over all final states and Boltzmann-averaging over initial states. In addition to being <u>direct</u> it is also <u>correct</u>, i.e., fully quantum mechanical and with no dynamical approximations.[2] The only assumption (other than the Born-Oppenheimer approximation, and thus that the dynamics evolves on a single potential energy surface) is that the non-interacting reactants are in a Boltzmann distribution of their internal states and relative translational energy.

Another important feature of this approach is that $C_f(t)$

decays to zero much more quickly than the time necessary for dynamical evolution all the way from reactants to products. This is analogous to the situation in classical mechanics where one begins trajectories on the dividing surface and runs them forward and backward in time to see which are reactive:[3,4] one needs to follow the classical trajectories only for short times in order to determine the rate constant (the net reactive flux), whereas one would need to follow them all the way to the reactant and product asymptotic regions in order to obtain the more detailed state-to-state scattering information. In fact, in the limit of classical mechanics $C_f(t)$ decays to zero in <u>infinitesimal</u> time (i.e., it is proportional to $\delta(t)$) if the reaction dynamics is <u>direct</u>, i.e., if there are no classical trajectories that cross the s=0 dividing surface more than once.[5] Quantum mechanically the situation is not quite so simple; even in the limit of direct dynamics a finite amount of time (of order $\hbar\beta \simeq$ 25 femtosec at T = 300°) is required for $C_f(t)$ to decay effectively to zero. To obtain the correct rate constant quantum mechanically, therefore, it is necessary to determine the (quantum) dynamics of the system at least for short times.

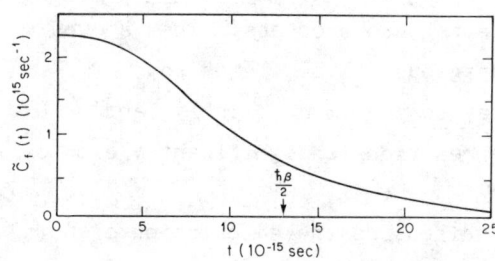

Figure 1. The flux-flux autocorrelation function for the 3-D H+H$_2$ reaction at T=300K.

Fig. 1 shows the reactive flux correlation function computed by Yamashita and Miller[6] for the H+H$_2$ → H$_2$+H reaction in 3-dimensions at room temperature (T = 300°). This is an example of direct dynamics, a case for which the classical approximation to $C_f(t)$ would be a delta function at t=0 and classical transition state theory thus exact. Fig. 1 shows how this is modified by quantum mechanics.

In passing it should also be noted that there are actually several different ways to define a reactive flux correlation function. If

$$C_f(t;\lambda) = \text{tr} \{\exp[(it/\hbar-\beta/2+\lambda)H] F \exp[(-it/\hbar-\beta/2-\lambda)H]F\} \quad (5)$$

then Yamamoto's[7] earlier definition of the reactive flux correlation function is

$$C_f^Y(t) \equiv \frac{1}{\beta} \int_{-\beta/2}^{\beta/2} d\lambda\, C_f(t;\lambda) \,. \quad (6)$$

Miller, Schwartz, and Tromp[1] noted that even though their $C_f(t)$, Eq. (1), and Yamamoto's,[7] Eq. (6), are different, the integrals of the two are the same, so that the same rate constants are obtained. (Actually, Miller et al.[1] noted that the integral of $C_f(t;\lambda)$ is independent of λ, so that the integral of $C_f(t;\lambda)$ yields the same rate constant for any value of λ.)

Tromp and Miller[8] have gone even further and generalized the definition of the flux operator in Eq. (2). They point out that the step function $h(s)$ can be replaced in Eq. (2) by any projector (projection operator) that projects onto products. They showed that this is particularly advantageous - i.e., leads to correlation functions which decay to zero more quickly, and without oscillations - when the reaction is significantly exo- or endo-thermic.

The remainder of the paper discusses the calculation of $C_f(t)$ by a path integral representation of the propagator/Boltzmann operators in Eq. (1).

SYSTEM-BATH HAMILTONIAN

Very often the reactive process of interest is characterized by a reaction coordinate s which is coupled to a number of harmonic modes. The generic form of this "system-bath" Hamiltonian is

$$H = \frac{P_s^2}{2m} + V_0(s) + \sum_k \left(\frac{P_k^2}{2m} + \frac{1}{2} m\omega_k^2 Q_k^2\right) - \sum_k Q_k f_k(s) . \quad (7)$$

The important practical feature is that in the $(P_k Q_k)$ degrees of freedom the Hamiltonian - or what is actually more relevant, the Lagrangian

$$L = \frac{m}{2} \dot{s}^2 - V_0(s) + \sum_k \left(\frac{1}{2} m \dot{Q}_k^2 - \frac{1}{2} m\omega_k^2 Q_k^2\right) + \sum_k f_k(s) Q_k^2 \quad (8)$$

- is a quadratic function of the coordinates Q_k.

A physically realistic Hamiltonian that is of this general form is the "reaction path" Hamiltonian of Miller, Handy, and Adams,[9] which has been used to describe a variety of dynamical processes in polyatomic molecular systems. The Lagrangian for this Hamiltonian is[6]

$$L = \frac{1}{2} \dot{s}^2 [1 + \sum_{k=1}^{F-1} Q_k B_{kF}(s) + \sum_{k,k'=1}^{F-1} Q_k Q_{k'} C_{k,k'}(s)]$$

$$- V_0(s) + \sum_{k=1}^{F-1} \frac{1}{2} \dot{Q}_k^2 - \frac{1}{2} \omega_k^2 Q_k^2 , \quad (9)$$

and though it is somewhat more complicated than that of Eq. (7), it still has the feature that the $\{Q_k\}$ coordinates appear only quadratically. (Here s is the reaction coordinate, the distance along the steepest descent path through the saddle point on the potential surface, and $\{Q_k\}$ are the local normal mode coordinates for motion orthogonal to the reaction path.)

Since the bath degrees of freedom, $(P_k Q_k)$, appear only quadratically, when the propagator/Boltzmann operators in the definition of the reactive flux correlation function, Eq. (1), are written in a Feynman path integral representation, the path integral over these bath degrees of freedom can be performed analytically, without approximation.[10] For the system-bath Lagrangian of Eqs. (8), and also (9), this gives[6,11]

$$C_f(t) = \left(\frac{\hbar}{2m}\right)^2 \left(\frac{\partial^2}{\partial s_N \partial s_0'} + \frac{\partial^2}{\partial s_0 \partial s_N} - 4\frac{\partial^2}{\partial s_0 \partial s_0'}\right)$$

$$\cdot \int_{s_N}^{s_0'} D\bar{s}(\tau) \int_{s_0}^{s_N} Ds(\tau) \, \exp\left[-\frac{1}{\hbar}\int_{it/\hbar}^{\frac{\hbar\beta}{2}} d\tau \, \frac{m}{2}\dot{\bar{s}}(\tau)^2 + V_0(\bar{s}(\tau))\right]$$

$$\cdot \exp\left[-\frac{1}{\hbar}\int_{-\frac{\hbar\beta}{2}}^{it/\hbar} d\tau \, \frac{m}{2}\dot{s}(\tau)^2 + V_0(s(\tau))\right] \cdot Z[s(\tau),\bar{s}(\tau)],$$

$$s_N = s_0' = s_0 = 0 \qquad\qquad (10)$$

where Z is the <u>partition functional</u> for the bath degrees of freedom. If the coupling terms $\{f_k(s)\}$ in Eq. (7) were zero, then Z would be simply the partition function for the (F-1) harmonic bath modes, but for non-zero coupling it is a functional of the two paths $s(\tau)$ and $\bar{s}(\tau)$ that must be path integrated over. Ref. 6, for example, gives the explicit expression for Z.

The important feature of Eq. (10) is that one is now left with only a one-degree-of-freedom quantum mechanical calculation, but it is a difficult problem because of the influence functional Z. Thus, it is not possible to convert this one degree of freedom path integral into a one dimensional Schrödinger equation; one must evaluate the path integral itself.

If the integrand of the path integral in Eq. (10) involved a purely <u>real</u> exponent - which it does if t=0 - then relatively well developed Monte Carlo methods could be used. Indeed, Monte Carlo path integrals for the Boltzmann operator $\exp(-\beta H)$ are now routinely calculated for relatively complex many-body systems.[12-17] However, this is in general not possible for t>0 because the integral is oscillatory. To circumvent this we have to date used an analytic continuation scheme.[1,6,11] Namely, t is taken to be purely imaginary,

$$it = \bar{\tau} \text{ (real)},\qquad(11)$$

so that the integrand in Eq. (10) is purely real and amenable to Monte Carlo integration. Values of $C_f(t) \equiv C_f(-i\bar{\tau})$ are calculated for various real values of $\bar{\tau}$ in the interval $-\frac{\hbar\beta}{2} < \bar{\tau} < \frac{\hbar\beta}{2}$, and these values are used to generate a Padé approximant which provides values of $C_f(t)$ for real t. Similar methods have also been used by Berne and his co-workers.[18]

These numerical analytic continuation methods work reasonably well for real values of t up to $\sim \frac{\hbar\beta}{2}$, but they become unstable for much larger times. Another approach, used by Behrman and Wolynes[19] is simply to use the real part of the complex exponential in the integrand as a Monte Carlo weighting function and to trust that the imaginary part is not too oscillatory for accurate Monte Carlo integration. This approach has given reasonable results also for short times, no longer than $\sim\hbar\beta/2$.

These short time results for the flux correlation function are sufficient for obtaining quantum corrections to transition state theory and thus constitute a significant advance. As Tromp and Miller[8] have discussed, the rate obtained from them can be used to obtain an approximate upper bound to the rate constant, so that one can invoke the variational aspect of classical transition state theory[20] and thus vary the location of the dividing surface to obtain the minimum (and thus best) possible rate constant. This all permits a quantum mechanical extension of transition state theory that is completely free of assumptions of separability of the reaction coordinate degree of freedom and <u>ad hoc</u> models for including tunneling corrections.

There are situations, however, where one needs to be able to evaluate the complex time propagator/Boltzmann operator $e^{-\beta_c H}$ for real times longer than $\hbar\beta/2$. This would be necessary, for example, to determine dynamical corrections to transition state theory, the quantum analog to classical trajectories that re-cross the dividing surface. The next section describes a very new approach by which it is indeed possible to carry out Monte Carlo evaluation of path integrals for real time about two orders of magnitude longer than $\hbar\beta/2$.

MONTE CARLO PATH INTEGRALS IN REAL TIME

Here we consider the complex time propagator for a one-dimensional system,

$$\langle x_N | e^{-\beta_c H} | x_0 \rangle = \left(\frac{mN}{2\pi\hbar^2\beta_c}\right)^{\frac{N}{2}} \int dx_1 \int dx_2 \ldots \int dx_{N-1}$$

$$\exp\left[\frac{-mN}{2\hbar^2\beta_c} \sum_{k=1}^{N} (x_k - x_{k-1})^2 - \frac{\beta_c}{N} \sum_{k=0}^{N} w_k V(x_k)\right], \quad (12a)$$

where

$$w_k = 1 \text{ for } k \neq 0, N$$

$$w_0 = w_N = \tfrac{1}{2}$$

and

$$\beta_c = \beta/2 + it/\hbar. \quad (12b)$$

(Since the approach we describe is a Monte Carlo method, it is changed little if an influence functional is added to the integrand in Eq. (12).) N, the number of "time slices", in Eq. (12) may be large, so that this is a many-dimensional integral even for one degree of freedom, thus making a Monte Carlo approach natural. As noted above, though, for real time dynamics (t>0) the integrand is oscillatory so that applications to date have only been possible for short times $\leq \hbar\beta/2$. In this section we describe a new approach[21] that is successful for t>>$\hbar\beta/2$. (Note that $\frac{\hbar\beta}{2}$ = 13 femtosec at T = 300°k, so that 1 psec corresponds to

$t/\frac{\hbar\beta}{2} = 80$.)

As a first step it is useful to make a linear transformation of the integration variables in Eq. (12a) to diagonalize the kinetic energy; this gives Coalson's[22] quasi-Fourier representation,

$$\langle x_N | e^{-\beta_c H} | x_0 \rangle = \langle x_N | e^{-\beta_c H_0} | x_0 \rangle e^{-i\phi(N-1)/2}$$

$$\cdot \int_{-\infty}^{\infty} da_1 \cdots \int_{-\infty}^{\infty} da_{N-1} \exp[-\pi e^{-i\phi} \sum_{k=1}^{N-1} a_k^2 - \Delta b \, e^{i\phi} \sum_{k=0}^{N} w_k V(x_k)], \quad (13)$$

where the coordinates $\{x_k\}$ are given in terms of the integration variables $\{a_k\}$ by

$$x_k = \frac{x_0(N-k) + x_N k}{N} + \sqrt{\frac{\pi \hbar^2 \Delta b}{mN}} \sum_{k'=1}^{N-1} a_{k'} \frac{\sin(\pi k k'/N)}{\sin(\pi k'/2N)} \quad (14)$$

with

$$\Delta b = |\beta_c|/N = [(\beta/2)^2 + (t/\hbar)^2]^{\frac{1}{2}}/N \quad (15a)$$

$$\phi = \arg(\beta_c) = \tan^{-1}(t/\frac{\hbar\beta}{2}), \quad (15b)$$

and where the first factor in Eq. (13) is the free particle matrix element

$$\langle x_N | e^{-\beta_c H_0} | x_0 \rangle = (\frac{m}{2\pi\hbar^2 \beta_c})^{\frac{1}{2}} \exp[-\frac{m}{2\hbar^2 \beta_c}(x_N - x_0)^2]. \quad (16)$$

The next step is to change integration variables so as to make the kinetic energy term in the exponent of the integral purely real and negative. New integration variables $\{c_k\}$ are thus defined by

$$c_k = a_k e^{-i\frac{\phi}{2}}, \qquad (17)$$

so that Eqs. (13) and (14) become

$$\langle x_N | e^{-\beta_c H} | x_0 \rangle = \langle x_N | e^{-\beta_c H_0} | x_0 \rangle$$

$$\times \int_{-\infty}^{\infty} dc_1 \ldots \int_{-\infty}^{\infty} dc_{N-1} \exp[-\pi \sum_{k=1}^{N-1} c_k^2] \exp[-\Delta b \, e^{i\phi} \sum_{k=0}^{N} w_k V(x_k)], \qquad (18)$$

where

$$x_k = \frac{x_0(N-k) + x_N k}{N} + e^{i\frac{\phi}{2}} \sqrt{\frac{\pi \hbar^2 \Delta b}{mN}} \sum_{k'=1}^{N-1} c_{k'} \frac{\sin(\pi k k'/N)}{\sin(\pi k'/2N)}. \qquad (19)$$

In writing Eq. (18) the integration path for the $\{c_k\}$ variables has been taken as the real axis $(-\infty, \infty)$, whereas Eq. (17) shows that the path originally is $(-\infty e^{-i\phi/2}, \infty e^{-i\phi/2})$; Eq. (18) thus assumes that the potential energy is sufficiently well-behaved that this distortion of the integration axis is valid.

Eq. (18) is now to be evaluated by Monte Carlo with the Gaussian factors $\exp[-\pi \sum_k c_k^2]$ (i.e., the kinetic energy) as weight function; i.e., the variables $\{z_k\}$,

$$z_k = \int_{-\infty}^{c_k} d\bar{c}_k \, e^{-\pi \bar{c}_k^2} \qquad (20)$$

are chosen as random numbers in the interval $(0,1)$ and the variables $\{c_k\}$ determined by Eq. (20). This is discussed more fully in earlier work.[23] The coordinates $\{x_k\}$ of Eq. (19), which appear in the potential energy factor of Eq. (18), are now complex, so it is necessary that the potential $V(x)$ be given as an analytic function of x which permits evaluation at complex x. The potential energy term in the exponent of Eq. (18) will thus in general be complex, so that one is still attempting to carry out Monte Carlo integration with an oscillatory integrand. To be successful, therefore, it must be true that the oscillations of the integrand are not too severe.

Although our ultimate interest in the present methodology is to evaluate reactive flux correlation functions, here we illustrate the basic features of the method by computing the diagonal propagator matrix element ($x_0 = x_N = 0$) of the complex β_c propagator. For a typical barrier potential, we use the Eckart function,

$$V(x) = V_0 \operatorname{sech}^2(\alpha x) \qquad (21)$$

with V_0 = 0.016 hartrees ($\approx$ 10 kcal/mole) and $\alpha = 2a_0^{-1}$. The mass of the particle is chosen to be that of an H atom, m = 1837 m_e. With these choices the imaginary frequency at the top of the barrier is $\approx$ 1832 cm^{-1}.

All calculations are for $\frac{\hbar\beta}{2}$ = 500 (in atomic units), corresponding to T $\approx$ 300°K, and for the present example the working expressions are Eqs. (18) and (19).

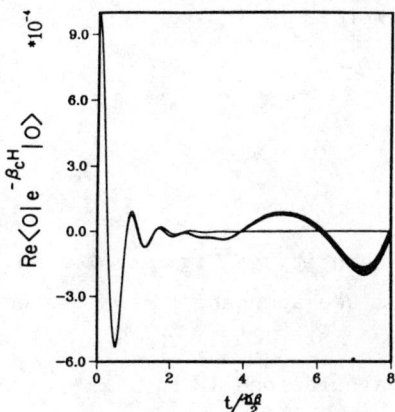

Figure 2. The real part of the matrix element $\langle 0|e^{-\beta_c H}|0\rangle$, $\beta_c = \beta/2 + it/\hbar$ for the Eckart potential barrier, eq. (21), as a function of time t (in units of $\hbar\beta/2$). N (the number of time slices) = 5. The width of the line is the usual Monte Carlo error estimate of the result.

Figure 2 shows the result for the real part of the diagonal

matrix element $\langle 0|e^{-\beta_c H}|0\rangle$ as a function of t for N=5 "time slices"; the imaginary part of the matrix element is similar. (The width of the broad lines in Figs. 2-4 gives the Monte Carlo error estimate.) For this value of N one obtains the correct result for $t/\frac{\hbar\beta}{2}$ up to ~2.

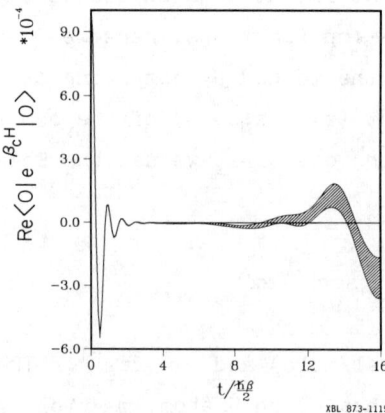

Figure 3. Same as Fig. 2 except N=11.

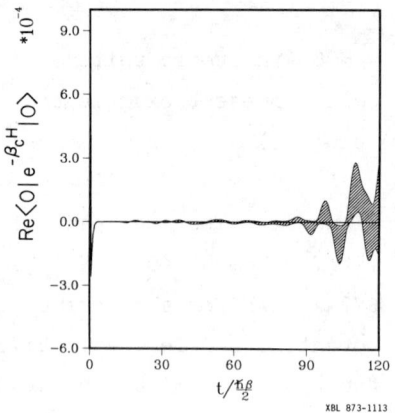

Figure 4. Same as Fig. 2 except N=101.

Figs. 3 and 4 show the same quantity as Fig. 2 N=11 and 101, respectively, and here the results are accurate for values up to $t/\frac{\hbar\beta}{2} \approx 6$ and ≈ 80.

The success shown by these calculations in Fig. 2-4 is quite spectacular; by a straight-forward Monte Carlo method, using only the kinetic energy as the Monte Carlo reference distribution, one obtains accurate results, with a reasonable number of Monte Carlo sample points, for real times much larger than $\frac{\hbar\beta}{2}$; for Fig. 4 the result is accurate for times up to t≈1 psec. Moreover, the important feature of the calculation is that it would be changed in no essential way if the integrand were augmented by a non-local influence functional that come from integrating out other degrees of freedom coupled to the present one. Alternatively, since the method is straight-forward Monte Carlo, it can be applied to systems with many degrees of freedom.

CONCLUDING REMARKS

For short times, of order $\hbar\beta$, there are a variety of methods for evaluating complex time propagator/Boltzmann operators and thus the reactive flux correlation function, e.g., direct Monte Carlo using only the real part of the kinetic energy part of the integrand of the path integral, or by numerical analytic continuation from Monte Carlo calculations of $C_f(t)$ at pure imaginary times. (For systems of only one or two degrees of freedom it is also possible to use the numerical matrix multiplicative algorithm or a basis set representative of the Hamiltonian.) Such calculations are sufficient to describe "direct" reaction dynamics, i.e., situations for which transition state theory is correct classically. One may thus say that determining the quantum dynamics for times of order $\hbar\beta$ defines a quantum transition state theory. This is very much in agreement with the semiclassical version of transition state theory[24] that involves periodic orbits ("instantons") through the transition state region.

Corrections to transition state theory, however, require longer time dynamics, and a new scheme has been described herein which allows Monte Carlo evaluation of the complex time propagator/Boltzmann operator for real times much longer than $\hbar\beta$. The essential trick, i.e., distorting the integration path to one along which the integrand is less oscillatory, is very reminiscent of the method of steepest descent.[25] After the contour distortion, though, the present approach utilizes the Monte Carlo integration method rather than the analytic approximation of the steepest descent method. Also, in the present approach the integration path has not been chosen to completely eliminate oscillations of the integrand but only to diminish them. This analogy to steepest descent suggests the possibility that cleverer contour distortions than the one we have used may be found to make the Monte Carlo integration even more efficient.

The complex paths that result from the contour distortion also remind one of the semiclassical approximation to propagator matrix elements which involve tunneling, for there the relevant classical paths are typically complex-valued.[26] Thus the contour distortion may be sampling paths close to these classical paths more efficiently.

ACKNOWLEDGMENT

The calculations reported herein were performed at the Berkeley theoretical chemistry computing center, supported by National Science Foundation grant CHE84-16345. This work was also supported by the Director, Office of Energy Research, Office of Basic Energy Sciences, Chemical Sciences Division of the U.S. Department of Energy under Contract Number DE-AC03-76SF00098.

REFERENCES

1. W. H. Miller, S. D. Schwartz, and J. W. Tromp, J. Chem. Phys. __79__, 4889 (1983);
2. W. H. Miller, J. Chem. Phys. __61__, 1823 (1974).
3. P. Pechukas, in W. H. Miller, Ed., __Dynamics of Molecular Collisions__ B, Plenum, N.Y., 1976, p. 269.

4. J. C. Keck, J. Chem. Phys. 32, 1035 (1960); Adv. Chem. Phys. 13, 85 (1967).
5. D. Chandler, J. Chem. Phys. 68, 2959 (1978).
6. K. Yamashita and W. H. Miller, J. Chem. Phys. 82, 5475 (1985).
7. (a) T. Yamamoto, J. Chem. Phys. 33, 281 (1960);
   (b) see also, S. F. Fischer, ibid. 53, 3195 (1970).
8. J. W. Tromp and W. H. Miller, J. Phys. Chem. 90, 3482 (1986).
9. W. H. Miller, N. C. Handy, and J. E. Adams, J. Chem. Phys. 72, 99 (1980). For a review and references to other papers, see W. H. Miller, J. Phys. Chem. 87, 3811 (1983).
10. R. P. Feynman and A. R. Hibbs, Quantum Mechanics and Path Integrals (McGraw-Hill, New York, 1965), p. 273 et seq.
11. R. Jaquet and W. H. Miller, J. Phys. Chem. 89, 2139 (1985).
12. (a) M. Parrinello and A. Rahman, J. Chem. Phys. 80, 860 (1984);
    (b) C. D. Jonah, C. Romero, and A. Rahman, Chem. Phys. Lett. 123, 209 (1986).
13. R. A. Kuharski and P. J. Rossky, (a) Chem. Phys. Lett. 103, 357 (1984);
    (b) J. Chem. Phys. 82, 5164 (1985).
14. (a) A. Nichols III, D. Chandler, Y. Singh, and D. Richardson, J. Chem. Phys. 81, 5109 (1984);
    (b) M. Sprik, M. L. Klein, and D. Chandler, J. Chem. Phys. 83, 3042 (1985).
15. (a) J. Bartholomew, R. Hall, and B. J. Berne, Phys. Rev. B 32, 548 (1985);
    (b) A. Wallquist and B. J. Berne, Chem. Phys. Lett. 117, 214 (1985);
    (c) A. Wallquist D. Thirumalai, and B. J. Berne, J. Chem. Phys. 85, 1583 (1986).
16. J. D. Doll, R. D. Coalson, and D. L. Freeman, Phys. Rev. Lett. 55, 1 (1985); J. D. Doll and D. L. Freeman, Science 234, 1356 (1986).

17. See also the entire issue J. Stat. Phys. $\underline{43}$, Nos. 5/6 (1986).
18. (a) D. Thirumalai and B. J. Berne, J. Chem. Phys. $\underline{79}$, 5029 (1983); $\underline{81}$, 2512 (1984);
    (b) D. Thirumalai, E. J. Bruskin, and B. J. Berne, J. Chem. Phys. $\underline{79}$, 5063 (1983);
    (c) D. Thirumalai and B. J. Berne, Ann. Rev. Phys. Chem. $\underline{37}$, 401 (1986).
19. E. C. Behrman and P. G. Wolynes, J. Chem. Phys. $\underline{83}$, 5863 (1985).
20. D. G. Truhlar, W. L. Hase, and J. T. Hynes, J. Chem. Phys. $\underline{87}$, 2664 (1983).
21. J. Chang and W. H. Miller, J. Chem. Phys., to be published.
22. R. D. Coalson, J. Chem. Phys. $\underline{85}$, 926 (1986).
23. W. H. Miller, J. Chem. Phys. $\underline{63}$, 1166 (1975).
24. W. H. Miller, J. Chem. Phys. $\underline{62}$, 1899 (1975).
25. P. M. Moore and H. Feshbach, <u>Methods of Theoretical Physics</u>, McGraw-Hill, N.Y., 1953, pp. 434-443.
26. W. H. Miller, Adv. Chem. Phys. $\underline{25}$, 69 (1974); Science $\underline{233}$, 171 (1986).

SOLID AND LIQUID MOLECULES AND MICROCLUSTERS

R. Stephen Berry
Oxford University, Oxford OX1 3QZ, U. K.*

ABSTRACT

By identifying conventional near-rigid molecules with solids and very nonrigid molecules with liquids, one is led to a quantum-statistical argument, based on densities of states, that some clusters should exhibit solid-like behavior at low temperatures, liquid-like behavior at high temperatures and an intermediate range of temperature within which the two "phases" coexist. The "freezing" temperature, the lower bound for thermodynamic stability of the liquid, and the "melting" temperature, the upper limit of thermodynamic stability of the solid, are sharp, according to this model, but unequal. In the intermediate range, the two forms coexist like chemical isomers. Molecular dynamic (MD) simulations of isoergic clusters and both MD and Monte Carlo (MC) simulations of isothermal clusters of argon show that clusters of certain sizes exhibit this behavior. However for the individual "phases" to be observable, certain relations among several time scales (including that of the observations) must be satisfied. Not all sizes of clusters meet these requirements; for those which do not satisfy them, experiments must show the clusters as slush. Consideration of small liquid clusters forces us to scrutinize what is meant by "liquid" at the molecular level.

INTRODUCTION

The notion that small clusters might exhibit distinguishable solid-like and liquid-like forms arose from classical isoergic molecular dynamic (MD) simulations of the 1970's, largely of argon represented by pairwise Lennard-Jones interactions[1]. Results of the simulations of that time, reviewed recently[2], indicated clearly that solid-like behavior could be expected at sufficiently low energies and that liquid-like behavior would appear at sufficiently high energies. "Sufficient" took on a meaning that depended on the number N of atoms

* Permanent address: University of Chicago, Chicago, Illinois 60637

© American Institute of Physics 1987

in the cluster. What occurs between the temperatures characterized by well-defined single "phases" was ambiguous from the evidence of that time. In particular, the two methods of simulation gave quite different caloric curves--curves of mean temperature (vibrational kinetic energy) vs. energy, or of mean energy vs. temperature; Monte Carlo calculations[3] showed monotonic curves of $<T(E)>$ vs. E whose positive slope is smaller in an intermediate region than in either the low-energy, solid-like region or the high-energy, liquid-like region. Whether the slopes of these curves (the inverse of the heat capacity) are continuous could not be distinguished from the simulation data. Molecular dynamics (MD) simulations[4] showed S-shaped curves of $<T(E)>$ vs. E, with distinctive regions of negative slope connecting a local maximum with a local minimum at a higher energy. This behavior appeared for N>6 and was interpreted as indicative of a first-order phase transition. It is shown schematically in Fig.1.

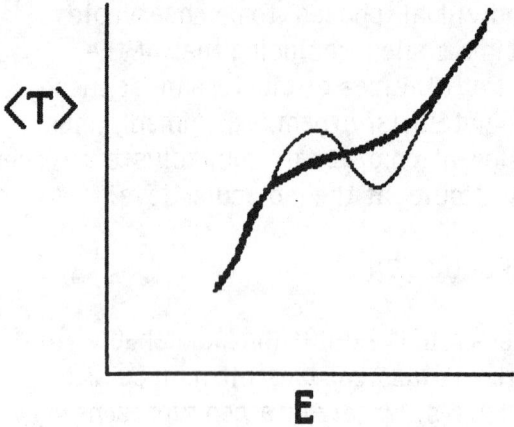

Figure 1. A schematic representation of the caloric curves derived from early Monte Carlo (heavy curve) and molecular dynamics (light curve) simulations of argon clusters. Different curves were derived for each value of N, the number of atoms in the cluster.

The striking results of these simulations inevitably raised the question of what the physical basis is for the existence of solid and the partition functions and thermodynamic quantities derived from their densities of states, the models need not be highly realistic, only good enough to describe the densities of states in the large.

3. Connect the states of the limiting cases into a correlation diagram by requiring that parity, total angular momentum and permutational symmetry be preserved for all intermediate conditions between the limits.

4. Quantify the abscissa of this correlation diagram in terms of a parameter of nonrigidity, γ, which serves a function much like a Landau order parameter; γ may be defined as a ratio of two specific spectroscopic intervals[8] so that $0<\gamma<1$ with the rigid limit at 0 and the nonrigid limit at 1.

5. Using the energy levels and degeneracies of the limiting cases, construct the partition functions $Q(T,\gamma)$ and, from them, the free energies $F(T,\gamma)$ for the limits for which we have models, which we identify as corresponding to $\gamma \approx 0$ and $\gamma \approx 1$.

6. Assume that most of the energy levels of the correlation diagram are monotonic and nonpathological functions of γ between the two limits.

These six steps and plausible models for the limiting cases are enough to lead us to the conclusions that a) we can rationalize the results of the 1970's simulations with a very spare, general representation of a cluster, and b) we are led to the prediction of sharp but unequal freezing and melting points for clusters. The latter prediction, we shall see, is a *permissive* prediction, not a proof of existence, in the sense that additional criteria must be satisfied if the predicted behavior is to be observable. However we can call upon the widely-accepted Gell-Mann's Law, which states that anything in nature that is not expressly forbidden will be found. Somewhere.

The logic following from the six steps just given goes this way; we only outline this since the full arguments are given elsewhere[2,7-9].

1. For virtually all reasonable models of the limiting cases, the density of states of the near-rigid, solid-like form is higher, in the lowest range of energies, than that of the nonrigid, liquid-like form. The weakest assumption about the zero-point energies of the two forms is that they are equal; with that assumption,

the rotational states of the solid are enough to account for the higher density. However any more realistic assumption requires that there is some energy of fusion, some energy required to promote the cluster from the solid to the liquid form; with this assumption, the solid may also have vibrational levels below the zero-point energy of the liquid.

2. The densities of states of both the solid-like and liquid-like limits increase with energy but the density of states of the liquid increases much faster, e.g. roughly as N^3 if both forms are treated as Einstein (completely degenerate) oscillator systems[7], so that at sufficiently high energies, the density of states of the liquid-like form becomes mush larger than that of the solid-like form.

3. Because of the relation between the densities of states of the two limits, the low-energy states have positive slope as functions of γ. Similarly, the overwhelming proportion of the high-energy states have negative slopes as functions of γ. Fig. 2 is such a correlation diagram, for the simplest example.

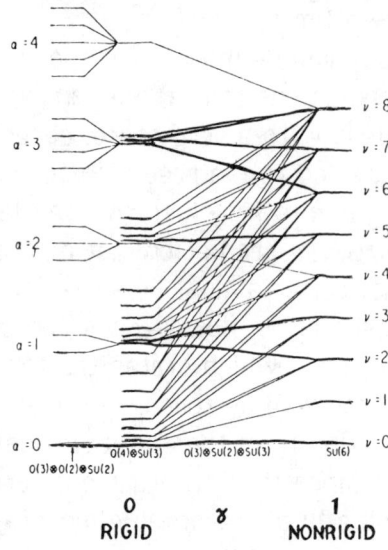

Fig. 2. A correlation diagram connecting the energies of the limiting rigid and nonrigid limits for a cluster of three identical atoms with an equilateral near-rigid structure.

4. Consequently the canonical partition function $Q(T,\gamma)$ is a monotonic, decreasing function of γ at low temperature but, at sufficiently high temperature, becomes a monotonic, increasing function of γ. Corresponding to this, the free energy $F(T,\gamma)$ is a monotonic, increasing function of γ for low T but for sufficiently high T is monotonic and decreasing. This means that at low temperatures, the free energy has only one minimum, which is at or near $\gamma=0$, so that at low temperatures, only the solid form is thermodynamically stable. Similarly, at sufficiently high temperatures, $F(T,\gamma)$ has a single minimum, near or at $\gamma=1$, so that only the liquid form is stable.

5. As the temperature is raised from the low range where only the solid is stable, the density of states near $\gamma=1$ increases faster than that near $\gamma=0$, so that $Q(T,0)$ increases slower with T than $Q(T,1)$, and, by the same token, $F(T,\gamma\approx 0)$ decreases slower with T than $F(T,\gamma\approx 1)$, and the curves of $F(T,\gamma)$ become more concave toward the γ-axis as T increases.

6. As the temperature increases, it reaches a value T_f at which the free energy $F(T,\gamma)$ develops a slope of zero at some value of γ at or near 1; that is, $[\partial F(T,\gamma)/\partial \gamma]_{T=T_f} = 0$. At temperatures above T_f, the curves of $F(T,\gamma)$ have a minimum at or near $\gamma=1$, so that for some range of temperatures above T_f, $F(T,\gamma)$ must have two minima. Therefore in this range, there should be *two* stable forms of the cluster, one corresponding to a solid, for which $\gamma\approx 0$, and the other, corresponding to a liquid, for which $\gamma\approx 1$.

7. As the temperature increases still further, it reaches a value T_{eq} for which the free energies at the two minima are equal: $F_{min}(T_{eq},\gamma\approx 0) = F_{min}(T_{eq},\gamma\approx 1)$. This equality insures that the two "phases" are present in equal amounts in a mixture containing both at equilibrium at this temperature.

8. Because $F(T,\gamma)$ becomes monotonic decreasing at sufficiently high T, and is a smooth function of both T and γ, there must be a temperature T_m at which the minimum near $\gamma=0$

turns into a point of zero slope; above T_m only the liquid-like form is thermodynamically stable. Below T_m, both solid-like and liquid-like forms are stable. Fig. 3 is a sketch of the behavior of $F(T,\gamma)$ for several values of T, from a T_1 for which only the solid is stable, through T_f, T_{eq}, an arbitrary T_4 for which both forms are stable but the liquid is predominant, T_m, and finally a T_6 for which only the liquid is stable.

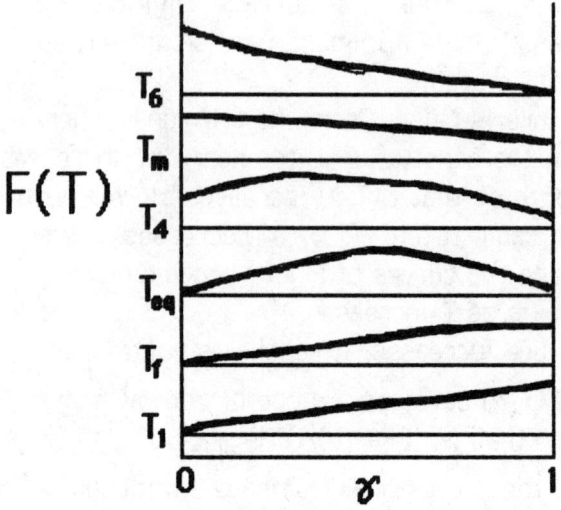

Fig. 3. A schematic representation of the free energy $F(T,\gamma)$ for several values of the temperature.

9. Hence our logic leads us to the inference that clusters of a given size N can be expected to have a sharp lower bound of temperature for stability of the liquid-like form and a sharp upper bound for the stability of the solid-like form. We can call these bounds, T_f and T_m, the freezing and melting temperatures, respectively, because no liquid exists below T_f and no solid exists above T_m in a system at equilibrium. Expressed this way, as the limits of existence for single phases, freezing and melting temperatures seem quite independent, with no necessary relation between them. Moreover we seem to find that clusters should have *sharp but unequal freezing and melting temperatures!*

The implications of this prediction are in some ways natural and not surprising, at least after the fact. Interpreted in light of this result, the identity of freezing and melting temperatures for bulk matter is a consequence of the enormous difference in total free energies of the two phases at any temperature even slightly away from T_{eq}. For a small cluster, in contrast, the difference $\Delta F(T,N)$ in total free energies of solid and liquid forms of an N-cluster may be a small fraction of $N \times \Delta E_b$, i.e. by a small fraction of the difference in total binding energies of the solid and liquid clusters, which need not be very different from kT at temperatures detectably different from T_{eq}. This means that the equilibrium ratio [solid]/[liquid], which is $K_{eq}=\exp[\Delta F(T,N)/kT]$, may well be a measurable quantity, rather than effectively infinite or infinitesimal, over a temperature range wide enough to be studied. In that range, we may think of the two forms as two coexisting phases, or as two isomers in equilibrium.

A concept useful for visualizing the two forms is that of the solid cluster as a tightly-bound, close-packed polyhedron or lattice, and the liquid as something produced by promoting at least one particle from the close-packed structure to an outer layer, leaving a vacancy. Then both the promoted particle and the hole are far more mobile than the particles of the closed-shell state before promotion. As a result (and we must examine this next "logical" step in more detail shortly), the cluster with the promoted particle is liquid-like and the closed-shell form is solid-like. In more quantum-mechanical terms, the solid-like form exists in a region of configuration space corresponding to a narrow, confining potential well, which in real systems is likely to be rather deep, where the energy levels are low-lying but widely separated. In the liquid-like region, the levels are closely spaced because the configuration space available to the nonrigid system is very large. The energies of even the lowest levels of the liquid-like system are, in realistic systems, likely to be well above the lowest levels of the solid-like system. (Our weakest-case model mentioned previously did not invoke this difference in energies.) This case is illustrated in Fig. 4.

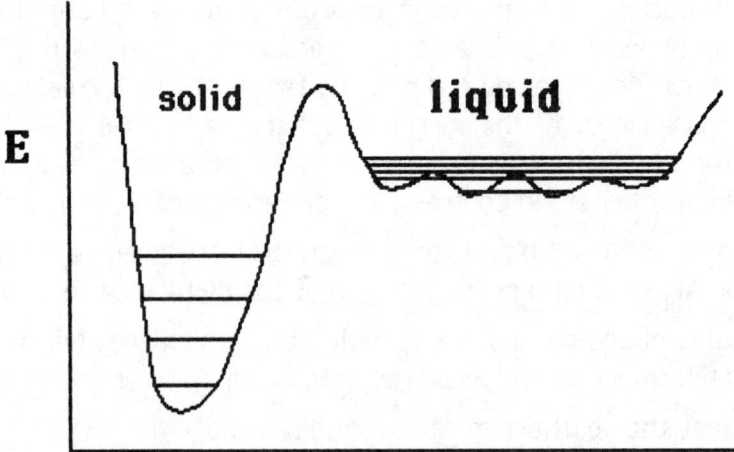

Fig. 4. A schematic representation of the effective potential of a cluster, showing regions of configuration space that correspond to solid-like and liquid-like behavior.

Not all clusters need have potentials that exhibit such distinguishable solid-like and liquid-like regions as that of Fig. 4, which corresponds rather well to Ar_{13} and some other clusters. The surface may have two or more kinds of minima analogous to different solid phases, as Fig. 5 shows; Ar_7, for example, has four distinct kinds of minima. In Fig. 5, the two solid-like minima are still fairly different from each other and from the liquid-like region but this need not be the case. The two kinds of regions could merge into each other without a clear distinction between different kinds of regions on the surface.

One otherpoint needs discussion here; that is the meaning of the nonrigidity parameter γ and the minima in $F(T,\gamma)$ that correspond to stable forms. A point of stability for one form of cluster corresponds to a specific representation around which the system's Hamiltonian can be developed to achieve a fairly accurate description of the rotational-vibrational-electronic state of the system. The minimum in $F(T,\gamma)$ at or near $\gamma=0$ corresponds to a Born-Oppenheimer representation with the Hamiltonian developed in a power series in $(m_e/M_{nuc})^{1/4}$ around a configuration at which the effective potential energy is a minimum.

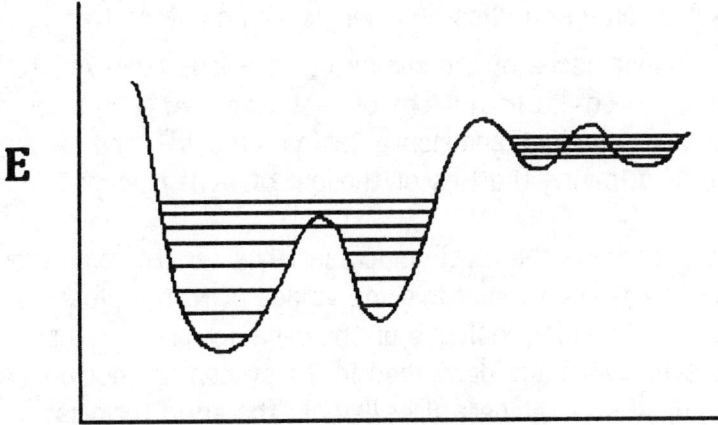

Fig. 5. A schematic representation of a potential surface for a cluster having two solid-like forms and a liquid-like form which is attainable only at temperatures well above that at which isomerization between the solid phases is allowed.

The minimum near $\gamma=1$ corresponds to an altogether different picture, a mean field expansion in which the description is, in lowest order, that of independent particles in the average field of all the others. A small-amplitude oscillator model is ineffective for a liquid, with its large-scale motions and soft modes. Yet both the liquid and the solid must ultimately be described by the same Hamiltonian. The two sets of states simply correspond to different sets of approximately good quantum numbers and therefore to different first approximations.

RECENT SIMULATIONS: RESULTS AND INSIGHTS

The prediction of a finite temperature range of coexistence is still not easy to test in the laboratory (but the next section mentions what has become available) so for this reason alone it would be worthwhile to do simulations. In fact simulations ought to be valuable for helping to prepare and design experiments as well. Some preliminary results oriented specifically toward phase changes and coexistence in clusters have been announced for alkali halides[10] and for silicon[11]; much more extensive results are available for simulated rare gas clusters. All to date are purely classical. These include detailed molecular dynamics studies of isoergic Ar_N clusters for $N=13$[12] and $N=7$[13], comparisons of argon clusters of sizes from Ar_6 to Ar_{33} by the same approach[14], simulations of isothermal Ar_{13} by both (Nosé-type) molecular dynamics and Monte Carlo methods[15], and quenching studies[16] following the line of thought of Stillinger and Weber[17].

A first result, perhaps the most important thus far, to come from the simulations is the finding that for some values of N, the clusters exhibit precisely the classical analogue of the behavior predicted by the quantum statistical analysis described in the preceding section. At low energies[12-14] or at low temperatures[15], only the solid form is found. At sufficiently high energies or temperatures, only the liquid form is found. For N=7, 9, 11, 13, 15 and 19, there are intermediate ranges of energy for which the two forms coexist in dynamic equilibrium. By this we mean the following. A molecular dynamics simulation is run for a cluster with no net translational or rotational energy; the vibrational temperature of the cluster is determined regularly, typically by averaging the kinetic energy over 500 steps, each of 10^{-14} sec. In the single-phase ranges of total energy, the distributions of temperature are unimodal; in the coexistence range, the temperature distribution is *bimodal*. According to several criteria, in the isoergic MD simulations, the forms present at low energies are solid-like, the forms present at high energies are liquid-like and in the intermediate range of energies, the high-temperature contribution to the bimodal distribution comes from hot, solid-like clusters and the low-temperature contribution, from cold, liquid-like clusters. These correspond of course to clusters in regions of deep potentials and of high-energy, shallow potentials, respectively, as indicated in Fig. 4. The isothermal studies have been done in detail only for N=13; for this system, the same behavior occurs but of course the distributions are those of the total energy rather than the temperature.

From the simulations one can develop $\langle T(E) \rangle$ or $\langle E(T) \rangle$, that is, the caloric function. For the unimodal regions of E or T, there is no question of how to take the desired mean. However this may be done in either of two ways for the regions in which the distributions are bimodal, for those cases exhibiting bimodality. One may divide the distributions into two Gaussians (and the distributions are indeed well fit by Gaussians), one around each of the peaks, and take two separate averages, or one may construct a single, long-time average over the entire distribution. When these two methods are used, one obtains results of the sort shown in Fig. 6 for Ar_{13}. By using the distributions about the two modes separately, we obtain smooth continuations of the solid-like and liquid-like branches, corresponding to a double-valued equation of state in the coexistence range of energy and temperature.

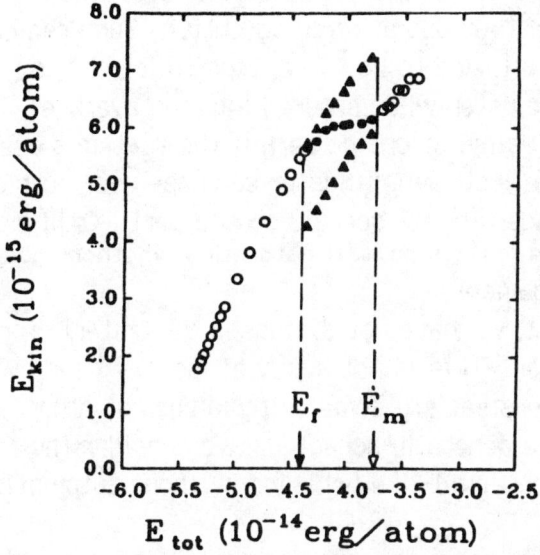

Fig. 6. The caloric function $\langle T(E) \rangle$ for Ar_{13}; open circles are used for single-phase regions; closed circles, for one long-time average at each T and triangles, for separated averages for each Gaussian peak in the bimodal region.

By using a single, long-time average at each energy, we obtain a monotonic curve but with a distinctive region of low slope connecting the simple, one-phase portions of the curve. With long enough runs, the MD simulations produce no regions in which $\partial \langle T(E) \rangle / \partial E < 0$; with short runs the S-shaped curves of Briant and Burton could be reproduced. When MC calculations are used with T as the independent variable and $\langle E \rangle$ the dependent function of T, the points so generated fall on the same curve, with the possible exception of points in the low-energy, low-temperature portion of the coexistence region. The (two) MC points generated in this region for Ar_{13} are very slightly lower than the points shown in Fig. 6, just about one diameter of the circles shown. Whether this is due to finite sampling errors or to a systematic consequence of the differences in the two distributions is not yet clear. The isoergic MD method of course yields trajectories that cannot penetrate into regions closed by saddles higher than the system's energy, so that the cluster may be prohibited from entering some regions where the potential is below the system's energy, especially regions of high but allowed potential. Such regions would, if the system were allowed to go there, tend to lower the mean temperature of the cluster when entered into the average. The isothermal MC or MD simulations do permit the system's energy to fluctuate so that the isothermal cluster can pass over barriers and reach regions forbidden to its isoergic counterpart. Only more detailed examination of the simulations will determine whether these small differences are fundamental.

Until this point, we have not discussed the criteria for liquid-like or solid-like behavior. This is not a trivial issue; it may well turn out that some clusters satisfy some criteria but not others, even among those that are generally accepted. We consider the following *possible* criteria for liquid-like behavior, without automatically accepting them:

1. passage from the well around one potential minimum to another on a time scale roughly comparable with the time scale for vibrations within a single well[7]--perhaps within a factor of 100 or 1000, rather than 10^8 or 10^{10} or more;
2. a high density of soft modes, with characteristic frequencies close to zero;
3. a diffusion coefficient well above zero;
4. a structure which, at every instant, appears amorphous or shows no significant order, long- or short-range;
5. an irregular, aperiodic or nonpolyhedral structure when the cluster is quenched to the potential minimum about which it is moving at any arbitrary moment;
6. a (Hausdorff) dimensionality for the hypersurface on which the system's phase point moves of approximately $6N-7$, corresponding to no constraints on the N-particle cluster except conservation of

momentum, energy and angular momentum.
We may be able to think of others as well. The corresponding criteria for solid-like behavior would be:

1'. passage among wells on a time scale much, much slower than the periods of vibration within a single well;
2'. a low- or zero-density of modes having very low frequencies;
3'. a diffusion coefficient that is very small or zero;
4'. a structure that is rarely far from a periodic or regular polyhedral form, so rarely that instantaneous snapshots display only small deviations from the recognizable, regular structure;
5'. a regular periodic or polyhedral structure when quenched to the nearest potential minimum;
6'. a (Hausdorff) dimensionality for the hypersurface on which the cluster's phase point moves which is equal to or only slightly larger than 3N-3 corresponding to one dimension for each independent harmonic oscillator.

If we consider conformability and fluidity as key characteristics of a liquid, then the first three criteria all should be accepted. The fourth is not so clear, particularly when we ask about liquid crystals. We shall not try to use criterion #4, although Briant and Burton did invoke it[4] with others in making their judgment that clusters can be liquid. The question of glassy, amorphous solids always enters if one tries to use this criterion. The fifth criterion, in our view, is not satisfactory. The quenching studies done by Stillinger and Weber[17] show unambiguously that some systems essentially always quench to nearly regular structures no matter how liquid-like or even dense-gas-like they are at the beginning of the quenching process. Put another way, we cannot expect a liquid to *avoid* regions in which the potential minimum corresponds to a regular structure. It may be that the liquid-like system does not, under some conditions, explore the regions of regular structures; the argument we introduced in connection with Fig. 4 would make that the case. However even in the situation represented by that Figure, if the energy of the system is greater than the barrier separating liquid-like and solid-like regions, the system is surely liquid-like, with the capability of exploring the entire configuration space. As for the last criterion, the answer to whether it is either valid or useful is still not known, but will be soon, we hope.

The first three criteria are well satisfied for the clusters we have examined by isoergic MD simulations and have found to exhibit bimodal temperature distributions: when the appropriate quantities, the passage rate from well to well[13], the density of soft modes[14,16] (from the Fourier transform of the velocity autocorrelation function) and the diffusion coefficient[14,16] (from the mean square displacement per unit time) are evaluated for each "phase" separately, they do indeed have the behavior familiar for that kind of phase. For cases that do not exhibit bimodal temperature distributions, there is of course no way of separating liquid-like and solid-like forms in what would be the coexistence region. There is, nevertheless, a criterion for *melting* associated with but not identical to the appearance of a bimodal temperature distribution, which also seems applicable as a signal of a transition of some kind for the cases with only unimodal temperature distributions. This is the sharp increase in δ, the root-mean-square bond length fluctuation[2,3,12,14], the old Lindemann criterion. Remarkably, the sharp increases in $\delta(N,T)$ with temperature or energy seem to occur from values of δ just below 10% of the mean nearest-neighbor distance; 10% was the value conjectured by Lindemann as the point above which melting would set in.

Although solid and liquid forms of Ar_{33} are unambiguously recognizable[14], no coexistence range has been resolved for this cluster. Larger clusters, over 200 argon atoms, clearly exhibit sharp changes that seem to be like first-order phase transitions between solid and liquid forms[18]. Very recently, MC calculations for an altogether different model[19] have also shown a range of energies and temperatures over which a solid-like and a liquid-like phase coexist, and show a bimodal distribution like those described above. This is a 10-state, two-dimensional Potts model for L×L lattices with L from 18 to 50. Furthermore it appears that for N of 55 or more, clusters of at least some sizes can exhibit liquid surfaces around solid cores[20]. However no generalizations can yet be made concerning the existence of clusters that are inhomogeneous with respect to phases.

TIME SCALES, DYNAMICS AND EQUILIBRIUM

The foregoing discussion alluded to the permissive character of the implication from the quantum statistical analysis that there could be a finite region of coexistence of liquid and solid clusters. The point was also made that clusters of only certain sizes exhibit such finite regions in the form of energy or temperature ranges within which the complementary dependent variable, temperature or energy respectively, has a bimodal distribution. Now we return to this issue and examine in more detail the conditions that must be met in order for us to observe coexisting phases in clusters--or, for that matter, in bulk matter.

If we accept criteria 1-4 and perhaps 6 as sufficient individually or together to recognize a liquid, then we implicitly accept the idea that being liquid has something to do with behavior *in time*, not merely with a property at an instant. The capacity of a liquid to conform certainly is related to its responses on the time scale of molecular movement. If, in trying to go swimming, I dive flat onto the surface of the water, I am quickly made aware that I can deliver an impulse to the water faster than it can respond by conforming; the water, under those conditions, delivers a noticeable impulse back to me! In short, it takes a finite amount of time for a substance to exhibit the properties we associate with "being liquid".

Accepting this condition leads us immediately to a set of related conditions on the relations among several time scales which are just what we seek. For a cluster to exhibit two phases in equilibrium, we have already found two conditions which are not related to time scales:

1. The stable form of the system at low temperatures must be solid, unlike the alkali trimers for example;
2. The system must be finite if the two forms are to exist in equilibrium at any temperature away from T_{eq}, the temperature at which the two forms have equal free energies.

Now we confront the problem of time scales to state the third criterion:

3. The system must persist in each "phase" for a time long relative to the time needed to establish equilibrium properties of that "phase" and our measurements to determine those properties

must require times short relative to the time spent in each "phase".

If the system stays for long intervals in each form, long enough to establish liquid-like and solid-like spectral densities and diffusion coefficients, but our measurements are much slower than the times spent in each form, we will only observe values of the properties that are averages over the two forms; we will have no indication of the presence of two forms in equilibrium. However we could test for this possibility by changing our method of observation to something faster, yet slow enough to reflect liquid-like properties if they were present. If, on the other hand, the first part of this condition is not met, and the system passes from solid-like to liquid-like forms too frequently for the development of two different velocity autocorrelation functions or two different mean square displacements, then no measurement, however fast, will distinguish two "phases". Such a system can only look to any observer like slush, something between solid-like and liquid-like. (We query whether the smooth solid-liquid transition reported in Ref.11 may be due to averaging over times longer than the duration spent in each phase; only long-time averages were used.)

Coexistence of two phases implies that a system can be sufficiently unconstrained to pass, given enough time, from one phase to the other. For small clusters, this happens on the time scale of 10-100 ps. Presumably it would require eons for 1 gm of ice to turn into 1 gm of water under conditions in which the two are in equilibrium and have equal free energies. Described in these terms, the cluster and the gram of water seem to differ only quantitatively but the quantities are so large that it is silly not to call the difference qualitative.

What we have, then, for clusters meeting the three conditions just stated, is a kind of phase change that falls outside the conventional classification by order. It is a phase change that we can relate to conventional phase changes because the width, in units of temperature, of the coexistence range, ΔT_c, becomes smaller as the cluster becomes larger--but ΔT_c is not a monotonic function of N. Details of specific structures make this width greater for such clusters as Ar_{13} and Ar_{19} than for clusters of neighboring sizes[2,14]. No general rule is yet known for the way T_m and T_f approach each other as functions of N.

EXPERIMENTAL EVIDENCE AND CONCLUSIONS

No experiments have yet been announced which are designed to test the proposition that finite clusters have a kind of phase equilibrium different from--but related to--the phase transitions of bulk matter. There is evidence from diffraction studies that clusters may be liquid or solid[21-23]. Evidence of liquid-like clusters has also come recently from photodissociation experiments[24,25]. The most recent and most relevant evidence on the subject is the analysis by Eichenauer and Le Roy[26,27] of studies of an infrared line shape and shift for a molecule of SF_6 in a cluster of argon atoms[28]. The ν_3 line broadens and shifts in a manner reminiscent of the behavior of nuclear magnetic resonance lines when they are observed through temperature ranges in which the molecule under study passes from one form to another: the line is insensitive to changes of temperature up to a point (or narrow region) at which it broadens considerably, and then, at a higher temperature again becomes sharp. Typically, the line starts at one end of the broad shape when the temperature is at one end of its transition range and ends at the other end of the broad shape when the temperature reaches the other end of the transition range. Eichenauer and Le Roy have carried out simulations to model this change of line shape[27]; the results are quite consistent with the existence of two "phases" in equilibrium, although, as they point out, this by no means proves the prediction.

The situation now stands at the point where experimental tests are much needed. Furthermore, we need to know the extent to which such coexistence phenomena may occur for clusters with binding forces quite different from the weak forces that hold rare gas atoms together. It becomes more interesting now to explore the degree to which phase coexistence and bimodality depends on the time scale of observations, which, in simulations, is replaced by the time scale of short-time averaging. And it would now be extremely useful to return to an analytic approach to infer values of the melting temperature from solvable models, with the larger goal of learning how the range of temperature of coexistence depends on the number of atoms in the cluster.

ACKNOWLEDGMENTS

The author wishes to acknowledge the very considerable efforts of his students and coworkers Drs. F. Amar, J. Jellinek and G. Natanson and T. L. Beck and H. T. Davis. He also wishes to thank Professors H. C. Andersen and R. J. Le Roy and Dr. D. Eichenauer for stimulating discussions, and the Physical Chemistry Laboratory of Oxford Unversity for its hospitality. The research reported here that was done by the author and his group at The University of Chicago was supported by a Grant from the National Science Foundation.

REFERENCES

1. D. J. McGinty, J. Chem. Phys. **58**, 4733 (1973); R. M. Cotterill, W. Damgaard Kristensen, J. W. Martin, L. B. Peterson and E. J. Jensen, Comput. Phys. Comm. **5**, 28 (1973); C. L. Briant and J. J. Burton, Nature **243**, 100 (1973); and J. K. Lee, J. A. Barker and F. F. Abraham, J. Chem. Phys. **58**, 3166 (1973); these seem to be the earliest papers to deal explicitly with simulations of solid and liquid forms of clusters. See Ref. 2 for a review of the work of this period.
2. R. S. Berry, T. L. Beck, H. L. Davis and J. Jellinek, Adv. Chem. Phys. (1987; in press).
3. c.f. R. D. Etters and J. B. Kaelberer, Phys. Rev. A **11**, 1068 (1975).
4. C. L. Briant and J. J. Burton, J. Chem. Phys. **63**, 2045 (1975).
5. G. Delacrétaz, G. Stein and L. Wöste, J. Chem. Phys. (in press); M. Broyer, G. Delacrétaz, P. Labastie, R. L. Whetten, J. P. Wolf and L. Wöste, Z. Phys. D **3**, 131 (1986); G. Delacrétaz, E. R. Grant, R. L. Whetten, L. Wöste and J. W. Zwanziger, Phys. Rev. Lett. **56**, 2598 (1986).
6. M. V. Berry, Proc. Roy. Soc. (London) A **392**, 45 (1984); previously, fractional quantum numbers in pseudorotation had been proposed by Longuet-Higgins in 1961; the phenomenon was examined further in 1963 by Herzberg and Longuet-Higgins and in terms of the need for a vector potential by Mead and Truhlar in a very elegant but rarely read 1979 paper; c.f. H. C. Longuet-Higgins,

Adv. Spectrosc. **2**, 429 (1961); G. Herzberg and H. C. Longuet-Higgins, Disc. Farad. Soc. **35**, 77 (1963) and C. A. Mead and D. G. Truhlar, J. Chem. Phys. **70**, 2284 (1979).
7. G. Natanson, F. Amar and R. S. Berry, J. Chem. Phys. **78**, 399 (1983).
8. R. S. Berry, J. Jellinek and G. Natanson, Chem. Phys. Lett. **107**, 227 (1984); ibid., Phys. Rev. A**30**, 919 (1984).
9. R. S. Berry, Proc. 1st NEC Symposium on Fundamental Approach to New Material Phases, Hakone, 1986 (Springer, New York, in press).
10. J. Luo, U. Landmann and J. Jortner, in *The Physics and Chemistry of Small Clusters*, P. Jena, ed. (Plenum, New York, in press, 1987); for other prior discussions of isomerization and nonrigidity in alkali halide clusters, see T. P. Martin, J. Chem. Phys. **69**, 2036 (1978), ibid. **72**, 3506 (1980), and T. P. Martin, Phys. Repts. **95**, 167 (1983).
11. E. Blaisten-Barojas and D. Levesque, in *The Physics and Chemistry of Small Clusters*, P. Jena, ed. (Plenum, New York, in press, 1987).
12. J. Jellinek, T. Beck and R. S. Berry, J. Chem. Phys. **84**, 2783 (1986).
13. F. Amar and R. S. Berry, J. Chem. Phys. **85**, 5943 (1986).
14. T. Beck, J. Jellinek and R. S. Berry, J. Chem. Phys. (in press, 1987).
15. H. L. Davis, J. Jellinek and R. S. Berry, J. Chem. Phys. (in press, 1987).
16. T. L. Beck and R. S. Berry, in *The Physics and Chemistry of Small Clusters*, P. Jena, ed. (Plenum, New York, in press, 1987).
17. F. H. Stillinger and T. L. Weber, Kinam **3A**, 159 (1981); ibid., Phys. Rev. A **25**, 978 (1982); ibid. A **28**, 2408 (1983); see also F. H. Stillinger and R. A. LaViolette, J. Chem. Phys. **83**, 6413 (1985).
18. N. Quirke, private communication; to be published.
19. M. S. S. Challa, D. P. Landau and K. Binder, Phys. Rev. B **34**, 1841 (1986).
20. V. V. Nauchitel and A.J. Pertsin, Mol. Phys. **40**, 1341 (1980).
21. J. Farges, M. F. de Faraudy, B. Raoult and G. Torchet, J. Chem. Phys. **59**, 3454 (1973); ibid., **78**, 5067 (1983); ibid., **84**, 3491 (1986).
22. B. G. De Boer and G. D. Stein, Surf. Sci. **106**, 84 (1981).
23. E. J. Valente and L. F. Bartell, J. Chem. Phys. **80**, 1451, 1458 (1984).
24. A. J. Stace, Chem. Phys. Lett. **99**, 470 (1983).
25. A. J. Stace, D. M. Bernard, J. J. Crooks and K. L. Reid, Mol. Phys. **60**, 671 (1987).
26. D. Eichenauer and R. J. Le Roy, Phys. Rev. Lett. **57**, 2920 (1986).
27. D. Eichenauer and R. J. Le Roy, J. Chem. Phys. (in press, 1987).
28. T. E. Gough, D. G. Knoght and G. Scoles, Chem. Phys. Lett. **97**, 155 (1983).

AIP Conference Proceedings

		L.C. Number	ISBN
No. 1	Feedback and Dynamic Control of Plasmas – 1970	70-141596	0-88318-100-2
No. 2	Particles and Fields – 1971 (Rochester)	71-184662	0-88318-101-0
No. 3	Thermal Expansion – 1971 (Corning)	72-76970	0-88318-102-9
No. 4	Superconductivity in d- and f-Band Metals (Rochester, 1971)	74-18879	0-88318-103-7
No. 5	Magnetism and Magnetic Materials – 1971 (2 parts) (Chicago)	59-2468	0-88318-104-5
No. 6	Particle Physics (Irvine, 1971)	72-81239	0-88318-105-3
No. 7	Exploring the History of Nuclear Physics – 1972	72-81883	0-88318-106-1
No. 8	Experimental Meson Spectroscopy –1972	72-88226	0-88318-107-X
No. 9	Cyclotrons – 1972 (Vancouver)	72-92798	0-88318-108-8
No. 10	Magnetism and Magnetic Materials – 1972	72-623469	0-88318-109-6
No. 11	Transport Phenomena – 1973 (Brown University Conference)	73-80682	0-88318-110-X
No. 12	Experiments on High Energy Particle Collisions – 1973 (Vanderbilt Conference)	73-81705	0-88318-111-8
No. 13	π-π Scattering – 1973 (Tallahassee Conference)	73-81704	0-88318-112-6
No. 14	Particles and Fields – 1973 (APS/DPF Berkeley)	73-91923	0-88318-113-4
No. 15	High Energy Collisions – 1973 (Stony Brook)	73-92324	0-88318-114-2
No. 16	Causality and Physical Theories (Wayne State University, 1973)	73-93420	0-88318-115-0
No. 17	Thermal Expansion – 1973 (Lake of the Ozarks)	73-94415	0-88318-116-9
No. 18	Magnetism and Magnetic Materials – 1973 (2 parts) (Boston)	59-2468	0-88318-117-7
No. 19	Physics and the Energy Problem – 1974 (APS Chicago)	73-94416	0-88318-118-5
No. 20	Tetrahedrally Bonded Amorphous Semiconductors (Yorktown Heights, 1974)	74-80145	0-88318-119-3
No. 21	Experimental Meson Spectroscopy – 1974 (Boston)	74-82628	0-88318-120-7
No. 22	Neutrinos – 1974 (Philadelphia)	74-82413	0-88318-121-5
No. 23	Particles and Fields – 1974 (APS/DPF Williamsburg)	74-27575	0-88318-122-3
No. 24	Magnetism and Magnetic Materials – 1974 (20th Annual Conference, San Francisco)	75-2647	0-88318-123-1
No. 25	Efficient Use of Energy (The APS Studies on the Technical Aspects of the More Efficient Use of Energy)	75-18227	0-88318-124-X

No.	Title	LCCN	ISBN
No. 26	High-Energy Physics and Nuclear Structure – 1975 (Santa Fe and Los Alamos)	75-26411	0-88318-125-8
No. 27	Topics in Statistical Mechanics and Biophysics: A Memorial to Julius L. Jackson (Wayne State University, 1975)	75-36309	0-88318-126-6
No. 28	Physics and Our World: A Symposium in Honor of Victor F. Weisskopf (M.I.T., 1974)	76-7207	0-88318-127-4
No. 29	Magnetism and Magnetic Materials – 1975 (21st Annual Conference, Philadelphia)	76-10931	0-88318-128-2
No. 30	Particle Searches and Discoveries – 1976 (Vanderbilt Conference)	76-19949	0-88318-129-0
No. 31	Structure and Excitations of Amorphous Solids (Williamsburg, VA, 1976)	76-22279	0-88318-130-4
No. 32	Materials Technology – 1976 (APS New York Meeting)	76-27967	0-88318-131-2
No. 33	Meson-Nuclear Physics – 1976 (Carnegie-Mellon Conference)	76-26811	0-88318-132-0
No. 34	Magnetism and Magnetic Materials – 1976 (Joint MMM-Intermag Conference, Pittsburgh)	76-47106	0-88318-133-9
No. 35	High Energy Physics with Polarized Beams and Targets (Argonne, 1976)	76-50181	0-88318-134-7
No. 36	Momentum Wave Functions – 1976 (Indiana University)	77-82145	0-88318-135-5
No. 37	Weak Interaction Physics – 1977 (Indiana University)	77-83344	0-88318-136-3
No. 38	Workshop on New Directions in Mossbauer Spectroscopy (Argonne, 1977)	77-90635	0-88318-137-1
No. 39	Physics Careers, Employment and Education (Penn State, 1977)	77-94053	0-88318-138-X
No. 40	Electrical Transport and Optical Properties of Inhomogeneous Media (Ohio State University, 1977)	78-54319	0-88318-139-8
No. 41	Nucleon-Nucleon Interactions – 1977 (Vancouver)	78-54249	0-88318-140-1
No. 42	Higher Energy Polarized Proton Beams (Ann Arbor, 1977)	78-55682	0-88318-141-X
No. 43	Particles and Fields – 1977 (APS/DPF, Argonne)	78-55683	0-88318-142-8
No. 44	Future Trends in Superconductive Electronics (Charlottesville, 1978)	77-9240	0-88318-143-6
No. 45	New Results in High Energy Physics – 1978 (Vanderbilt Conference)	78-67196	0-88318-144-4
No. 46	Topics in Nonlinear Dynamics (La Jolla Institute)	78-57870	0-88318-145-2
No. 47	Clustering Aspects of Nuclear Structure and Nuclear Reactions (Winnepeg, 1978)	78-64942	0-88318-146-0
No. 48	Current Trends in the Theory of Fields (Tallahassee, 1978)	78-72948	0-88318-147-9

No.	Title		
No. 49	Cosmic Rays and Particle Physics – 1978 (Bartol Conference)	79-50489	0-88318-148-7
No. 50	Laser-Solid Interactions and Laser Processing – 1978 (Boston)	79-51564	0-88318-149-5
No. 51	High Energy Physics with Polarized Beams and Polarized Targets (Argonne, 1978)	79-64565	0-88318-150-9
No. 52	Long-Distance Neutrino Detection – 1978 (C.L. Cowan Memorial Symposium)	79-52078	0-88318-151-7
No. 53	Modulated Structures – 1979 (Kailua Kona, Hawaii)	79-53846	0-88318-152-5
No. 54	Meson-Nuclear Physics – 1979 (Houston)	79-53978	0-88318-153-3
No. 55	Quantum Chromodynamics (La Jolla, 1978)	79-54969	0-88318-154-1
No. 56	Particle Acceleration Mechanisms in Astrophysics (La Jolla, 1979)	79-55844	0-88318-155-X
No. 57	Nonlinear Dynamics and the Beam-Beam Interaction (Brookhaven, 1979)	79-57341	0-88318-156-8
No. 58	Inhomogeneous Superconductors – 1979 (Berkeley Springs, W.V.)	79-57620	0-88318-157-6
No. 59	Particles and Fields – 1979 (APS/DPF Montreal)	80-66631	0-88318-158-4
No. 60	History of the ZGS (Argonne, 1979)	80-67694	0-88318-159-2
No. 61	Aspects of the Kinetics and Dynamics of Surface Reactions (La Jolla Institute, 1979)	80-68004	0-88318-160-6
No. 62	High Energy e^+e^- Interactions (Vanderbilt, 1980)	80-53377	0-88318-161-4
No. 63	Supernovae Spectra (La Jolla, 1980)	80-70019	0-88318-162-2
No. 64	Laboratory EXAFS Facilities – 1980 (Univ. of Washington)	80-70579	0-88318-163-0
No. 65	Optics in Four Dimensions – 1980 (ICO, Ensenada)	80-70771	0-88318-164-9
No. 66	Physics in the Automotive Industry – 1980 (APS/AAPT Topical Conference)	80-70987	0-88318-165-7
No. 67	Experimental Meson Spectroscopy – 1980 (Sixth International Conference, Brookhaven)	80-71123	0-88318-166-5
No. 68	High Energy Physics – 1980 (XX International Conference, Madison)	81-65032	0-88318-167-3
No. 69	Polarization Phenomena in Nuclear Physics – 1980 (Fifth International Symposium, Santa Fe)	81-65107	0-88318-168-1
No. 70	Chemistry and Physics of Coal Utilization – 1980 (APS, Morgantown)	81-65106	0-88318-169-X
No. 71	Group Theory and its Applications in Physics – 1980 (Latin American School of Physics, Mexico City)	81-66132	0-88318-170-3
No. 72	Weak Interactions as a Probe of Unification (Virginia Polytechnic Institute – 1980)	81-67184	0-88318-171-1
No. 73	Tetrahedrally Bonded Amorphous Semiconductors (Carefree, Arizona, 1981)	81-67419	0-88318-172-X

No. 74	Perturbative Quantum Chromodynamics (Tallahassee, 1981)	81-70372	0-88318-173-8
No. 75	Low Energy X-Ray Diagnostics – 1981 (Monterey)	81-69841	0-88318-174-6
No. 76	Nonlinear Properties of Internal Waves (La Jolla Institute, 1981)	81-71062	0-88318-175-4
No. 77	Gamma Ray Transients and Related Astrophysical Phenomena (La Jolla Institute, 1981)	81-71543	0-88318-176-2
No. 78	Shock Waves in Condensed Matter – 1981 (Menlo Park)	82-70014	0-88318-177-0
No. 79	Pion Production and Absorption in Nuclei – 1981 (Indiana University Cyclotron Facility)	82-70678	0-88318-178-9
No. 80	Polarized Proton Ion Sources (Ann Arbor, 1981)	82-71025	0-88318-179-7
No. 81	Particles and Fields –1981: Testing the Standard Model (APS/DPF, Santa Cruz)	82-71156	0-88318-180-0
No. 82	Interpretation of Climate and Photochemical Models, Ozone and Temperature Measurements (La Jolla Institute, 1981)	82-71345	0-88318-181-9
No. 83	The Galactic Center (Cal. Inst. of Tech., 1982)	82-71635	0-88318-182-7
No. 84	Physics in the Steel Industry (APS/AISI, Lehigh University, 1981)	82-72033	0-88318-183-5
No. 85	Proton-Antiproton Collider Physics –1981 (Madison, Wisconsin)	82-72141	0-88318-184-3
No. 86	Momentum Wave Functions – 1982 (Adelaide, Australia)	82-72375	0-88318-185-1
No. 87	Physics of High Energy Particle Accelerators (Fermilab Summer School, 1981)	82-72421	0-88318-186-X
No. 88	Mathematical Methods in Hydrodynamics and Integrability in Dynamical Systems (La Jolla Institute, 1981)	82-72462	0-88318-187-8
No. 89	Neutron Scattering – 1981 (Argonne National Laboratory)	82-73094	0-88318-188-6
No. 90	Laser Techniques for Extreme Ultraviolt Spectroscopy (Boulder, 1982)	82-73205	0-88318-189-4
No. 91	Laser Acceleration of Particles (Los Alamos, 1982)	82-73361	0-88318-190-8
No. 92	The State of Particle Accelerators and High Energy Physics (Fermilab, 1981)	82-73861	0-88318-191-6
No. 93	Novel Results in Particle Physics (Vanderbilt, 1982)	82-73954	0-88318-192-4
No. 94	X-Ray and Atomic Inner-Shell Physics – 1982 (International Conference, U. of Oregon)	82-74075	0-88318-193-2
No. 95	High Energy Spin Physics – 1982 (Brookhaven National Laboratory)	83-70154	0-88318-194-0
No. 96	Science Underground (Los Alamos, 1982)	83-70377	0-88318-195-9

No.	Title		
No. 97	The Interaction Between Medium Energy Nucleons in Nuclei – 1982 (Indiana University)	83-70649	0-88318-196-7
No. 98	Particles and Fields – 1982 (APS/DPF University of Maryland)	83-70807	0-88318-197-5
No. 99	Neutrino Mass and Gauge Structure of Weak Interactions (Telemark, 1982)	83-71072	0-88318-198-3
No. 100	Excimer Lasers – 1983 (OSA, Lake Tahoe, Nevada)	83-71437	0-88318-199-1
No. 101	Positron-Electron Pairs in Astrophysics (Goddard Space Flight Center, 1983)	83-71926	0-88318-200-9
No. 102	Intense Medium Energy Sources of Strangeness (UC-Sant Cruz, 1983)	83-72261	0-88318-201-7
No. 103	Quantum Fluids and Solids – 1983 (Sanibel Island, Florida)	83-72440	0-88318-202-5
No. 104	Physics, Technology and the Nuclear Arms Race (APS Baltimore –1983)	83-72533	0-88318-203-3
No. 105	Physics of High Energy Particle Accelerators (SLAC Summer School, 1982)	83-72986	0-88318-304-8
No. 106	Predictability of Fluid Motions (La Jolla Institute, 1983)	83-73641	0-88318-305-6
No. 107	Physics and Chemistry of Porous Media (Schlumberger-Doll Research, 1983)	83-73640	0-88318-306-4
No. 108	The Time Projection Chamber (TRIUMF, Vancouver, 1983)	83-83445	0-88318-307-2
No. 109	Random Walks and Their Applications in the Physical and Biological Sciences (NBS/La Jolla Institute, 1982)	84-70208	0-88318-308-0
No. 110	Hadron Substructure in Nuclear Physics (Indiana University, 1983)	84-70165	0-88318-309-9
No. 111	Production and Neutralization of Negative Ions and Beams (3rd Int'l Symposium, Brookhaven, 1983)	84-70379	0-88318-310-2
No. 112	Particles and Fields – 1983 (APS/DPF, Blacksburg, VA)	84-70378	0-88318-311-0
No. 113	Experimental Meson Spectroscopy – 1983 (Seventh International Conference, Brookhaven)	84-70910	0-88318-312-9
No. 114	Low Energy Tests of Conservation Laws in Particle Physics (Blacksburg, VA, 1983)	84-71157	0-88318-313-7
No. 115	High Energy Transients in Astrophysics (Santa Cruz, CA, 1983)	84-71205	0-88318-314-5
No. 116	Problems in Unification and Supergravity (La Jolla Institute, 1983)	84-71246	0-88318-315-3
No. 117	Polarized Proton Ion Sources (TRIUMF, Vancouver, 1983)	84-71235	0-88318-316-1

No.	Title		
No. 118	Free Electron Generation of Extreme Ultraviolet Coherent Radiation (Brookhaven/OSA, 1983)	84-71539	0-88318-317-X
No. 119	Laser Techniques in the Extreme Ultraviolet (OSA, Boulder, Colorado, 1984)	84-72128	0-88318-318-8
No. 120	Optical Effects in Amorphous Semiconductors (Snowbird, Utah, 1984)	84-72419	0-88318-319-6
No. 121	High Energy e^+e^- Interactions (Vanderbilt, 1984)	84-72632	0-88318-320-X
No. 122	The Physics of VLSI (Xerox, Palo Alto, 1984)	84-72729	0-88318-321-8
No. 123	Intersections Between Particle and Nuclear Physics (Steamboat Springs, 1984)	84-72790	0-88318-322-6
No. 124	Neutron-Nucleus Collisions – A Probe of Nuclear Structure (Burr Oak State Park - 1984)	84-73216	0-88318-323-4
No. 125	Capture Gamma-Ray Spectroscopy and Related Topics – 1984 (Internat. Symposium, Knoxville)	84-73303	0-88318-324-2
No. 126	Solar Neutrinos and Neutrino Astronomy (Homestake, 1984)	84-63143	0-88318-325-0
No. 127	Physics of High Energy Particle Accelerators (BNL/SUNY Summer School, 1983)	85-70057	0-88318-326-9
No. 128	Nuclear Physics with Stored, Cooled Beams (McCormick's Creek State Park, Indiana, 1984)	85-71167	0-88318-327-7
No. 129	Radiofrequency Plasma Heating (Sixth Topical Conference, Callaway Gardens, GA, 1985)	85-48027	0-88318-328-5
No. 130	Laser Acceleration of Particles (Malibu, California, 1985)	85-48028	0-88318-329-3
No. 131	Workshop on Polarized ^{3}He Beams and Targets (Princeton, New Jersey, 1984)	85-48026	0-88318-330-7
No. 132	Hadron Spectroscopy–1985 (International Conference, Univ. of Maryland)	85-72537	0-88318-331-5
No. 133	Hadronic Probes and Nuclear Interactions (Arizona State University, 1985)	85-72638	0-88318-332-3
No. 134	The State of High Energy Physics (BNL/SUNY Summer School, 1983)	85-73170	0-88318-333-1
No. 135	Energy Sources: Conservation and Renewables (APS, Washington, DC, 1985)	85-73019	0-88318-334-X
No. 136	Atomic Theory Workshop on Relativistic and QED Effects in Heavy Atoms	85-73790	0-88318-335-8
No. 137	Polymer-Flow Interaction (La Jolla Institute, 1985)	85-73915	0-88318-336-6
No. 138	Frontiers in Electronic Materials and Processing (Houston, TX, 1985)	86-70108	0-88318-337-4
No. 139	High-Current, High-Brightness, and High-Duty Factor Ion Injectors (La Jolla Institute, 1985)	86-70245	0-88318-338-2

No. 140	Boron-Rich Solids (Albuquerque, NM, 1985)	86-70246	0-88318-339-0
No. 141	Gamma-Ray Bursts (Stanford, CA, 1984)	86-70761	0-88318-340-4
No. 142	Nuclear Structure at High Spin, Excitation, and Momentum Transfer (Indiana University, 1985)	86-70837	0-88318-341-2
No. 143	Mexican School of Particles and Fields (Oaxtepec, México, 1984)	86-81187	0-88318-342-0
No. 144	Magnetospheric Phenomena in Astrophysics (Los Alamos, 1984)	86-71149	0-88318-343-9
No. 145	Polarized Beams at SSC & Polarized Antiprotons (Ann Arbor, MI & Bodega Bay, CA, 1985)	86-71343	0-88318-344-7
No. 146	Advances in Laser Science–I (Dallas, TX, 1985)	86-71536	0-88318-345-5
No. 147	Short Wavelength Coherent Radiation: Generation and Applications (Monterey, CA, 1986)	86-71674	0-88318-346-3
No. 148	Space Colonization: Technology and The Liberal Arts (Geneva, NY, 1985)	86-71675	0-88318-347-1
No. 149	Physics and Chemistry of Protective Coatings (Universal City, CA, 1985)	86-72019	0-88318-348-X
No. 150	Intersections Between Particle and Nuclear Physics (Lake Louise, Canada, 1986)	86-72018	0-88318-349-8
No. 151	Neural Networks for Computing (Snowbird, UT, 1986)	86-72481	0-88318-351-X
No. 152	Heavy Ion Inertial Fusion (Washington, DC, 1986)	86-73185	0-88318-352-8
No. 153	Physics of Particle Accelerators (SLAC Summer School, 1985) (Fermilab Summer School, 1984)	87-70103	0-88318-353-6
No. 154	Physics and Chemistry of Porous Media—II (Ridge Field, CT, 1986)	83-73640	0-88318-354-4
No. 155	The Galactic Center: Proceedings of the Symposium Honoring C. H. Townes (Berkeley, CA, 1986)	86-73186	0-88318-355-2
No. 156	Advanced Accelerator Concepts (Madison, WI, 1986)	87-70635	0-88318-358-0
No. 157	Stability of Amorphous Silicon Alloy Materials and Devices (Palo Alto, CA, 1987)	87-70990	0-88318-359-9
No. 158	Production and Neutralization of Negative Ions and Beams (Brookhaven, NY, 1986)	87-71695	0-88318-358-7

No. 159	Applications of Radio-Frequency Power to Plasma: Seventh Topical Conference (Kissimmee, FL, 1987)	87-71812	0-88318-359-5
No. 160	Advances in Laser Science–II (Seattle, WA, 1986)	87-71962	0-88318-360-9
No. 161	Electron Scattering in Nuclear and Particle Science: In Commemoration of the 35th Anniversary of the Lyman-Hanson-Scott Experiment (Urbana, IL, 1986)	87-72403	0-88318-361-7